WILLYS K. SILVERS

The Coat Colors of Mice

A Model for Mammalian Gene Action and Interaction

With 66 Illustrations and 3 Color Plates

Springer-Verlag
New York Heidelberg Berlin

Willys K. Silvers
Department of Human Genetics
University of Pennsylvania
School of Medicine
Philadelphia, Pennsylvania 19104

Library of Congress Cataloging in Publication Data

Silvers, Willys Kent, 1929–
The coat colors of mice.

Bibliography: p.
Includes index.
1. Mice—Genetics. 2. Color of animals. 3. Mammals
—Genetics. I. Title.
QL737.R638S535 599′.3233 78-24533

Printed in the United States of America.

9 8 7 6 5 4 3 2 1

ISBN 0-387-90367-4 Springer-Verlag New York
ISBN 3-540-90367-4 Springer-Verlag Berlin Heidelberg

To Herman B. Chase

The enthusiasm he fostered made this a labor of love.

Preface

Many investigators seem to be fascinated by the coat colors of the mammals with which they work. This seems to be the case particularly for those utilizing isogenic strains of mice, not only because such strains display widely different phenotypes, but because scientists, by definition, are an inquisitive lot and it is sometimes difficult for the uninitiated to comprehend how such phenotypes are produced. This bewilderment becomes even more apparent if the investigator happens to be involved in breeding studies and a number of attractively colored animals, quite different from the original stocks, appear. Thus I can recall numerous occasions when my colleagues, frequently working in areas completely unrelated to any aspect of genetics, have come to me with an attractively pigmented animal or, more likely, with a population of segregating coat color types (usually because they have not tended their animals properly and have ended up with a cage full of F_2s displaying a number of different colors). How, they ask, do such colors come about?

While in some cases it is easy to take chalk in hand and explain what has been going on (segregating) and why, in other cases it is virtually impossible. It is extremely difficult because while the interactions of many coat-color factors obey the simple laws of heredity and of predictable gene interactions, others do not. Thus one is faced with having to say "it is just an empirical fact that . . ." a statement which proves nothing other than that the empirical fact is "ignorance."

Indeed, I will always remember how one such "empirical fact" revealed my ignorance at the worst possible time—my preliminary examination for the Ph.D. In my day, prelims at the University of Chicago were conducted "one on one," i.e., each student spent one hour with *each* faculty member. When my turn to spend an hour with my mentor, Professor Sewall Wright, came, I was a nervous wreck. I was not nervous because Professor Wright had a reputation for being difficult in such situations, on the contrary he is a most compassionate individual, going out of his way to make things as pleasant as they can be under such trying circumstances. I was nervous because I knew that Professor Wright was just as likely to hit me with population genetics or biometry, two areas with which I have never felt at home, as he was to spend most of the time on the inheritance of coat colors in guinea pigs. I was relieved when he looked at me and said "Silvers, what do you know about the coat-color genetics of guinea pigs?" I replied that I knew more about the colors of mice, but would do my best if he would rather talk about guinea pigs (but what a relief; guinea pigs and not mathamatical equations!). He said he was going to write down a number of genotypes and I was to tell him what the animals looked like. Well we started off and I was doing splendidly, basing all my answers not on personal experience, for I had very little, but on some "book knowledge" about what each coat-color determinant was supposed to do. Finally, Professor Wright (I think with a twinkle in his eye, although I'm not sure) wrote down a genotype, looked at me, and said "Silvers, what do you think this animal looks like?" A fast calculation told me that on the basis of the performance of each allele before me the animal should be a pale cream, and I stated just that. Professor Wright, not wanting to tell me I was stupid but at the same time wanting to impress upon me the fact that I was wrong, looked at me and said "No, Silvers, not quite. It happens to be an intense brown!" Yes, Professor Wright had caught me flat-footed. I had not worked with guinea pigs, and he happened to have chosen a genotype which didn't abide completely by the "rules."

In this book many examples of such exceptions will be noted. While, on the one hand, they can make life difficult (especially on preliminary examinations), on the other such exceptions often lead to more important rules.

How I wish I could state that the subject matter which follows will serve as a *simple* guide for explaining to the uninitiated the whys and wherefores of coat color inheritance. However, I have no delusions about its success as a "mouse watcher's color-guide"; for this it is not. What I do hope it accomplishes, however, is to impress upon the reader that all the basic principles of genetics—the predictable and unpredictable interaction of simple Mendelian factors and of polygenes; the influence of specific modifiers and of the genetic background as a whole; dominance, codominance, semidominance, and epistasis, etc.—are all intimately involved in making mice the colors they are.

I am indebted to many of my colleagues who in different ways helped me during the course of this effort. To all of them I wish to express my gratitude. Special appreciation is due Tim Poole who brought me back into the pigment field and who encouraged me to embark on this endeavor; George C.

McKay, Jr. and Priscilla W. Lane who, upon request, provided me with many of the colored pictures; William Fore who reproduced many of the colored slides; Paul Mortensen who did many of the drawings and helped with the photography; Rita J.S. Phillips whom I continually bothered but only because each unquiry produced so much useful information; Joan Staats who time and time again promptly supplied me with information from *Mouse News Letter*s I did not have; Maureen Geibler who typed the final draft; Herman Chase, David Gasser, and Tim Poole who very kindly read through and commented on the final draft; and Mark Licker of Springer-Verlag who more than fulfilled his part of the bargain. I also wish to acknowledge the National Institutes of Health for supporting all of my pigment cell research and for permitting me to use my current grant (CA-18640) to help defray many of the expenses in preparing what follows. Finally, I would like most to thank my family—Abigail, Deborah, and Kent—not only for putting up with me, but for having to spend many hours listening to Francis Albert Sinatra who, in my opinion, but not theirs, made all the tedious tasks associated with this effort *almost* enjoyable.

Willys K. Silvers

Contents

Symbols, Names, and (Chromosomal Assignments) of Reported Mutant Genes Affecting Coat Color in the House Mouse

Agouti series (2)
- A^{y} Lethal yellow
- A^{vy} Viable yellow
- A^{iy} Intermediate yellow
- A^{sy} Sienna yellow
- A^{w} White-bellied agouti
- A^{i} Intermediate agouti
- a^{td} Tanoid
- a^{t} Black and tan
- a Nonagouti
- a^{e} Extreme nonagouti
- a^{m} Mottled agouti
- a^{u} Agouti umbrous
- a^{da} Nonagouti with dark agouti belly
- a^{l} Nonagouti lethal
- a^{x} Lethal nonagouti
- A^{s} Agouti-suppressor
- a^{6H}

ash Ashen (9)

b series (4)
- b Brown
- b^{c} Cordovan
- B^{lt} Light
- B^{w} White-based brown

Bf Buff (5)

- bg Beige (13)
- bg^{J} Beige-Jackson
- bg^{slt} Beige (slate)

Bs Belly spot

Bst Belly spot and tail

- bt Belted (15)
- bt^{J} Belted-Jackson

bt-2 Belted-2

c (albino) series (7)
- c Albino
- c^{ch} Chinchilla
- c^{i} Intense chinchilla
- c^{e} Extreme dilution
- c^{h} Himalayan
- c^{p} Platinum
- c^{m} Chinchilla-mottled

d (dilute) series (9)
- d Dilute

d^l Dilute-lethal
d^s Slight dilution
d^{15} Dilute-15

da Dark (7)

Dfp Dark foot pads

dp Dilution-Peru

Extension series (8)
E^{so} Sombre
E^{tob} Tobacco darkening
e Recessive yellow

ep Pale ear (19)

f Flexed-tailed (13)

Fk Fleck

Flk Freckled (14)

Ga Greying with age

gl Grey-lethal (10)

gm Gunmetal (14)

gr Grizzled (10)

le Light ear (5)

Li Light

ln Leaden (1)

ls Lethal spotting (2)

m Misty (4)

$M(c^m)$ Modifier of c^m

md Mahoganoid (16)

mg Mahogany (2)

mh Mocha (10)

Microphthalmia series (6)
mi Microphthalmia
mi^{Crc} Microphthalmia-Clinical Research Center
mi^{bw} Black-eyed white
mi^{ew} Eyeless white
mi^{rw} Red-eyed white
mi^{sp} Microphthalmia-spotted
mi^{ws} White-spot
mi^{x} Microphthalmia-*x*
Mi^{b} Microphthalmia-brownish
Mi^{or} Microphthalmia-Oak Ridge
Mi^{wh} White

Mottled series (X)
Mo Mottled
Mo^{bl} Blotchy
Mo^{br} Brindled
Mo^{dp} Dappled
Mo^{vbr} Viable-brindled
To Tortoiseshell

Ms Mosaic (X)

mu Muted (13)

nc Nonagouti curly (16)

Och Ochre (4)

p (pink eyed) series (7)
p Pink-eyed dilution
p^{bs} *p*-black-eyed sterile
p^{cp} *p*-cleft palate
p^{d} Dark pink-eye
p^{dn} *p*-darkening
p^{m1} Pink-eyed mottled-1
p^{m2} Pink-eyed mottled-2
p^{r} Japanese ruby
p^{s} *p*-sterile
p^{un} Pink-eyed unstable
p^{x} *p*-extra dark
p^{6H} —
p^{25H} —

pa Pallid (2)

pb Piebald

pe Pearl (13)

Pew Pewter (X)

Ph Patch (5)
Ph^e Patch-extended

pl Platino

Rn Roan (14)

rs Recessive spotting (5)

ru Ruby-eye (19)

ru-2 Ruby-eye-2 (7)
$ru\text{-}2^r$ Ruby-eye-2 (Oak Ridge)
$ru\text{-}2^{hz}$ Haze
$ru\text{-}2^{mr}$ Maroon

Rw Rump-white (5)

s Piebald (14)
s^l Piebald-lethal

sea Sepia (1)

si Silver (10)

Steel series (10)
Sl Steel
Sl^{cg} Cloud-grey
Sl^{con} Contrasted
Sl^d Steel-Dickie
Sl^{gb} Grizzle-belly
Sl^m Steel-Miller
Sl^{so} Sooty
sl^{du} Dusty

Slt Slaty (14)

Sp Splotch (1)
Sp^J Splotch-Jackson
Sp^d Delayed splotch

te Light-head

tp Taupe (7)

U Umbrous

uw Underwhite (15)

Va Varitint-waddler (12)
Va^J Varitint-waddler-Jackson

W (*dominant spotting*) *series* (5)
W Dominant spotting
W^x Dominant spotting-*x*
W^a Ames dominant spotting
W^b Ballantyne's spotting
W^f *W*-fertile
W^j Jay's dominant spotting
W^{pw} Panda-white
W^v Viable dominant spotting
W^s Strong's dominant spotting
W^{sh} Sash
W^e Extreme dominant spotting

Y^m Yellow mottling (X)

Chapter 1

Introduction

Interest in coat-color genetics is almost as old as the science of genetics itself, for it was only shortly after the rebirth of Mendelism, at the beginning of the century, that W.E. Castle and his students, as well as others, initiated studies on the inheritance of specific coat colors in guinea pigs, rats, rabbits, and mice. Although these investigators were completely unaware of the anatomical basis of pigmentation, not to mention its biochemistry, their studies clearly established that the production of coat-color patterns involved a local interaction of specific gene products which was relatively unaffected by systemic or environmental factors.

It remained for subsequent investigators to produce the evidence that melanogenesis is the sole prerogative of specialized branched or dendritic cells, now known as *melanocytes*, of neural crest origin (Rawles, 1940, 1947, 1948), and that these cells function as unicellular melanin-secreting glands in the epidermis (see Billingham and Silvers, 1960). This elucidation of the cellular basis of pigment formation set the stage for extensive studies on the physiological genetics of pigmentation. These studies are directed toward answering the important question of how genes which influence pigmentation produce their effects, and it is precisely this question which forms the principal subject matter of this book.

Since the influence or influences of a specific genetic locus can be established only on the basis of variations (*alleles*) from the "wild type" that have been produced by mutations, it is obvious that the subject matter which follows initially depended upon the occurrence, recognition, and description of mutations, along with their preservation. Mutations give rise to alleles which not only identify the locus but which, on the basis of their effects, tell us something about the kind of activity with which the locus is involved.

In the house mouse (*Mus musculus* L.) more genes have been identified which affect coat color than any other trait. Moreover, this number has increased enormously over the past 25 years. During this period the number of coat-color determinants has risen from 32 to more than 130 and the number of loci involved has increased from approximately 20 to more than 50. While there have been a number of reviews concerned either entirely or in part with these genes (Little, 1958; Billingham and Silvers, 1960; Silvers, 1961; Deol, 1963, 1970a; Foster, 1965; Wolfe and Coleman, 1966; Searle, 1968a; Quevedo, 1969a,b, 1971) none approaches, either in scope or detail, the masterful treatment which Grüneberg gave this subject in Chapters 4 and 5 of the 1952 edition of his book *The Genetics of the Mouse*. Indeed, one of the reasons this book was initiated was because it was believed the time had come to bring these chapters up to date.[1]

The subject has been attacked in a fashion similar to Grüneberg's. Thus, for the most part, each locus is considered separately. While this seemed to be the most logical approach, it has the disadvantage of often failing to emphasize the fact that each phenotype results from the cumulative effect of a multiplicity of determinants. For this reason I would like to call the reader's attention to the classic studies of E.S. Russell (1946, 1948, 1949a,b); as far as I am aware these are the only ones which attempt to define each phenotype in terms of the actions and interactions of all the participating factors.

Many coat-color determinants have pleiotropic effects which cannot, at least at present, be related to their influence on pigmentation; these too are considered, if only briefly, in the hope that they may provide important clues for future investigations. Much of the information relating to these other effects, and indeed much of the detailed information concerning the pigmentary influences of specific alleles, are to be found in the notes at the end of each chapter. This format was adopted not to deemphasize the information presented in these notes but only to enhance the continuity of the text.

I. The Coat of the Mouse and Its Development

Any consideration of coat color requires some appreciation of the intimate relationship that exists between the melanocyte and the hair in which its product, the *melanin granule* (or *mature melanosome*), is deposited. Therefore, before embarking on the main theme of this effort, it is necessary to comment briefly on the coat of the mouse and its development (for more extensive treatments see Dry, 1926; Chase et al., 1951; Chase, 1954, 1958; Chase and Silver, 1969).

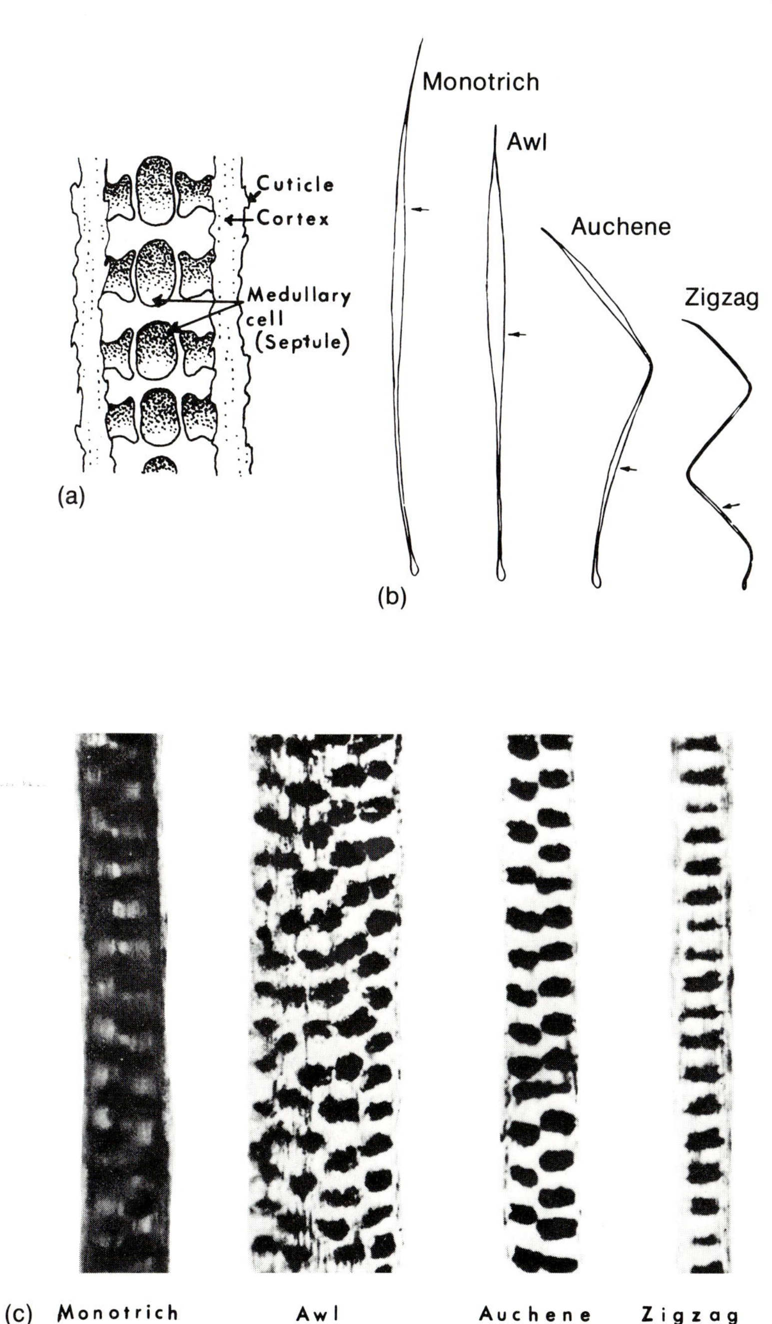

Figure 1-1. (a) Diagram of a longitudinal section of a hair. (b) The four main hair types drawn to scale (approx., ×9). Arrows indicate where photographs were taken for (c) (approx., ×400). (b) and (c) from H.B. Chase and H. Rauch, 1950. Reproduced with permission of the authors and The Wistar Institute Press.

The coat of the mouse consists of two kinds of hairs, the overhairs and the underhairs. There are three types of overhairs (which together make up about 20% of the total number of hairs), the guard hairs (or monotrichs) and the awls, which have no constrictions, and the auchenes which have a single constriction.[2] The underhairs or zigzags constitute the remaining and predominant hair population. These fibers are shorter than the other hair types and usually have three flat constrictions, the segments following each being angulated against each other (and hence zigzagged). Because of their small size these underhairs play only a minor role in determining the overall color of the animal (Deol, 1970a).

In fine structure all hairs are essentially similar, consisting of a wide central medulla surrounded by a narrow cortex which, in turn, is surrounded by a thin cuticle. The tips and bases of all the hairs are solid and deficient in medullary cells, and these cells may also be absent at the constrictions. Other than in these regions, the medullary cells are arranged transversely, separated from one another by areas devoid of melanin granules. In the overhairs these medullary cells may form rows of three, four, or even five *septules*, whereas in zigzags there is only a single row of *septa*. Although pigment granules are normally present both in the cortex and in the medulla of the shaft, most of the pigment occurs in the medulla. The four main hair types and some of their characteristics are shown in Figure 1-1.

The hair itself is formed in the hair follicle which begins as an epidermal invagination into the dermis. The dermis forms a thickening immediately beneath, and the blind end of the invagination comes to surround it partly. The dermal thickening develops into the hair papilla, and the surrounding part of the invagination forms the hair bulb (see Figure 1-2).

The melanocytes of the mature hair follicle are highly dendritic cells found

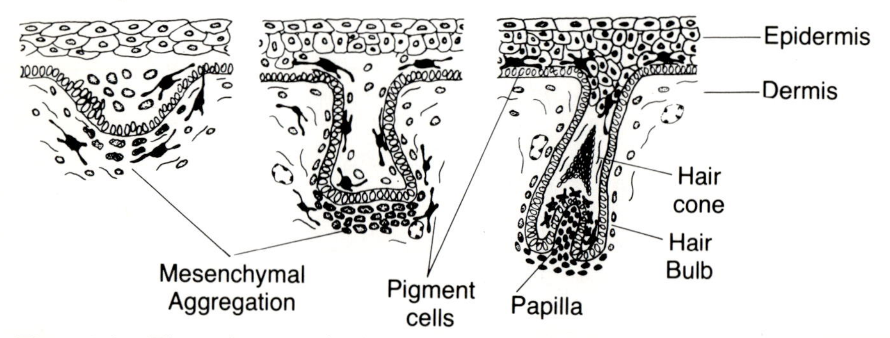

Figure 1-2. The early stages in the development of the hair follicle and hair. Each hair is formed in a follicle which arises embryologically as a thickening and inpushing of the epidermis. Subsequently, the base of the follicle thickens forming a cylinder on top of a papilla of mesodermal origin. By multiplication of this matrix of epidermal cells the hair shaft is formed, while the outer wall of the follicle forms a sheath round the base of the hair. Based on Searle (1968a).

in the hair bulb which their embryonic precursors (*melanoblasts*) enter as each follicle is formed (Figure 1-2).[3] These cells, which are derived from the neural crest,[4] secrete pigment granules (mature melanosomes) into the hair cortex and medulla as they develop.[5] This deposition of granules continues during the entire growing phase of the hair (about 17 days). This phase is known as *anagen* and has been further divided into six substages (see Chase et al., 1951). When matrix cell proliferation in the bulb (the stage known as *catagen*) ceases, the pigment cells also cease producing pigment and generally no further cells enter the shaft as medulla. This stage is followed by the resting or *telogen* stage. In young animals this phase may last 10 days or less, whereas in older mice it may persist for months before a new wave of hair growth begins. Nevertheless, new cycles can be initiated by plucking telogen hairs.

Notes

[1]Nevertheless, the remarkable book by A.G. Searle, *Comparative Genetics of Coat Colour in Mammals*, deserves special mention. Because this comprehensive and well-organized monograph concerns the coat-color determinants of all mammals, it serves in many ways as a "companion" to this effort.

[2]According to Grüneberg (1969; see also Dry, 1926) roughly 1% of the dorsal hairs of mice are guard hairs. Among the remaining overhairs, awls usually predominate and auchenes are rare.

[3]In addition to the melanocytes of the hair bulb, populations of pigmented melanocytes also occur in the ear and tail skin, but not usually in the hairy (trunk) skin of adult mice. Nevertheless, the epidermis of the hairy skin of *newborn* mice displays numerous dopa-positive melanocytes which increase during the first few days of life but then decline and virtually disappear by 3–4 weeks of age (Rovee and Reams, 1964; Quevedo et al., 1966; Hirobe and Takeuchi, 1977a). These cells, however, persist in a nonpigmented state (Reynolds, 1954; Quevedo and J. Smith, 1968) and reappear, i.e., synthesize melanin, when the mice are exposed to ultraviolet light (Quevedo and McTague, 1963; Quevedo et al., 1967).

[4]The neural crest becomes differentiated from the lateral margins of the neural plate when the neural tube develops during early ontogeny. It appears initially in the mid-brain region shortly before the neural folds close; subsequently it is found more and more caudally. Its development in more posterior regions is, typically, from the dorsal portion of the neural tube, where cells undergo rearrangement and rapid proliferation. From the resultant rather loosely connected mass, cells migrate in two directions: laterally toward the dorsal portion of the somite and ventrally along the sides of the neural tube. It is from the former group of neural crest cells that melanoblasts arise (Searle, 1968a).

[5]In contrast to the melanocytes of the hair bulbs and skin etc., those of the retina are formed from the outer cells of the optic cup. These melanocytes are histologically very distinct from those of neural crest origin and form a simple epithelial layer of hexagonal cells on the outer surface of the retina from which fine cytoplasmic processes run inward between the retinal rods (Searle, 1968a).

Chapter 2

The Agouti and Extension Series of Alleles, Umbrous, and Sable

Although there are a great number of loci which influence the synthesis of melanin in mice,[1] there are only two loci which control the *nature* of the pigment formed. Thus the *agouti* and *extension* series of alleles determine the relative amount and distribution of yellow pigment (phaeomelanin) and black or brown pigment (eumelanin) in the hairs of the coat.

I. The Agouti Locus

The agouti locus (chromosome 2) is particularly interesting because it is a coat-color determinant which acts via the hair follicle. It also appears to be a complex locus. The number of alleles which have been described as being part of this complex—17 in all—and the number of reported mutations (see Chapter 12, Section IV), also exceeds that recorded for any other locus concerned with the synthesis of melanin. Some of these mutations are listed by Grüneberg (1952, 1966a), Wallace (1954, 1965), L. Russell and Maddux (1964), and Dickie (1969a).

A. Alleles

1. Lethal Yellow (A^y)

Lethal yellow (A^y) represents the top dominant of the agouti series in the sense that it does not matter with which of the series it is heterozygous, the phenotype is either a rich yellow or orange color (Plate 1–A) or, on some backgrounds, a "sooty yellow" or "sable" color (see Section III). This allele, which represents an old mutant of the mouse fancy, was described first by Cúenot (1905) who recognized the fact that it was lethal when homozygous (see also Castle and Little, 1910; Ibsen and Steigleder, 1917). However, while Cúenot believed that an A^y-bearing egg could not be fertilized by an A^y-bearing sperm, this is not the case. A^y/A^y embryos are formed but display characteristic abnormalities at the morula or blastocyst stage (Kirkham, 1919; Robertson, 1942a) and die early on the sixth day of gestation after the trophectoderm of the blastocyst has come into contact with the uterine epithelium, i.e., before implantation is complete (see Eaton and M.M. Green, 1962, 1963; Eaton, 1968; Pedersen, 1974; Calarco and Pedersen, 1976). Although according to Eaton and M.M. Green (1963) death is due to a lag in the differentiation of the trophoblast giant cells, preventing the normal interaction between embryo and endometrium necessary for successful implantation, the primary lesion responsible for the failure of giant cells to differentiate from the trophectoderm is not known.[2] Robertson (1942b) demonstrated that A^y/A^y embryos which develop in a nonyellow mother, a feat accomplished by mating a/a females bearing A^y/a ovarian isografts with A^y/a males, proceed to a somewhat more advanced stage of implantation, attaining about twice the number of cells, than A^y homozygotes which develop in yellow mothers. Indeed, there is some evidence that this maternal influence on the development of A^y homozygotes holds for A^y heterozygotes as well (G. Wolff and Bartke, 1966).

The A^y allele is also of interest because it often is associated with stimulation of normal body growth (Heston and Vlahakis, 1961a), obesity (Dickerson and Gowen, 1946, 1947; Grüneberg, 1952), sterility (M.C. Green, 1966a), and a diabetes-like syndrome (Hummel et al., 1972). Moreover, this allele has been associated with an increased susceptibility to both spontaneous (Heston and Deringer, 1947) and induced (Heston, 1942) pulmonary tumors, to spontaneous hepatomas in males and spontaneous mammary tumors in females (Heston and Vlahakis, 1961a), to induced skin tumors (Vlahakis and Heston, 1963), and to spontaneous reticular neoplasms (Deringer, 1970).[3] The fact that all of these pleiotropic effects are unique to A^y, or to animals of yellow phenotype, i.e., these effects are not associated with any of the *nonyellow* agouti series alleles, is undoubtedly significant both in terms of the structure of the locus and its mode of action. Indeed, the observation that A^y differs in so many ways from the other members of the series led Sir Ronald Fisher to suggest that it may represent a deletion covering more than one locus (Wallace, 1954).

2. Viable Yellow (A^{vy})

Viable yellow (A^{vy}) arose spontaneously in the C3H/HeJ strain (Dickie, 1962a) and although it may produce a phenotype similar to A^y, it is not lethal. The most characteristic feature of the A^{vy} allele is the variation in phenotypes associated with it (G. Wolff, 1971). Some homozygous viable yellow animals when weaned are a reasonably clear yellow, indistinguishable from $A^{y}/-$, but many become sooty with successive molts (Dickie, 1962a); some (homozygotes and heterozygotes) have a peculiar mottled pattern (Plate 1–B; see also Figure 7-6b), first observed when they are 4 or 5 days old, which may vary from a small black patch on an otherwise yellow coat to a complete intermixture of small black and yellow patches (Dickie, 1962a); still others display a marked visual resemblance to agouti mice even though, on close examination, disparities are apparent (Galbraith and G. Wolff, 1974).[4] The frequency of this "pseudoagouti" phenotype seems to be strain dependent (G. Wolff, 1971). Inasmuch as genetically identical littermates often display these variations in pigmentation, and since A^{vy} animals of agouti-like phenotype often produce yellow offspring with variable degrees of mottling (Dickie, 1962a; G. Wolff and Pitot, 1973), it is apparent that a considerable amount of the variation in pigmentation displayed by these mice must be a consequence of nonhereditary factors (G. Wolff and Pitot, 1973) (see note 9 and Chapter 7, Section VI). Nevertheless, taken together, the expression of the A^{vy} gene seems to strike a precarious balance between the A^{y} and A alleles (Galbraith and G. Wolff, 1974).

Although not lethal A^{vy} is associated with many of the other effects of A^{y}. Moreover, these pleiotropic effects, which include an influence on normal and neoplastic growth, fat deposition, and hormonal and enzymatic levels (Heston and Vlahakis, 1968; G. Wolff, 1965, 1970a,b,c; G. Wolff and Flack, 1971; G. Wolff and Richard, 1970; G. Wolff and Pitot, 1972a,b, 1973), are particularly interesting because they appear to be correlated with the amount of yellow in the coat (G. Wolff, 1965, 1971). For example, homozygous and heterozygous A^{vy} mice which are predominantly yellow are more likely to become obese than agouti viable yellow animals (G. Wolff, 1965; Dickie, 1969a), and A^{vy} mice which are of agouti phenotype are less likely than mottled yellow mice to develop hepatomas (G. Wolff and Pitot, 1972b). Clearly, for some reason, A^{vy} mice of agouti phenotype are metabolically more similar to a/a mice than they are to mottled A^{vy} animals (G. Wolff, 1965; G. Wolff and Pitot, 1973) (see note 23).

3. Intermediate Yellow (A^{iy})

Intermediate yellow (A^{iy}) originated in the C3H/HeJ strain. The original mutant (A^{iy}/A) was a very sooty yellow with a slightly lighter belly, which is the characteristic phenotype of most animals bearing this mutation (Dickie, 1969a). Intermediate yellow mice are not as mottled with dark areas as some A^{vy} mice, but like A^{vy} they occasionally resemble the wild (agouti) phenotype. The ears of A^{iy} mice appear darker than those of lethal or viable yellow animals. Both homozygotes and heterozygotes, which look alike, become obese as adults. However, the maximum weight of these animals is lower than $A^{vy}/-$ or $A^{y}/-$ genotypes. As in the case for A^{vy}, A^{iy} animals of agouti phenotype do not become as obese as, or display the very mild hyperglycemia of, phenotypically yellow A^{iy} mice (Dickie, 1969a).[5]

4. Sienna Yellow (A^{sy})

Sienna yellow (A^{sy}) occurred in the C57BL/6J strain. Although little information is available on this allele, A^{sy} heterozygotes appear to be a very dark or sooty yellow (displaying yellow hairs with black tips) whereas homozygous A^{sy} mice are phenotypically a clearer yellow. A^{y}/A^{sy} heterozygotes are viable and fertile (Dickie, 1969b).

5. White (or Yellow) Bellied Agouti (A^{w})

It should be noted that this allele has also been referred to as A^{L} (light-bellied agouti). Mice displaying this phenotype have a typical agouti dorsum, i.e., a unique pattern of pigmentation characterized by a subapical yellow band on an otherwise black (or brown) hair. This black–yellow–black pattern results from a rapid shift from deposition of eumelanin to deposition of phaeomelanin and back again to eumelanin in the hair shaft. This pattern has been described in detail by Werneke (1916), Dry (1928), Dunn (1936), Kaliss (1942), E. Russell (1949b), and Galbraith (1964). The belly of A^{w} mice is white, cream, or tan (yellow) as a consequence of the fact that, depending on the genetic background, the hairs originating on the ventrum are either frequently nonpigmented, possess yellow pigment, or are predominantly yellow with black bases (Silvers, 1958b). A^{w} is dominant to A and all lower alleles.

6. Agouti (A)

This is the so-called wild type coloration[6] characterized by an agouti dorsum,[7] identical in all respects to the A^{w} dorsum described above, but a darker than A^{w} ventrum (Plate 1–C). Although the hairs of the ventrum are predominantly banded, some have yellow tips and black bases.

7. Intermediate Agouti (A^{i})

Intermediate agouti (A^{i}) arose spontaneously in the C57BL/6J strain. The belly of the original deviant (proven to be A^{i}/a) was dull yellow. There were some agouti hairs on the sides of the body but the subterminal yellow band was almost absent from the dorsal hairs, so that the back had a very dark appearance. As A^{i}/a heterozygotes age, the dorsum lightens so that the animals resemble $A^{w}/-$ mice (Dickie, 1969a). A^{i}/A^{i}, A^{i}/A, and A^{i}/a^{td} genotypes are indistinguishable from light-bellied agouti (Dickie, 1962b; M.C. Green, 1966a). Young A^{i}/a^{t} animals, like A^{i}/a, have a dark back and light belly with agouti hairs along the sides (Dickie, 1969a).

8. Tanoid (a^{td})

Tanoid (a^{td}) also represents a mutation in the C57BL/6J strain. The deviant animal had a tan (yellow) belly, typical of that found on black-and-tan (a^t) mice (see **9**) and banded "agouti" hairs among the predominantly black hairs on the sides of the body. The dorsum was very dark but with the first molt the number of "agouti" hairs increased, spreading over the animal's back so that the animal resembled a dark modification of "agouti" (A or A^w) (Loosli, 1963). Although precise information is lacking, tanoid animals must, at least at some stage of life, bear a striking similarity to young A^i/a^t and A^i/a mice. The tanoid allele is dominant to nonagouti (a) and black-and-tan (a^t) and A^w/a^{td} and A/a^{td} heterozygotes are phenotypically indistinguishable from $A^w/-$ animals.

9. Black-and-tan (a^t)

This phenotype (Plate 1–D) was first described by Dunn (1928) who obtained a strain of black-and-tan mice from an English fancier, and demonstrated that an allele of yellow, agouti, and nonagouti was responsible for the pigment pattern. Black-and-tan mice have a black dorsum and yellow or cream belly. As in the case of A^w the color of the ventrum varies with the genetic background; in some stocks the small amount of pigment in ventral hairs is phaeomelanin, whereas in other stocks almost all of the ventral hairs are yellow with black bases (Silvers and E. Russell, 1955). Some yellow pigment also is found in hairs originating on and behind the ears of a^t mice. This allele is particularly interesting because it is recessive to A on the dorsum, but dominant to A on the ventrum. Thus the A/a^t heterozygote is phenotypically indistinguishable from light-bellied agouti ($A^w/-$).

10. Nonagouti (a)

Nonagouti (a) represents another old mutant of the mouse fancy. This allele obtains its name from the fact that a/a mice display nonbanded (eumelanotic) hairs. a/a mice are black (Plate 1–E) and $a/a,b/b$ animals are brown (Plate 1–F). Although a/a hairs are almost exclusively pigmented with eumelanin, the hairs originating on and behind the ears as well as the hairs around the genital papilla and mammae, are yellow (at least in part).

11. Extreme Nonagouti (a^e)

Extreme nonagouti (a^e) was found among the descendants of an irradiated mouse (Hollander and Gowen, 1956). It is recessive to all the other alleles of the agouti series and is characterized as *completely* eumelanotic—i.e., no yellow hairs occur on or behind the ears or around the nipples and perineum.

12. Mottled Agouti (a^m)

Mottled agouti (a^m) is an allele at the agouti locus which originated in radiation experiments at Oak Ridge (L. Russell, 1964, 1965). This allele is similar to A^{vy} in that it induces a mottled coat, many a^m/a animals possessing agouti (actually, as described above for A^{vy}, "pseudoagouti")[8] and nonagouti patches of fur freely intermingled at their edges (L. Russell, 1964, 1965). Moreover, as in the case of A^{vy}, a^m produces a continuum of phenotypic variability. Unlike A^{vy}, however, this variability encompasses phenotypes of the lower agouti series alleles. Thus a^m/a^m mice vary in color from agouti to completely black, mimicking in a few cases the phenotypes of $a/-$ and a^e/a^e animals (G. Wolff, 1971). The fact that A^{vy} is associated with phenotypes ranging from agouti up to yellow, while a^m produces phenotypes ranging from agouti down to extreme nonagouti, is especially interesting in that these two alleles can produce almost the entire spectrum of agouti locus phenotypes. This has been nicely emphasized by G. Wolff (1971) who has established a stock, the so-called VYm stock, in which both of these alleles are maintained (see Figure 2-1 and Chapter 7, Section VI).[9]

In addition to the above alleles, three others have recently been recovered from radiation experiments at Harwell (R.J.S. Phillips, 1976); these are agouti umbrous (a^u), nonagouti with dark agouti belly (a^{da}), and nonagouti lethal (a^l).

13. Agouti Umbrous (a^u)

Agouti umbrous (a^u) resembles A^s (see **17**) in its effect but does not crossover with a. a^u homozygotes have a dark agouti dorsum and ventrum as well as dark agouti pinna hairs. a^u/a animals also display an umbrous back and dark agouti belly but their pinna hairs are yellow. a^u/a^t heterozygotes have an umbrous back, tan (yellow) belly, and yellow ear hairs (R.J.S. Phillips, 1976) (see note 10).

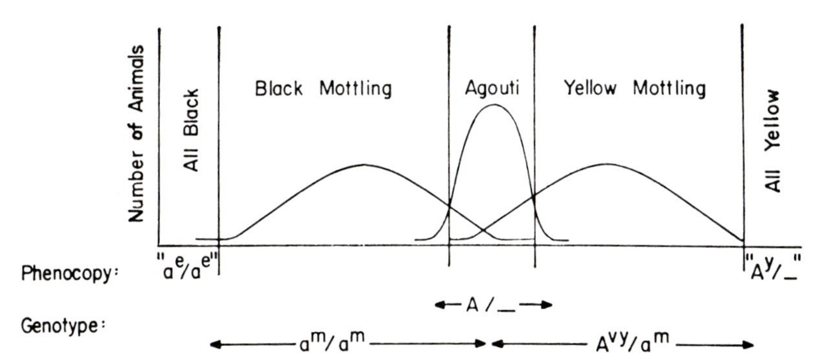

Figure 2-1. Schematic representation of phenotypic variability of coat-color patterns among A^{vy}/a^m and a^m/a^m mice in Wolff's VYm stock. The $A/-$ genotype does not segregate in this stock, but has been included since its expression includes a slight amount of phenotypic variability in the form of small black or yellowish areas in the coat-color pattern of some inbred A/A mice. The all-black mice and agouti animals which occur in the VYm stock are phenocopies of the a^e/a^e and $A/-$ coat-color phenotypes. From G. Wolff (1971). Reproduced with permission of the author and The University of Chicago Press.

14. Nonagouti with Dark Agouti Belly (a^{da})

This allele when homozygous, or when heterozygous with *a*, produces either a nonagouti or extreme umbrous back, dark agouti belly, and yellow pinna hairs (R.J.S. Phillips, 1976).

15. Nonagouti Lethal (a^{l})

Nonagouti lethal (a^{l}) is presumably a prenatal lethal when homozygous, but it differs from a^{x} (see **16**) in that phenotypically a^{l}/a mice are indistinguishable from a/a genotypes. a^{e}/a^{l} mice resemble a^{e}/a^{e} animals, i.e., they have black pinna hairs. The interaction of this new lethal with A^{y} has not been reported (R.J.S. Phillips, 1976) (see note 10).

The above 15 alleles have all been ascribed to the agouti locus and although some investigators believe that many of the phenotypes associated with these alleles result from the expression of more than one locus (see section B), there is no concrete evidence to *prove* this assertion. Nevertheless, there are two mutations which are considered members of the "agouti" series which have been shown to be very closely linked; these are a^{x} and A^{s}.

16. Lethal Nonagouti (a^{x})

Lethal nonagouti (a^{x}) is also a radiation-induced mutation and one which is lethal *in utero* when homozygous. a^{x}, which is recessive to A^{y}, A^{w}, A, and a^{t} but dominant to *a*, is characterized by a slightly paler belly than that found in a/a animals. This mutation shows about 0.5% recombination with A^{y} (L. Russell et al., 1963). A^{y}/a^{x} mice are perfectly viable and preliminary studies indicate that the lethal effect in a^{x} homozygotes is different from that seen in A^{y} homozygotes (Papaioannou and Mardon, 1978) (see note 10).

17. Agouti-Suppressor (A^{s})

This determinant is of special interest not only because it displays 0.6% crossing over with A^{w} and a^{t} (R.J.S. Phillips, 1966a), but because it recently has been shown to have resulted from an inversion (Evans and R.J.S. Phillips, 1978) and therefore almost certainly reflects some sort of position effect.[10] It occurred among the offspring of an X-irradiated (C3H $\times$ 101)F_1 male (R.J.S. Phillips, 1961a) and was originally believed to be a recessive allele of the agouti locus and similar to A^{i} (R.J.S. Phillips, 1959, 1960). When homozygous, A^{s} reduces the amount of phaeomelanin normally produced by whatever agouti alleles are present. As described by R.J.S. Phillips (1966a): "The yellow pinna hairs found with all alleles except a^{e} become black (or brown on a homozygous brown background etc.); with a^{t}/a^{t} the back and belly are black; with A^{w}/A^{w} the size of the normal yellow band and the percentage of banded hairs are reduced giving a very dark agouti effect shading to almost complete nonagouti on the top of the head (with A^{w}/a^{t} this "nonagouti" patch is rather larger, but the two genotypes are

Table 2-1 Phenotypes of Agouti-Suppressor (A^s) in Combination with Other Agouti Locus Alleles[a]

Genotype	Corresponding phenotype		
	Back	Belly	Pinna hairs
A^sA^w/A^sA^w	Dark agouti[b]	Dark agouti[b]	Black
A^sA^w/A^sa^t	Dark agouti[b]	Dark agouti[b]	Black
A^sa^t/A^sa^t	Nonagouti	Nonagouti	Black
$A^sA^w/{+}A^w$	Normal agouti	Yellow	Yellow
$A^sa^t/{+}A^w$	Normal agouti	Yellow	Yellow
$A^sA^w/{+}A$	Normal agouti	Normal agouti	Yellow
$A^sa^t/{+}A$	Normal agouti	Normal agouti	Yellow
$A^sA^w/{+}a^t$	Umbrous[c]	Yellow	Yellow
$A^sA^w/{+}a$	Umbrous[c]	Dark agouti[b]	Yellow
$A^sA^w/{+}a^e$	Umbrous[c]	Dark agouti[b]	Black
$A^sa^t/{+}a^t$	Nonagouti	Yellow	Yellow
$A^sa^t/{+}a$	Nonagouti	Nonagouti	Yellow
$A^sa^t/{+}a^e$	Nonagouti	Nonagouti	Black

[a]From R.J.S. Phillips (1966a).
[b]Dark agouti animals have less yellow banding in their agouti areas.
[c]Umbrous animals: the center back is nonagouti, the flanks are dark agouti.

not positively distinguishable). The belly of the A^sA^w/A^sA^w animal is also dark agouti; the A^sa^t/A^sa^t animal is undistinguishable from a^e/a^e."

The fascinating aspect of the behavior of A^s is that it shows a *cis–trans* position effect (E. Lewis, 1961), i.e., when heterozygous it affects only the agouti allele to which it is linked. Thus, for example, $A^sa^t/{+}A^w$ is phenotypically indistinguishable from light-bellied agouti whereas $A^sA^w/{+}a^t$ animals have an umbrous back, characterized by a nonagouti mid-dorsum and dark agouti flanks, and a yellow belly; $A^sa^t/{+}a^e$ mice are phenotypically identical with extreme nonagouti, while $A^sa^e/{+}a^t$ animals look like normal black-and-tan mice (see Table 2-1). In other respects A^s behaves in a normal Mendelian fashion (R.J.S. Phillips, 1966a).[11]

B. Structure of the Locus (Simple vs Complex)

As the above description of the phenotype(s) associated with each agouti locus allele testifies, the locus is a complicated one (see Table 2-2). In general, as one proceeds from top dominance (A^y) to bottom recessive (a^e) there is an increase in the amount of eumelanin, indicating that the synthesis of less black is dominant to the production of more. This tendency to produce yellow pigment appears also to be greater in ventral than in dorsal hair

Table 2-2. Phenotypes of Various Combinations of Some Agouti Locus Alleles[a,b,c]

	A^y	A^{iy}	A^w	A^i	A	a^{td}	a^t	a^x	a	a^e
A^y	Lethal	YYY	YYY	—	YYY	—	YYY	YYY	YYY	YYY
	A^{iy}	Viable	—	—	Sooty yellow	—	—	—	YYY–BBY[d]	—
		A^w	BYY	—	BYY	BYY	BYY	BYY	BYY	BYY
			A^i	BYY	BYY	—	—	—	UYY	—
				A	BBY	BYY	BYY	BBY	BBY	BBY
					a^{td}	U[D]YY	—	—	U[D]YY	—
						a^t	NYY	NYY	NYY	NYY
							a^x	Lethal	NPY	—
								a	NNY	NNY
									a^e	NNN

[a]Each phenotype, where known, is indicated by three symbols referring to back belly and pinna hairs respectively.
[b]From R.J.S. Phillips (1966a).
[c]B, banded (agouti); U, umbrous; U[D], umbrous combined with dark-agouti; Y, yellow; N, nonyellow, unbanded hairs (i.e., black on black background); P, as for N but paler.
[d]Varies in phenotype from A^y-like to wild type.

follicles and is even greater in the hairs of the pinna. However, there are some important exceptions to these generalizations and these have convinced some investigators that the agouti locus is composed of several very closely linked "mini-loci," each of which controls the proportion of yellow and black pigment for different parts of the body, e.g., dorsum, ventrum, ears, etc. (Pincus, 1929; Keeler, 1931; Wallace, 1954, 1962, 1965). Indeed, according to Wallace (1965), the situation to be envisaged is similar to the pseudoallelic system proposed by Fisher to account for the Rh blood group antigens in man (Race and Sanger, 1962).

Let us consider the evidence supporting this complex locus theory. Probably the best evidence is that inasmuch as a^x and A^s have already been demonstrated to be closely linked to A^y, and A^w and a^t, respectively, and there is some evidence that A^y is pseudoallelic to a just as it is to a^x (L. Russell et al., 1963), there is precedence for an even more extensive series of closely linked determinants. Further support for a complex locus stems from the observation that as one proceeds down the agouti locus series, the change from yellow to black occurs in one place in the series for one part of the pelage and in another for a different area. Thus phaeomelanin disappears from the mid-dorsum at compound A^i/a (but not in A^i/A^i), from the lateral dorsum at a^t/a^t and its lower compounds, from the ventrum at a/a and its lower compounds, and from the pinna and genital ridge at a^e/a^e. Although these changes may merely reflect the fact that certain regions of the body, e.g., the ventrum and pinna, present a milieu more favorable for the synthesis of phaeomelanin, they also are compatible with the notion that several very closely linked loci are involved, one controlling the proportion of yellow-black pigment for one part of the body, and one for another region.[12] For example, as pointed out by Wallace (1965), if one "mini-locus," D, controlled dorsal banding, and another, V, controlled ventral banding, with dominance of the yellow component in each case, the genotypes A/a^t and A^w/a, which have the same phenotype, light-bellied agouti, would become Dv/dV and DV/dv, respectively. Such an interpretation would explain the observation that the series A^w:A:a^t represents a grading yellow to black dorsally but not ventrally.[13]

Proponents of the complex locus hypothesis argue also that it is difficult to envisage on the simple locus theory how two alleles, A^{vy} and a^m (and perhaps A^{iy} should also be included), which are relatively high and low in the dominance hierarchy, produce mottling, whereas it does not appear in any of the intermediate compounds (Wallace, 1965). According to the multiple locus theory, mottling is attributed to yet another closely linked locus—or possibly two further loci (Wallace, 1965).

Mutations to light (yellow) belly are much more frequent than mutations to other agouti locus genotypes both in the laboratory (Isherwood et al., 1960; Dickie, 1969a)[14] and in the wild (Wallace, 1954) and the advocates of the pseudoallelism theory believe that this too is consonant with their position. They argue that it is more likely that a locus closely linked to the *a* locus has a much higher mutation rate than to suppose that some agouti locus alleles differ so drastically from the others in terms of their stability.

Supporters of the complex locus hypothesis believe also that the occurrence of certain phenotypes from known matings may represent crossovers rather than gene mutations. For example, Wallace (1954) reported that a nonagouti (a/a) mouse was produced from an A/a^t ♀ × a/a ♂ mating. While it is conceivable that a mutation to a occurred from A or a^t, it is also possible that the unexpected phenotype resulted from a crossover. If yellow belly is designated as W, a gene closely linked to a, and nonyellow belly as w, its recessive allele, then Wallace's mating can be designated Aw/aW ♀ × aw/aw ♂ and a simple crossover could yield an aw ovum (see Keeler, 1931; Grüneberg, 1966a).

Finally, the advocates for the complex locus hypothesis believe that evolutionary considerations support their contention. They find it difficult to understand why if one locus is involved, two different genotypes, A/a^t and $A^w/-$, should have as a result of selection the same phenotype (Wallace, 1954).

In spite of all the evidence supporting the complex locus theory, there are some observations which argue against it. On numerous occasions a/a matings have produced $A^w/-$ offspring (Little and Hummel, 1947; Bhat, 1949; Hoecker, 1950; Dickie, 1969a) (see note 13). While according to the simple locus concept this merely requires a single mutation, albeit one which "skips a step," i.e., $a \rightarrow A^w$ (skipping a^t), the complex locus interpretation requires two simultaneous mutations, one from $a \rightarrow A$ and the other from $w \rightarrow W$. Nevertheless, two such simultaneous mutations in a single animal have been documented (Dickie, 1969)[15] and one could conceive that if two loci are involved, they are so closely associated that a mutating agent affecting one would frequently affect the other (Wallace, 1954). Still more difficult to reconcile with the two locus hypothesis is R.J.S. Phillips' (1966a) observation that A^s affects the belly of A^w mice differently than it does the ventrum of a^t animals; the belly of A^sA^w homozygotes is a dark agouti whereas the ventrum of A^sa^t homozygotes is nonagouti. According to the complex locus interpretation, this would indicate that the yellow belly of A^w and a^t mice are not equivalent and would necessitate a locus for banding and a locus for yellow belly rather than one for dorsal and one for ventral banding (R.J.S. Phillips, 1966a).

Any hypothesis involving two loci would also be expected to produce some phenotypes with the ventrum darker than the dorsum or with plain hair on the dorsum and banded hair on the belly, and these have never been observed. Indeed, the relationship of back, belly, and ears is such that if the ears are dark, so is the belly and if the ventrum is dark, so is the dorsum (R.J.S. Phillips, 1966a).

Finally it should be noted that the mutation to extreme nonagouti first occurred in an agouti stock (Hollander and Gowen, 1956) which, according to the complex locus theory, would require at least three distinct mutations: one affecting the dorsum, one affecting the ventrum, and one affecting the ears![16]

Clearly, the structure of the agouti locus remains to be resolved but it probably will not be until the specific biochemical processes involved in producing the various agouti locus phenotypes are determined.

C. Eumelanin and Phaeomelanin: Structure, Ultrastructure, and Biosynthesis

Regardless of the chromosomal structure of the agouti locus it is apparent that it determines whether eumelanin, phaeomelanin, or both of these pigments are synthesized in the melanocytes of the hair bulb. Moreover, it should be emphasized that these two kinds of pigment are appreciably different from each other when examined under the electron (Moyer, 1966; Sakurai et al., 1975) or conventional microscope (E. Russell, 1946, 1948, 1949a,b). When viewed under the latter, yellow granules from either $A^y/-$ or $A/-$ hairs are uniformly round and of approximately the same intensity and size, whereas black granules are much more variable in their characteristics (see Figure 4-6) with three different grades of color, four recognizably different shapes, and a wide variation in size (see Table 2-3) (E. Russell, 1949a). The total pigment volume in black hairs also is about five times higher than in yellow hairs (E. Russell, 1948) and whereas eumelanin is insoluble in almost all solvents and resistant to chemical treatment, phaeomel-

Table 2-3. Granule Properties of Phaeomelanin and Eumelanins[a]

Attribute	Color series		
	Yellow	Black-fuscous	Brown
Total pigment volume, highest	281	1332	480
Upper limit, medullary number	47 ± 2	98 ± 2	95 ± 4
Upper limit, cortical number	39 ± 7	219 ± 7	160 ± 8
Color intensity	Aniline yellow	Black, black-fuscous, fuscous	Carob brown Mummy brown
Shape	Round	Long-oval, oval, round, shred	Round, shred
Mean greater diameter (μm)	0.83–0.66	1.44–0.62	0.98–0.61

[a]From E.S. Russell (1949a).

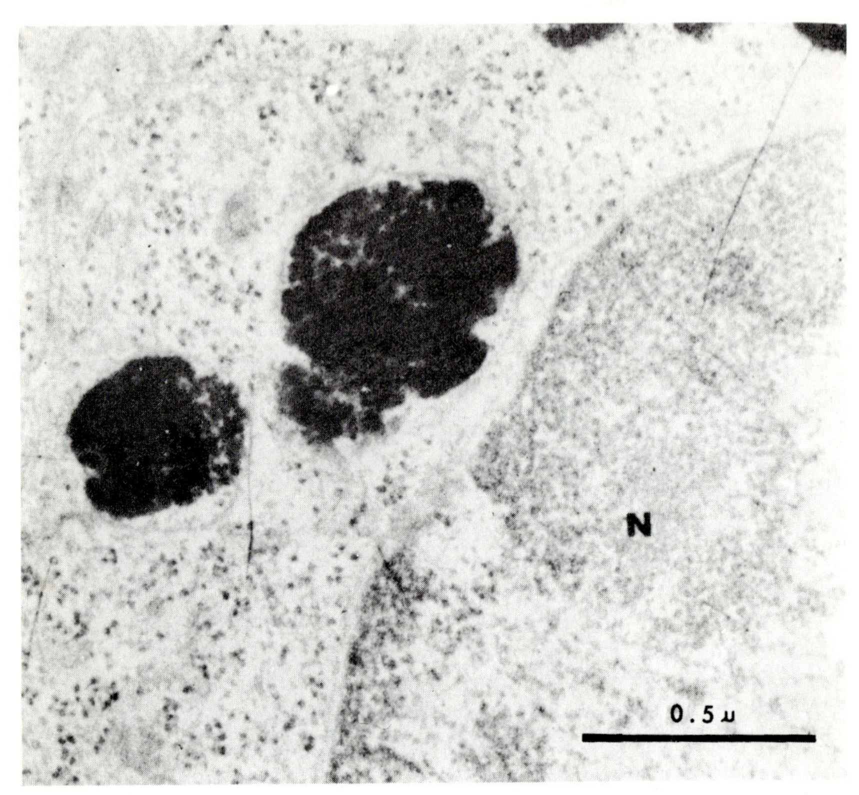

Figure 2-2. Melanin granules in a melanocyte from a hair follicle of a yellow mouse (C57BL/6–A^y). Note the lack of an organized matrix and the incomplete "unit membranes" bounding the granules. The nucleus of the melanocyte is indicated by N. From F.H. Moyer (1966). Reproduced with permission of the *American Zoologist*.

anin is soluble in dilute alkali (Jimbow et al., 1976; see also Ikejima and Takeuchi, 1978).

When examined under the electron microscope, yellow and black granules again display striking differences which suggest that the ontogeny of yellow granules is quite distinct from granules producing eumelanin (Moyer, 1966; Sakurai et al., 1975). One characteristic feature of the ultrastructure of yellow granules is that they do not possess any organized matrix (Figure 2-2) so that although pigment is laid down in discrete areas, it is deposited randomly on a tangled mat of exceedingly fine fibers which are almost translucent to the electron beam (Moyer, 1966). According to Moyer (1966), "there is no ordered aggregation of fibers nor is there any organized cross-linking. As the phaeomelanin accumulates, the areas of deposition spread and fuse until finally the granule assumes a very dense homogeneous appearance. Very occasionally the areas of melanin deposition in the granules are found in linear arrays reminiscent of intermediate stages in the ontogeny of eumelanin granules [see Chapter 3; Section I, E and F], but even then no organized matrix is visible."

Of particular relevance is the ultrastructural study by Sakurai and his associates (1975) on *A*/*A* melanocytes. These investigators observed that hair follicle melanocytes in transition from black to yellow possess *both* eumelanosomes and phaeomelanosomes (Figure 2-3). This not only consti-

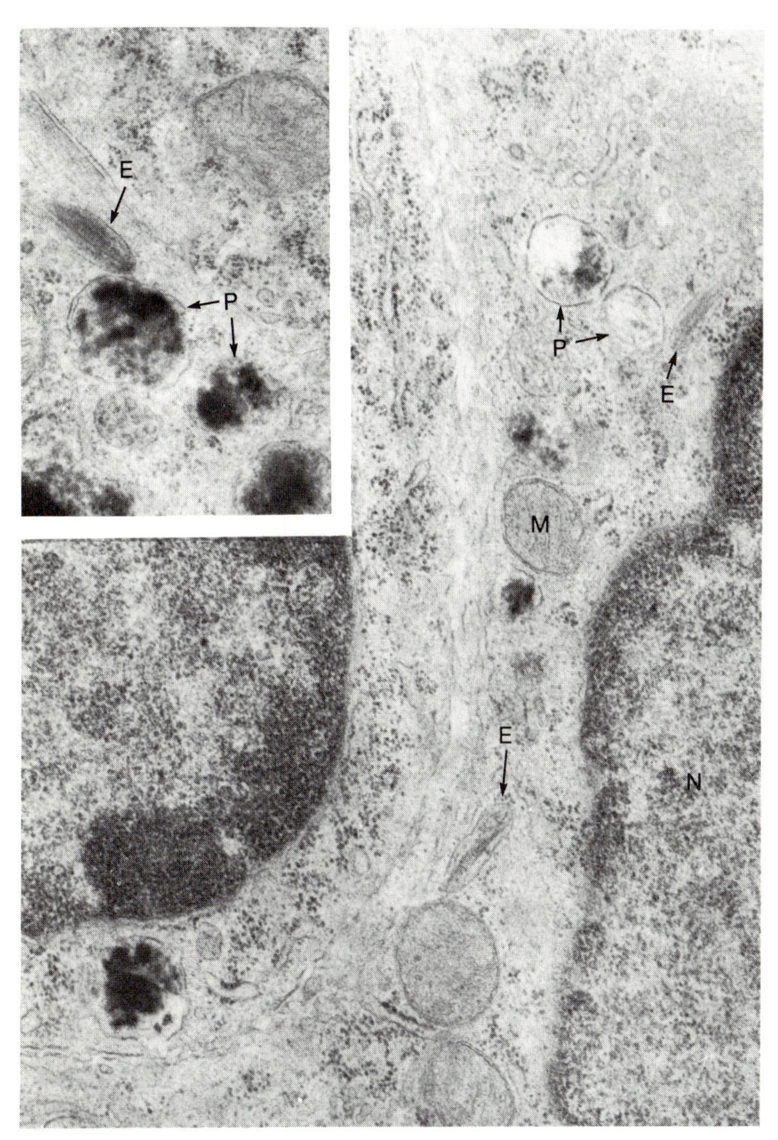

Figure 2-3. Eumelanosomes (E) and phaeomelanosomes (P) occurring within the same hair bulb melanocytes of a 3-day-old agouti mouse. N; nucleus. M; mitochondrian. Approx., ×28,000. From T. Sakurai et al., 1975. Reproduced with permission of the authors and Academic Press Inc.

tutes the best evidence to date that the shift from eumelanin to phaeomelanin synthesis, or vice versa, occurs within a single cell, but it indicates that one population of cells is responsible for all the patterns of pigmentation associated with the *a*-locus (see also Galbraith, 1964; Geschwind et al., 1972).

In regard to the biosynthesis of these two melanins (Figure 2-4), the prevailing contention has been that eumelanin is a protein conjugate formed by the coupling of a quinonoid polymer, indole-5,6-quinone, with protein to form eumelanin granules. Tyrosine is the natural precursor of eumelanin, and a single, copper-containing, enzyme complex, tyrosinase, is involved in the first two steps of the conversion of this colorless amino acid to melanin: step one, the hydroxylation of tyrosine to 3,4-dihydroxyphenylalanine (dopa) and step two, the oxidation of dopa to dopaquinone. According to this scheme, eumelanin is a homopolymer of indole-5,6-quinone units, formed from intramolecular rearrangements of oxidative products of dopaquinone, linked through a single bond type (Fitzpatrick and Lerner, 1954; Lerner, 1955; Fitzpatrick et al., 1958; Lerner and Case, 1959; Fitzpatrick and Kukita, 1959; Foster, 1965). However, there is also some evidence that eumelanin may actually be a heteropolymer, or a random polymer derived from the linkage of many different indoles, including 5,6-dihydroxyindole (Nicolaus and Piattelli, 1965), or that it is a highly irregular, three-dimensional polymer composed of several types of monomers joined by different covalent bonds (Blois et al., 1964; Jimbow et al., 1976).

Although at one time there was evidence to suggest that phaeomelanin might be derived from tryptophan, i.e., that an ortho-aminophenol derived from tryptophan might be oxidized by dopaquinone to produce a yellow pigment under the influence of a genetically controlled switch mechanism (Fitzpatrick et al., 1958; Fitzpatrick and Kukita, 1959; Foster, 1965), yellow pigment, too, is almost certainly a product of tyrosine metabolism.

Support for the notion that tryptophan might be involved in phaeomelanin synthesis stemmed primarily from the efforts of Foster (1951) who found that oxygen uptake in pulverized yellow skin was stimulated considerably when it was incubated with tryptophan (see also Nachmias, 1959). Moreover, the production of a yellow pigment accompanied this reaction. While it was known that incubating yellow skin in tyrosine produced a black pigment, it was believed that yellow skin possessed an inhibitor of tyrosinase since when *a*/*a* skin was mixed with A^y/a skin the latter appeared to inhibit the tyrosinase activity normally present in the *a*/*a* tissue (Foster, 1951). Nevertheless, all attempts to demonstrate directly the formation of yellow pigment from tryptophan have failed. Thus while Markert (1955) and Coleman (1962) were unable to show that ^{14}C-labeled tryptophan was incorporated into melanin, these investigators found that ^{14}C-labeled tyrosine was included. The study by Coleman (1962) is particularly pertinent since he employed mice of various genotypes. He observed that if labeled tryptophan was injected into yellow, black, or brown baby mice it was not incorporated into any of these pigments, whereas labeled tyrosine occurred in all

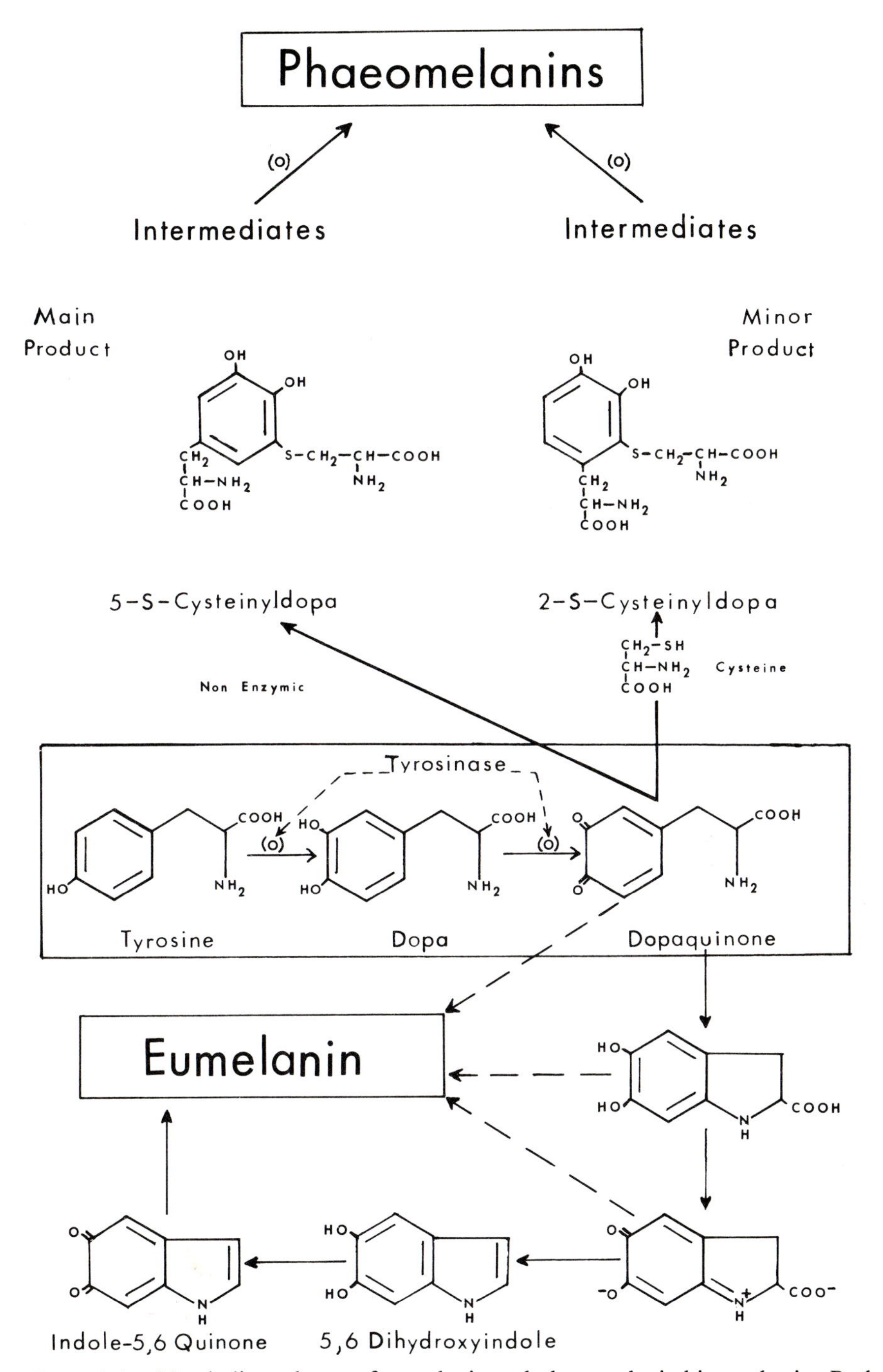

Figure 2-4. Metabolic pathway of eumelanin and phaeomelanin biosynthesis. Both begin with the enzymatic (tyrosinase) conversion of tyrosine to dopa to dopaquinone. Adapted from K. Jimbow et al. (1976).

of them, albeit at a somewhat reduced rate in yellow, as compared with black and brown skin. Coleman observed also that skin slices prepared from yellow animals likewise failed to incorporate [^{14}C]tryptophan, but did incorporate [^{14}C]tyrosine at about one-third the normal rate observed for nonagouti skin (see Table 3-1).[17] These results strongly imply that tryptophan plays no role in the synthesis of melanin and that tyrosine is the common precursor of both eumelanin and phaeomelanin. The fact that tyrosine is incorporated at a reduced rate into yellow pigment may indicate that the normal sequence of events which lead to the synthesis of eumelanin by hair bulb melanocytes is permanently (e.g., in the case of A^y) or periodically (e.g., in the case of A) interrupted, or somehow diverted, resulting in a smaller polymer with altered physical properties (Wolfe and Coleman, 1966). Indeed, the little that is known about the synthesis of phaeomelanin suggests that it is formed by a modification of the eumelanin pathway, involving the interaction of cysteine with dopaquinone (Jimbow et al., 1976).[18]

D. Role of Sulfhydryls in the Expression of Agouti Locus Alleles

In 1963 Cleffmann reported that if pieces of skin from late embryos or from young mice of different agouti locus constitutions, including $A^y/-$, were grown *in vitro*, hair growth continued and *eumelanin* was produced. From this Cleffmann concluded that all melanocytes synthesize eumelanin under standard *in vitro* conditions. However, if sulfhydryl (SH) compounds such as glutathione were added to the nutrient medium all melanocytes, regardless of their age or genotype (provided of course that they were not from albino donors), could be induced to produce phaeomelanin. The minimum concentration of SH compounds which were required to induce the *in vitro* synthesis of phaeomelanin varied with the agouti locus genotype of the explant. a/a and $a^t/-$ pigment cells required a significantly higher concentration of SH compounds throughout their entire period *in vitro* than $A^y/-$ melanocytes. On the other hand, $A^w/-$ and $A/-$ melanocytes displayed cyclic variations in their requirements of SH compounds. During that stage of the hair growth cycle when eumelanin was normally synthesized, a high level of SH compounds, equivalent to that required by a/a cells, was necessary to induce phaeomelanin synthesis, whereas during the period when the yellow band was being formed significantly fewer SH compounds were needed. Even more intriguing was the fact that Cleffmann claimed that the *in vitro* synthesis of phaeomelanin was not confined to the melanocytes of the hair bulb. The same *in vitro* conditions which promoted yellow pigment production in the hair follicle induced dermal melanocytes to synthesize phaeomelanin. This was suprising because this is *not* the case *in vivo*. Thus $A^y/-$ mice produce phaeomelanin *only* in a hair bulb environment; in other regions of the animal where melanin is produced, e.g., the eye, ear, tail, harderian gland etc., only eumelanin occurs (Markert and Silvers, 1956). Cleffmann also observed that pigment cells which are synthesizing phaeomelanin take up more SH compounds and less melanin precursors ([^{14}C]tyrosine, [^{14}C]dopa) than eumelanin-producing cells and that these SH compounds are incorporated to a greater extent into phaeomelanin than into eumelanin (Cleffmann, 1964). Because these results helped account for how the agouti locus operates, i.e., it somehow controlled the amount of SH compounds in the melanocyte, a number of hypotheses on *a*-locus gene expression have been based upon these findings (Foster, 1965; R.J.S. Phillips, 1966a; G. Wolff, 1963, 1971; Geschwind et al., 1972). Unfortunately, however, all of these hypotheses must be considered with caution since the original observations of Cleffmann await confirmation. Thus Knisely and his associates (1975) observed that neither A^y/a explants synthesized eumelanin under standard *in vitro* conditions, nor did A/A explants, in the eumelanin phase of pigment production, synthesize phaeomelanin when cultured in a concentration of glutathione which, according to Cleffmann (1963), should have induced them to do so.[19] Galbraith and Patrignani (1976) have likewise failed to corroborate Cleffmann's (1964) contention that yellow and black melanocytes differ according to their SH metabolism. They found that "yellow and black melanocytes, regardless of genotype, possess equivalent amounts of histochemically detectable sulfhydryl compounds."

E. Action of Genes at the Agouti Locus

The fact that in $A^y/-$ mice phaeomelanin appears to occur only in hair follicles (see Silvers, 1957),[20] as well as the observation that yellow and black granules occur in the same cell in A/A hair bulbs (Galbraith, 1964; Geschwind et al., 1972; Sakurai et al., 1975), strongly suggest that the genes at the agouti locus act via the follicular environment. However, these findings do not constitute proof that this is the case. Proof requires evaluating the behavior of melanocytes of one agouti-locus genotype in hair follicles comprised of cells of a different agouti-locus constitution. Fortunately this situation can be achieved experimentally by transplanting histocompatible skin from near term or newborn donors to newborn recipients (Silvers, 1963). This procedure takes advantage of the fact that when such grafts are made to neonatal recipients, some host melanoblasts migrate across the graft boundary and become established in the developing hair follicles of the graft (Reed and Sander, 1937; Reed, 1938, Reed and Henderson, 1940). Thus employing this technique it was found that when intensely pigmented black mice of the genotype $a/a;C/c^e$ (produced by crossing mice of an inbred color stock, $A^y/a;c^e/c^e$ with C57BL/6, $a/a;C/C$ animals) were grafted shortly after birth with neonatal skin from black-eyed white mice of the genotype $A^y/a;c^e/c^e$, some intensely colored *yellow* hairs appeared among the nonpigmented hairs within the border of the graft. Conversely, when neonatal $a/a;c^e/c^e$ skin was transplanted to yellow-pigmented ($A^y/a;C/c^e$) recipients, some intensely colored *black* hairs were found among the predominantly pale brown-pigmented hairs of the ($a/a;c^e/c^e$) graft (Silvers and E. Russell, 1955). Other experiments, utilizing the same technique, demonstrated that when agouti (but albino) skin was transplanted to either intensely pigmented yellow or black recipients, some hairs pigmented with the typical agouti pattern, i.e., black with a yellow band, arose among the albino hairs of the graft (Silvers, 1958b).

While all of these results were consistent with the hypothesis that it was the agouti locus constitution of the receiving hair follicle which determined the kind of melanin synthesized by the invading cell, nevertheless, because all the grafts, including those which were nonpigmented, possessed indigenous populations of melanocytes, one had to rule out the possibility that these indigenous populations were not being "turned on" by some chemical infiltrate of host origin. This possibility was eliminated by repeating some of these experiments with grafts from white spots. Inasmuch as white-spotting results from an absence of melanocytes (see Chapter 3, Section II, A), such grafts do not possess any indigenous population that can be "turned on." The results employing these grafts not only confirmed the previous findings (Silvers and E. Russell, 1955; Silvers, 1958a,b) but they indicated that the number of pigmented hairs within the border of "white spots" was usually greater than in transplants known to contain an indigenous melanoblast population. Evidently, in the latter case the melanoblasts of host origin have to compete with those of the graft for occupancy in hair bulbs.

Additional experiments with $A^w/-$ and $a^t/-$ transplants demonstrated that the expression of *a*-locus genic activity depended not only upon the genotype of the follicular environment, but also upon the location of this environment on the integument. Thus when potentially intensely pigmented melanocytes of *any* agouti locus constitution invaded a dorsal, nonpigmented, a^t/a^t graft they produced intensely pigmented black hairs, whereas when these cells were introduced into ventral follicles of the same genotype, some all-yellow hairs as well as some yellow hairs with black bases (characteristic of the ventral hairs of many intensely pigmented black-and-tan genotypes) resulted (Silvers, 1958b).[21] These observations, together with the finding that $a^t/-$ and $A^w/-$ melanocytes were likewise able to respond completely to the agouti locus genotype of the receiving hair follicle, implied that in $a^t/-$ and $A^w/-$ mice ventrality and dorsality of location were not important per se but that, in conjunction with their genotype, they presented different follicular environments which influenced the expression of the melanocyte (Silvers, 1961). The results of these transplantation experiments are summarized in Table 2-4.[22]

Once it was established that the genes at the agouti locus produce their effect by altering the follicular milieu the stage was set to locate more precisely the activity of these alleles. Inasmuch as hair follicles have both an epidermal and dermal component—the epidermal ectoderm forms the epithelial

Table 2-4. Summary of Results following Incorporation of Melanocytes of One Agouti Locus Genotype into Dorsal (D) or Ventral (V) Hair Follicles of a Different Agouti Locus Genotype

Genotype of host	Phenotype of host	Genotype of graft	Phenotype of graft	Tract origin of graft	Colors of graft hairs pigmented by host melanocytes
$a/a;C/c^e$	Black	$A^y/a;c^e/c^e$	White	D or V	Yellow
$A^y/a;C/c^e$	Yellow	$a/a;c^e/c^e$	Pale Brown	D or V	Black
$a^t/a;C/c^e$	Black-and-tan	$A^y/a;c^e/c^e$	White	D or V	Yellow
$a^t/a;C/c^e$	Black-and-tan	$a/a;c^e/c^e$	Pale brown	D or V	Black
$A^w/a;C/c^e$	Yellow-bellied-agouti	$A^y/a;c^e/c^e$	White	D or V	Yellow
$A^w/a;C/c^e$	Yellow-bellied-agouti	$a/a;c^e/c^e$	Pale brown	D or V	Black
$a/a;C/C$	Black	$a^t/a^t;Mi^{wh}/Mi^{wh}$	White[a]	D	Black
$a/a;C/C$	Black	$a^t/a^t;Mi^{wh}/Mi^{wh}$	White[a]	V	Yellow and black[b]
$A^y/a;C/C$	Yellow	$a^t/a^t;Mi^{wh}/Mi^{wh}$	White[a]	D	Black
$A^y/a;C/C$	Yellow	$a^t/a^t;Mi^{wh}/Mi^{wh}$	White[a]	V	Yellow and black[b]
$a/a;C/c$	Black	$A^w/A^w;c/c$	White	D	Agouti and some black[c]
$a/a;C/c$	Black	$A^w/A^w:c/c$	White	V	Yellow with black base[d]
$A^y/A^w;C/c$	Yellow	$A^w/A^w;c/c$	White	D	Agouti and some black[c]
$A^y/A^w;C/c$	Yellow	$A^w/A^w;c/c$	White	V	Yellow and black[d]

[a]Absence of melanocytes.
[b]Both yellow hairs with black bases and some all-yellow hairs occurred. These are characteristic of the hairs found on the ventrum of many a^t/a^t mice.
[c]Characteristic of the hairs found on the dorsum of A^w/A^w mice.
[d]Characteristic of the hairs found on the ventrum of many A^w/A^w mice.

portion of the follicle and the mesoderm the dermal papilla—attention was focused on which of these components was responsible for the agouti pattern. Mayer and Fishbane (1972) and Poole (1974) treated small pieces of skin, derived from the side of the trunk of 13- to 17-day-old a/a and A^w/A^w (or A/A) embryos, with trypsin so that the epidermis and dermis could be separated and then recombined and allowed to differentiate in the testis of a histocompatible host or in a chick embryo. They found that when a/a dermis was recombined with A^w/A^w (or A/A) epidermis the hairs which differentiated did *not* display the agouti pattern. On the other hand, the reciprocal combination of agouti dermis and nonagouti epidermis always resulted in typical agouti hairs. These findings indicated that the genotype of the epidermal (ectodermal) component had no influence on the kind of pigment synthesized and that *a*-locus activity was mediated via the dermis. This conclusion had to be revised, however, when it was observed that recombinations of A^y/a epidermis with $+/+$, a/a, or a^e/a^e dermis produced hairs completely pigmented with phaeomelanin (Poole, 1974, 1975). Indeed, yellow pigmented hairs were produced even in recombinations of "young" (13 day) embryonic yellow epidermis with "old" (17 day) a/a or a^e/a^e dermis: dermis in which dermal papillae were already present.

Further investigations demonstrated that only in $A^y/-$ skin does *both* the epidermis and dermis induce the synthesis of phaeomelanin.[23] Thus the regional pigmentation pattern displayed by black-and-tan mice, like the agouti pattern, is determined by the dermis regardless of whether it is of ventral or dorsal origin (Poole and Silvers, 1976a). Experimental recombinations of 14- to 16-day-old embryonic a^t/a^t ventral dermis and a^t/a^t dorsal epidermis give rise to yellow-pigmented hairs, whereas the reciprocal combination of dorsal dermis and ventral epidermis develop hairs pigmented with eumelanin. Black hairs are likewise produced when a^t/a^t dorsal dermis is combined with a/a ventral epidermis or when a/a ventral dermis is associated with a^t/a^t ventral epidermis. This latter finding is especially revealing because it demonstrates that the epidermis of a^ta^t ventral skin, unlike that of A^y/a skin, does not promote the synthesis of phaeomelanin (Poole and Silvers, 1976a). It appears, therefore, that although yellow and black-and-tan animals have similar ventral phenotypes, only the A^y allele can act via the dermal and epidermal components of the follicle to induce the formation of yellow hairs. A summary of the pigmentation of hairs obtained by recombining dermis and epidermis between a^t/a^t, a/a dorsal and ventral, and A^y/a ventral embryonic skin is presented in Table 2-5. Obviously the combination of *both* dermal ventrality and the a^t allele is necessary for the development of the black-and-tan ventral phenotype.

In spite of all the effort which has been directed toward investigating the action of genes at the agouti locus, the specific effect of this locus remains unknown. It was once proposed that the banding of agouti hairs was caused by the rate of hair growth, i.e., an elevated mitotic rate in the follicular ma-

Table 2-5. Pigmentation of Hairs Obtained by Recombining Dermis and Epidermis between Black-and-Tan (a^t/a^t), Nonagouti (a/a) Dorsal and Ventral, and Yellow (A^y/a) Ventral Embryonic Skin

Type of recombination		
Dermis	Epidermis	Pigmentation of hair
a^t/a^t ventral	a^t/a^t dorsal	Yellow
a^t/a^t dorsal	a^t/a^t ventral	Black
a/a ventral	a^t/a^t dorsal	Black
a^t/a^t dorsal	a/a ventral	Black
a/a ventral	a^t/a^t ventral	Black
a^t/a^t ventral	a/a ventral	Yellow
a/a dorsal	A^y/a ventral	Yellow
a^t/a^t dorsal	A^y/a ventral	Yellow

trix somehow promoted phaeomelanin synthesis (Cleffmann, 1953, 1960, 1963). While this possibility was consistent with the observation that the agouti pattern could be altered by the administration of the mitotic inhibitor colchicine during the period of yellow band formation (Galbraith, 1964), as well as with the fact that there was an association between the initiation of yellow banding in A/A mice with an increased mitotic rate in their hair follicles, this notion has had to be discarded. Thus the very meticulous analysis by Galbraith (1971) has revealed clearly that a/a hair bulbs (as well as A^y/a bulbs) display an increased mitotic rate at the same phase of hair growth when the yellow band is forming in A/A animals. Indeed, genetically yellow and genetically black bulbs possess virtually identical mitotic rates at all stages of hair growth. Moreover, if there was a causal relationship between increased mitotic activity and yellow pigment production one would expect the cessation of phaeomelanin production in A/A hair bulbs to be associated with a *decreased* mitotic rate and this, too, does not occur. In fact, the rate of cell division during the return to eumelanin synthesis is, in most instances, higher than during band formation (Galbraith, 1971).[24]

Galbraith and Arceci (1974) have also shown that A^y/a and a/a hair bulbs do not differ with respect to their melanocyte populations and so it is most unlikely that a fluctuation in the number of pigment cells during hair growth is responsible for the agouti pattern.[25] Actually the only factor which has been shown to be related to the synthesis of phaeomelanin in agouti mice is bulb mass. As A/A hair bulbs increase in size there is a loss in the capacity of their melanocyte populations to synthesize yellow pigment; no phaeomelanin is produced by A/A melanocytes in hair bulbs that exceed 110 μm in diameter (Galbraith, 1969). Thus neither guard hairs nor the vibrissae show the agouti pattern.

While, as the above testifies, an abundant amount of information has been collected on the action of *a*-locus genes, the crucial experiments defining the primary action of this locus have yet to be reported.

II. The Extension Series of Alleles

It has been known for some time that there are a number of mammals (e.g., rabbit, guinea pig, dog) which have a series of alleles that extend or restrict the amount of eumelanin, with an opposite effect on phaeomelanin (see Searle, 1968b). It is therefore not surprising that the mouse has a comparable series. What is surprising, however, is that this has only relatively recently been established. Thus, while as long ago as 1912 Hagedoorn reported a recessive mutation that caused yellow coat color, because the strain carrying this mutant soon became extinct, it was not until sombre was described in 1961, and recessive yellow was reported again in 1968, that an "extension series of alleles" for the mouse finally was confirmed. This series of alleles is now known to be located on chromosome 8 (Falconer and Isaacson, 1962).

A. Sombre (E^{so})

1. Origin and Influence on Pigmentation

This dominant mutation[26] has an interesting history in that it appeared within 3 days in two unrelated C3H litters, in each case from dams that had been drenched with dibutylphthalate, an antiectoparasite agent. Indeed, because of this situation it initially was suspected that these deviants represented "phenocopies" and not true mutations (N. Bateman, 1961).

Aside from the fact that on the C3H (A/A) background on which E^{so} occurred E^{so}/E heterozygotes have dark ears and mammae, they resemble nonagouti (a/a) mice. Often when nonagoutis are mature their flanks are flecked with yellowtipped hairs and their ventrums are grey. These same features occur in $A/A;E^{so}/E$ mice where, according to Bateman, they are even more conspicuous. Moreover, both $a/a;E/E$ and $A/A;E^{so}/E$ mice have considerable numbers of "yellowed" perineal hairs right from their first pelage.

$A/A;E^{so}/E^{so}$ mice are strikingly black (Plate 1–G) and are easily recognized from E^{so} heterozygotes from as early as 12 days of age. However, at least on the C3H background, they fall short of being *completely* black, as they display a few "yellowed hairs" on the perineum. In fact, as Bateman notes, it is only this feature which spoils their perfect mimicry of extreme nonagouti (a^e/a^e). Viability and fertility are normal in both heterozygotes and homozygotes.

2. Interaction with Other Coat Color Determinants

Insofar as the interaction of E^{so} with other mutants is concerned, Bateman remarks that $A^{w}/A;E^{so}/E$ heterozygotes, unlike $A^{w}/A;E/E$ animals, do not have light bellies; presumably they are indistinguishable from $A/A;E^{so}/E$ genotypes.

G. Wolff and his colleagues (1978) have compared viable sombre ($A^{vy}/A;E^{so}/E$) with agouti sombre ($A/a^{e};E^{so}/E$) mice and although they found both of these genotypes to be black[27] they nevertheless could usually be distinguished on the basis of body weight and ventral pigmentation. Sombre A^{vy} males reached a mean body weight of 44.2 $\pm$0.6 g at 18 weeks, whereas sombre A males had a mean weight of 30.7 $\pm$ 0.4 g at the same age. It thus appears that E^{so} is epistatic to A^{vy} only with regard to coat color.[28] This is important because it indicates that the abnormally regulated synthesis and/or deposition of fat caused by A^{vy} (and A^{y}) is in no way related to phaeomelanin synthesis (G. Wolff et al., 1978).

With regard to the ventral pigmentation of $A^{vy}/A;E^{so}/E$ and $A/a^{e};E^{so}/E$ animals Wolff and his associates observed that, in general, it was more dilute in mice of the former genotype. Microscopic examination of plucked hair samples showed that the most noticeable difference occurred in the monotrichs. In agouti sombre mice these hairs are heavily and uniformly pigmented throughout their length, whereas those of viable yellow sombre animals are not heavily pigmented at the tip and tend to display over their entire length a somewhat uneven distribution of eumelanin granules. The bases of these monotrichs also frequently lack pigmentation.[29]

The coat of both viable yellow sombre and agouti sombre mice appear to possess a faint yellow cast in certain regions of the body. These include the ventral neck region, the chest, the femoral region of the forelimb, the base of the ear, and the perianal region. Nevertheless, microscopic examination of plucked hairs from these areas failed to reveal any phaeomelanin granules, the hairs being either devoid of pigment or dilute black. Since hairs from these various regions are much finer than those of the main pelage, Wolff and his associates believe that this property, in combination with their dilutely pigmented (or unpigmented) condition, is responsible for their faint yellow appearance.

One of the most interesting observations in this study was that many, but not all, zigzags from agouti sombre (and probably to a lesser extent from viable yellow sombre) mice lack pigment or contain relatively little in exactly the same region which is pigmented yellow in $A/-;E/E$ animals, i.e., in the apical region of the shaft which precedes the first constriction. Consequently, although no yellow pigment is detected in these mice, the sparse pigmentation which takes its place yields a pattern similar to the agouti pattern.[30] This "banding" pattern is important because it indicates that eumelanin synthesis is inhibited at definite periods during hair growth regardless of whether phaeomelanin synthesis occurs or not (G. Wolff et al., 1978).

B. Tobacco Darkening (E^{tob})

1. Origin and Influence on Coat Color

This allele of the extension series received its name from the fact that it originated in the tobacco-mouse (*Mus poschiavinus* Fatio),[31] where it is responsible for darkening of the coat. According to von Lehmann (1973), as a consequence of this gene young ($A/A;E^{tob}/E^{tob}$) tobacco-mice have a nonagouti back and "cannot be distinguished from black mice until the 8th week; later the flanks become agouti." When tobacco mice are crossed with nonagouti black ($a/a;B/B$) house mice the F_1 ($A/a;E^{tob}/E$) display a wild type pelage but the dorsum may be darkened. Crosses to yellow (A^y) or agouti (A) house mice produce yellow or agouti mice with darkened backs, though less so with agouti. The expression of E^{tob} in subsequent brother × sister generations derived from the (house × tobacco mouse)F_1 indicates that, when homozygous, E^{tob} is epistatic over nonagouti. It therefore appears that E^{tob}, like E^{so}, is incompletely dominant over E. This is substantiated further by the observation that when $A/a;E^{tob}/E$ animals are backcrossed to tobacco mice ($A/A;E^{tob}/E^{tob}$) half of the offspring are dark ($A/-;E^{tob}/E^{tob}$) and half are grey with some darkening of the back ($A/-;E^{tob}/E$) (von Lehmann, 1973). In a subsequent report von Lehmann (1974) presents evidence that tobacco darkening is a member of the E-series.[32]

C. Recessive Yellow (e)

1. Origin and Influence on Coat Color

This mutation occurred in Hauschka's C57BL/6 subline and is very likely a recurrence of Hagedoorn's mutant. Although the coat of adult e/e mice is a clear yellow and, according to Hauschka and his associates (1968), somewhat less orange than the coat of $A^y/a;E/E$ animals of the YBR/Wi strain, prior to weaning it displays some "dorsally concentrated umbrous sootiness" which diminishes with successive molts. In fact, this sootiness is so intense that it is difficult to distinguish $a/a;e/e$ from $a/a;E/E$ when the hairs first emerge through the skin of infant animals (see Poole and Silvers, 1976b).[33] Like A^y/a mice, the eyes of $a/a;e/e$ animals are black.

2. Recessive Yellow and Lethal Yellow: Similarities and Differences

Histological examination of the yellow granules in the coat of e/e mice revealed that they were much the same as the granules in $A^y/-$ animals. Nevertheless, the cortex of e/e hairs appeared to contain significantly less pigment than the cortex of $A^y/-$ hairs (Hauschka et al., 1968).[34]

In both e/e and A^y/a mice phaeomelanin can be extracted from the hairs by the same treatment; immersion in 10% potassium hydroxide for 15 minutes (Hauschka et al., 1968). In both yellow genotypes all extrafollicular melanocytes also produce only *eumelanin*. Thus Lamoreux and Mayer

(1975) observed that the melanocytes located in the dermis of the skin, leg muscles, choroid, harderian gland, and meninges of the brain and spinal cord of young *e/e* mice all were black. Moreover, although exact counts of melanocytes were not made, these investigators estimated that their numbers were similar in these locations in *a/a;e/e* and *a/a;E/E* genotypes. The melanocytes of *e/e* ear skin likewise possess eumelanin but here, as in A^y/– ear skin, they are not, for some unknown reason, as numerous as in *a/a;E/E* ear skin (Poole and Silvers, unpublished).

Among the numerous differences between recessive and lethal yellow is the fact that mice homozygous for the former are viable and, unlike A^y*/a;E/E* animals, breed well, do not become obese, and, at least on the basis of current evidence, are no more susceptible to tumors than *a/a;E/*– genotypes. Furthermore, while leaden (*ln/ln*) has been reported to be epistatic to *e/e* (Hauschka et al., 1968), this is not the case with A^y/–. (For other differences see notes 33 and 35; Section II, D; Chapter 3, Section II, C; Chapter 4, note 9; and Chapter 12, Section III.)

As with sombre it appears that while *e/e* is epistatic to A^y/– in terms of its effect on pigmentation (Searle and Beechey, 1970), it probably does not interfere with the other manifestations of the lethal yellow allele. Thus while the coats of A^y*/a;e/e* mice are identical with those of *a/a;e/e* animals they become as obese as A^y*/a;E/*– genotypes (Lamoreux, 1973).[35]

3. Site of Gene Action

Inasmuch as recessive yellow is the only coat-color determinant besides those of the agouti series which promotes the synthesis of phaeomelanin, it was important to determine its site of action. Does it act via the follicular milieu, as is the case for the agouti series of alleles, or does it act within the melanocyte? Two independent investigations have been carried out to answer this question and the results of both are in accord with the premise that it is the *e/e* genotype of the melanoblast which determines its phaeomelanin-synthesizing activity; nevertheless, these cells produce yellow pigment only in a hair follicle environment.

In the first of these studies (Lamoreux and Mayer, 1975) grafts composed of 9-day-old neural tube, including neural crest cells, and 11-day-old neural crest-free embryonic skin were placed in the coelom of the chick and allowed to differentiate for 15 days. It was found that when *a/a;e/e* neural tube was grafted with *a/a;E/E* skin, the hairs of recovered grafts were pigmented yellow, whereas *a/a;e/e* skin produced black hairs when combined with *a/a;E/E* neural tube.

In a simultaneously conducted series of experiments, Poole and Silvers (1976b) determined the kind of pigment produced by various dermal–epidermal combinations of *e/e* and *E/E* embryonic skin when allowed to differentiate in the testes of adult hosts for 3 weeks. They found that when 13- to 15-day-old *a/a;E/E* epidermis (which possesses numerous melanoblasts) was combined with *a/a;e/e* dermis (which is either melanoblast free or contains very few melanoblasts), that the hairs which formed were completely black. On the other hand, in the reciprocal combination, consisting of black (*a/a;E/E*) dermis and recessive yellow (*a/a;e/e*) epidermis, all the hairs were pigmented in the recessive yellow pattern, i.e., they were composed predominantly of phaeomelanin with eumelanin at the tips. Poole and Silvers also determined the behavior of dermal–epidermal recombinants that differed at *both* the extension and agouti loci. This was accomplished by recombining recessive yellow (*a/a;e/e*) and agouti (*A/A;E/E*) dermis and epidermis. They found that when *a/a;e/e* dermis was combined with *A/A;E/E* epidermis all the hairs which developed were *black*, i.e., the *A/A;E/E* melanoblasts in the epidermis had responded to the *a/a* dermis. However, the reciprocal combination of *A/A;E/E* dermis and *a/a;e/e* epidermis developed hairs pigmented with the recessive yellow pattern (since in this case the hairs were pigmented by *a/a;e/e* melanocytes).

Since *a/a;e/e* melanocytes did not respond to any influence from either an *a/a* or *A/A* dermis, a final series of experiments was aimed at introducing these cells into nonagouti black (but nonpigmented) hair follicles. This was accomplished by combining 12-day-old embryonic recessive yellow dermis (which still possessed a population of melanoblasts) with *a/a;c/c* (albino) epidermis of the same age. Although, as a consequence of the fact that this epidermis too possessed melanoblasts, not all the hairs which developed in these grafts were pigmented, i.e., some were populated only by amelanotic melanocytes (see Chapter 10, Section I, D), nevertheless, those hairs which were pigmented all displayed the recessive yellow pattern. The results of these various dermal–epidermal recombinations are summarized in Table 2-6.

It therefore seems evident that even though *e/e* melanocytes produce

Table 2-6. Pigmentation of Hairs Obtained by Recombinations of Dermis and Epidermis between Recessive Yellow (*a/a;e/e*) and Black (*a/a;E/E*), Agouti (*A/A;E/E*), or Albino (*a/a;c/c;E/E*) Embryonic Skin

Type oı recombination		Age (days) of donor embryos	Pigmentation of hair
Dermis	Epidermis		
a/a;e/e	*a/a;E/E*[a]	13–15	Black
a/a;E/E	*a/a;e/e*[a]	13–15	Yellow
A/A;E/E	*a/a;e/e*[a]	13–15	Yellow
a/a;e/e	*A/A;E/E*[a]	13–15	Black
a/a;e/e[a]	*a/a;c/c;E/E*[a]	12	Yellow and white

[a]Origin of melanocytes.

yellow pigment only in hair follicles, recessive yellow, unlike the alleles at the agouti locus, acts autonomously within the melanoblast. What it is that is unique about the follicular environment which turns on phaeomelanin synthesis in *e/e* pigment cells is not known, but perhaps it is related to the fact that melanocytes in hair follicles synthesize melanosomes at a much more rapid rate, for inclusion into the growing hair, than extra-follicular cells (Lamoreux and Mayer, 1975).

D. The Effect of Melanocyte-Stimulating Hormone (MSH) on Phaeomelanin Synthesis

Geschwind and his colleagues (Geschwind, 1966; Geschwind and Huseby, 1966, 1972; Geschwind et al., 1972) have studied the effect of α-MSH on the coat color of various genotypes. These efforts stemmed from the observation that the coat of an agouti mouse which had been maintained on stilbestrol for 11 months, and which had developed a pituitary tumor, was much darker than normal. Transplantation of the tumor to other agouti mice also maintained on the same stilbestrol-containing diet likewise produced darkening of their coats and microscopic examination of dorsal hairs from these mice revealed that the yellow band had been obliterated completely. It was subsequently found that the injection of α-MSH into otherwise untreated agouti mice produced the same effect and attempts were made to modify the phenotypes of other genotypes either directly by raising circulating MSH levels with injections of the hormone or indirectly by transplanting the tumor. However, these treatments only influenced the expression of the agouti locus where the most obvious effect was to convert the pigmentation to the nonagouti type. Thus in *a/a* animals homozygous for either brown (*b*), albinism (*c*), dilute (*d*), or pink-eye (*p*), MSH had no effect.[36] Its most dramatic influence was on the coat of the A^y/– mouse; in regions which had been plucked 5–8 days previously or which were in active growth at the time of hormone treatment, only intensely black or brown colored hairs, in A^y/–;*B*/– and A^y/–;*b/b* genotypes, respectively, emerged (see Figures 2-5 and 2-6).[37] This influence of the hormone on phaeomelanin was also recognizable in *A*/–;*p/p* animals where normally the hairs are basically grey with subterminal yellow bands which gives an overall yellow appearance to the coat. Following the administration of MSH, however, growing hairs were totally grey and the coat was indistinguishable from *a/a*;*p/p* genotypes.

In spite of the fact that MSH converted phaeomelanin to eumelanin synthesis when the former resulted from the action of agouti locus alleles, this was *not* the case when the yellow pigment was a manifestation of recessive yellow (*e/e*). Thus the color of recessive yellow mice was completely unaltered either by MSH administration (Figure 2-5) or by the presence of the tumor. Recessive yellow mice heterozygous for lethal yellow, i.e., A^y/*a*;*e/e*, also continued to produce phaeomelanin when exposed to MSH. This

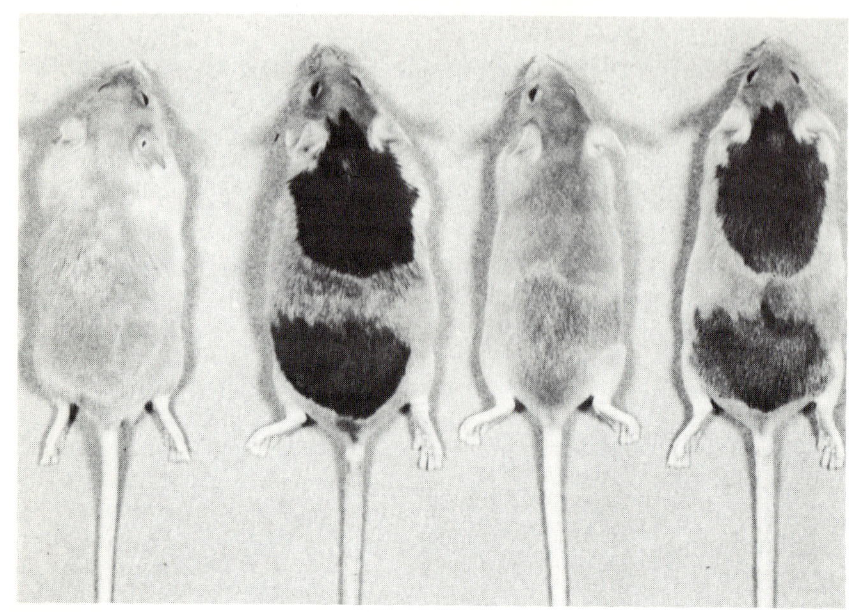

Figure 2-5. From left to right: recessive yellow (*e*/*e*) mouse which had been partially plucked, and 5 days later had received a series of 10 daily injections of MSH; lethal yellow (A^y) mouse similarly treated; A^y mouse which had been plucked but which received medium rather than MSH; A^y mouse similarly treated but which received MSH only on days 8 and 9 after plucking. All mice were 90 days old and were photographed 20 days after plucking. From I.I. Geschwind et al. (1972). Reproduced with the permission of the authors and Academic Press, Inc.

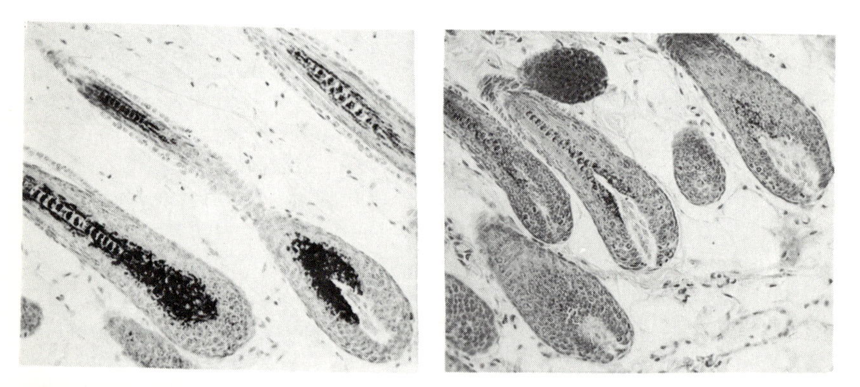

Figure 2-6. On the left is a histological section of the dorsal skin from an A^y mouse that had received a single inoculation of MSH 8 days after plucking. It was sacrificed 48 hours later. Note the presence of black pigment in the hair bulbs and shafts. On the right is a similar section taken from the dorsal skin of an A^y mouse that was similarly treated but received medium, alone. Although in this black and white photograph the pigment appears dark, it was yellow both grossly and when viewed through the microscope. Approx., ×150. From I.I. Geschwind et al., 1972. Reproduced with permission of the authors and Academic Press Inc.

provides further evidence that the pigmentation of these genotypes is identical with that of $a/a;e/e$ animals. Moreover, in contrast to the site of action of agouti locus alleles, since the E-locus acts within the melanoblast, it also indicates that the influence MSH has on converting phaeomelanin synthesis to eumelanin production is mediated via the follicular milieu.[38]

III. "Umbrous" and "Sable" Determinants

Although we have already considered sombre (E^{so}) and tobacco darkening (E^{tob}), alleles of the *extension locus* which cause darkening of the coat, there are a number of other unrelated determinants which also produce similar effects. Thus, as mentioned previously, yellow ($A^{y}/-;B/-$) mice are often characterized, especially in outbred populations, by variable degrees of sootiness. In some animals the sootiness is confined to a mid-dorsal streak, in others this streak is wider, covering the entire back and sometimes the flanks, so that only the belly is phenotypically "yellow." This situation is due to the admixture of hairs possessing significant amounts of eumelanin to the yellow fur. If these hairs are relatively few, the animal appears to be dingy yellow; with increasing numbers, however, the dorsum becomes darker until, at the extreme, an animal with a dark sable back and yellow belly is produced.[39] As in the case of E^{so} and E^{tob}, these modifications are not limited to animals of yellow genotype as the amount of phaeomelanin may also be drastically reduced in black agouti ($A/-;B/-$) and even in nonagouti animals. This dark form of agouti has been termed "umbrous"[40] (Barrows, 1934; Mather and North, 1940).

A. Polygenic Factors

The initial studies on the inheritance of factors which darken the coat color of mice were conducted more than 60 years ago. In 1916 Little attempted to alter the agouti pattern by crossing the darkest heterozygous agouti animals selected from a "sooty yellow" stock to nonagouti mice from the same stock. Both the agouti and nonagouti animals became darker as the selected crossing progressed (Little, 1916). These findings indicated that the darkly pigmented phenotypes resulted from the cumulative action of a number of genes and that high concentrations of these genes could make even nonagouti mice look darker. This was subsequently substantiated by Dunn (1920a) who demonstrated that the responsible factors were inherited independently of the agouti series. Dunn (1920a) also firmly established that both the sable and umbrous phenotypes had an essentially similar hereditary basis. If the genes responsible for producing the sable phenotype in $A^{y}/-$ mice were incorporated into $A/-$ animals, their backs were almost black, and agouti hairs with yellow tips were confined to the flanks and belly. Dunn's investigations also revealed that the expression of both umbrous and sable varied between wide limits and that the darker grades produced the more variable offspring. This indicated that the factors responsible for the darkening were either largely dominant in their effects or very strongly cumulative (Robinson, 1959).[41]

Barrows (1934) too studied the umbrous phenotype. Unlike Dunn, he concluded that it depended upon the presence of *a* in the heterozygous condition along with one or more dominant or semidominant modifying "umbrous genes" whose effect was to make nonagouti partially dominant to agouti.

B. Mendelian Mutations

1. Umbrous (*U*)

Although, as the above work testifies, sable and umbrous phenotypes often result from the cumulative effect of an unknown number of genes, as already noted (Section II) they can also be produced by single Mendelian factors. The first of these was described by Mather and North (1940) and was appropriately designated "umbrous" (*U*). Umbrous (linkage not known) is a semidominant which appears also to interact with *a*.

Contrary to usual belief, *A* is not invariably fully dominant to *a*, even in the absence of *U*; the $A/a;u/u$ phenotype displays a slight darkening along the medial line as compared with $A/A;u/u$. The difference is difficult to perceive in living mice, but becomes more obvious from comparisons of skins, although some overlapping occurs (Mather and North, 1940). This slight darkening effect which *a* has when heterozygous with *A* is further augmented by the presence of *U*. Thus as a consequence of the interaction of *U* and *a* the following six agoutis may be ranked for increasing umbrous: $A/A;u/u < A/a;u/u = A/A;U/u < A/A;U/U = A/a;U/u < A/a;U/U$.

While the basis for this interaction is not known, Mather and North (1940), like Barrows (1934), stress that *U* may modify the dominance of agouti over nonagouti. In fact a significant part of their paper is devoted to discussing this possibility with regard to various theories of dominance. Although they may be correct, the alternative and simpler explanation, that *a* is itself an "umbrous gene" acting in concert with *U*, should not be overlooked. This possibility is attractive especially since, as noted above, even in the absence of *U*, A/a mice may be slightly darker than A/A animals.

Mather and North (1940; Grüneberg, 1952) noted that a sable color of low intensity occurred in $A^y/a;U/U$ mice, but reported that umbrous had no effect upon nonagouti (a/a) individuals. This, however, is not the case as $a/a;U/U$ animals are significantly darker than $a/a;u/u$ mice with a phenotype bearing a close resemblence to extreme nonagouti (a^e/a^e) and sombre ($E^{so}/–$) (Poole and Silvers, unpublished). Indeed, the difference between $a/a;U/U$

and *a/a;u/u* animals is so striking that it is difficult to understand how it was overlooked.

More recently Robinson (1959) described a similar "umbrous" gene segregating in a stock of fancy orgin but no tests were made to determine whether this mutation was the same as, or allelic with, Mather and North's. This is unfortunate since it is not known whether Robinson's failure to detect any obvious differences between umbrous animals of A^y/a^t versus A^y/A^w or A^w/a^t versus A^w/A^w genotypes was due to the fact that his mutant was different from the one previously described, or to the fact that umbrous may not interact with a^t the way it does with *a*, or to the fact that prepared skins were not critically matched.

2. Mahogany (*mg*)

In 1960 Lane and M.C. Green described a recessive umbrous gene which they named mahogany (*mg*). The mutation appeared in agouti mice of unknown origin and the animals fit the descriptions of umbrous as given by Barrows and by Mather and North. "The central dorsal hairs were considerably darker than those of normal agouti mice, the yellow ticking characteristic of agouti being considerably reduced. The darkening was present but less intense in the lateral hairs. The ventral hairs were almost solid grey with no yellow ticking. In addition the ears and tail appeared more deeply pigmented than those of normal wild-type mice" (Lane and M.C. Green, 1960). Mahogany proved to be detectable also in nonagouti mice; *a/a;mg/mg* mice are coal black with no lightly pigmented hairs evident behind the ears or around the perineum. The tail, feet, and ears are also darker than in normal nonagouti animals (Lane and M.C. Green, 1960). In fact, in general appearance nonagouti mahognay mice fit the description of the nonagouti mice selected for dark modifiers by Little (1916), mice homozygous for extreme nonagouti, and nonagouti mice homozygous for *U*. Mahogany proved to be on chromosome 2, about 12 cM from agouti.[42]

3. Mahoganoid (*md*), Nonagouti Curly (*nc*) and Dark (*da*)

Three other umbros mutations have been reported and briefly described. Mahoganoid (*md*), another recessive (chromosome 16),[43] occurred in the C3H/HeJ strain and is identical to mahogany (*mg*) on both agouti and nonagouti backgrounds (Lane, 1960a). Nonagouti curly (*nc*), which is probably an allele of *md*,[44] is likewise recessive and was recognized in an F_2 animal from a mutagenesis experiment using caffeine. The deviant was nonagouti with black pinna hairs and curly whiskers. *nc*, like E^{so}, is epistatic to A^w so that $nc/nc;A^w/A^w$ mice look like a^e/a^e animals but have curly whiskers and a slightly plush coat (R.J.S. Phillips, personal communication).[45] *nc* homozygotes also have a reduced viability (R.J.S. Phillips, 1963, 1971).

Finally, dark (*da*) arose in the CBA/Fa strain. Animals homozygous for this recessive mutation are smaller than normal with reduced fertility. When combined with A or A^y, dark produces a phenotype in which the yellow pigment on the back is replaced by black so that both look nonagouti except on the flanks. The darkening of the back of $A/-;da/da$ decreases as the animals become older. This umbrous gene, which was reported by Falconer (1956a), has been assigned to chromosome 7. Unfortunately, it may be extinct.

C. "Reds"

Before concluding this section it should be mentioned that there are certain yellows of the fancy which are known under the name of "reds" (Grüneberg, 1952). Such animals are genetically brown ($A^y/-;b/b$) and the intensity of their "red" coat is probably dependent on intensifiers, analogous to, but distinct from, the "umbrous" genes (Dunn, 1916). A detailed analysis of this phenotype has not yet been made.

Notes

[1]Throughout this book the wild-type allele is designated sometimes by + or, for the sake of clarity, sometimes by its alphabetical symbol. Thus $+ = B, C, D, P, S, w$ etc.

[2]Cizadlo and Granholm (1978a) have histologically examined embryos derived from $A^y/a \times A^y/a$ and $a/a \times A^y/a$ matings at 105 hours postcoitum. Their observations confirm the suggestion of Calarco and Pedersen (1976) that the cells of the inner cell mass are more susceptible to the effects of A^y/A^y than are trophoblast cells, although both populations eventually succomb. Their findings are also in accord with Eaton and M.M. Green's (1962, 1963) conclusion that failure of trophoblast giant cell proliferation in A^y/A^y embryos is the cause of the lethality, and with Pedersen (1974) and Calarco and Pedersen's (1976) conclusion that the effects of A^y homozygosity occur over an extended period of time. Thus individual blastomeres may become arrested as early as the 4- to 8-cell stage with "normal" development proceeding in the remaining cells until the late blastocyst stage. Finally, their observations are also consistent with Pedersen and Spindle's (1976) report that presumed homozygous A^y blastocysts do not develop as fast and possess fewer cells than normal littermates when cultured from cleavage stages. (See also Cizadlo and Granholm (1978b) and Granholm and P. Johnson (1978).)

[3]Heston and Vlahakis (1961b) have shown that the effect which A^y has in increasing the incidence of methylcholanthrene-induced pulmonary tumors seems to be related to its effect on the growth rate. Thus they found that by restricting the food intake of yellow mice they were able to reduce their susceptibility to these tumors to the level of nonyellow (a/a)—but otherwise genetically identical—mice. From experiments based on the graft-versus-host reactivity of spleen cells, Gasser and Fischgrund (1973) observed that the immune responsiveness of A^y/a (and A^{vy}/a) mice may be impaired and suggest that this may be involved in their higher tumor susceptibility.

[4]Some of the pigment patterns displayed by the hairs of "agouti" A^{vy}/a phenotypes are strikingly different from the standard agouti pattern. These aberrant pigment patterns are especially apparent among the awls (Galbraith and G. Wolff, 1974).

[5]This association of the yellow phenotype with obesity in both A^{iy} and A^{vy} genotypes led Dickie (1969a) to suggest that those environmental factors which stimulate phaeomelanin synthesis may block or inhibit a pathway used in normal fat metabolism (but see Section II, A, 2).

[6]Although *A* is usually regarded as the wild-type allele of this series by many authors, according to Grüneberg (1952) this is an oversimplification. Thus he notes that the Western European house mouse (*M. m. domesticus* Rutty) usually is grey-bellied, but sporadic cases of A^w are by no means rare, and there are entire populations in which this allele has replaced *A* (Clarke, 1914). Moreover, according to Schwarz and Schwarz (1943) all wild subspecies of *M. musculus* are A^w, and the dark belly of *A* is the hallmark of commensalism. It is therefore evident that one can equally justify A^w and *A* as the wild-type allele of the series.

[7]Agouti (and white-bellied-agouti) mice also possess some all-black (or in *b*/*b* animals, brown) hairs on the dorsum. Thus, according to Galbraith (1969), of the four main hair types which comprise the dorsal pelage, zigzags are invariably banded, auchenes are usually banded, and monotrichs are never banded. Awls, on the other hand, may be either banded or totally black.

[8]"Pseudoagouti" a^m/a mice are indistinguishable from "pseudoagouti" A^{vy}/a animals except by breeding tests.

[9]Very recently G. Wolff (1978) has published the results of extensive breeding experiments which indicate that the phenotype expressed by viable yellow (A^{vy}) and mottled agouti (a^m) genotypes is significantly influenced by the agouti locus genotype and genetic background of the dam. Thus in a reciprocal cross between a C57BL/6 (*a*/*a*) subline and strain AM/Wf, which is homozygous for a^m, he found that 29.5% of the offspring of the C57BL/6 mothers were of the "pseudoagouti" phenotype whereas no animals displaying this phenotype were produced by the AM dams. Wolff also found in two different reciprocal crosses using four different inbred strains, that the proportion of A^{vy}/a offspring of "pseudoagouti" phenotype differed according to the strain of the mother. He also found in $A^{vy}/a \times a/a$ reciprocal matings that, regardless of strain, mottled yellow A^{vy}/a females produced significantly fewer "pseudoagouti" A^{vy}/a young than did black (*a*/*a*) females. Although Wolff does not consider the possibility that his results may reflect some maternal influence on the selection of phenoclones, in the light of Mintz's observations (see Chapter 7), this seems quite possible.

[10]Although A^s was suspected of being an inversion because it suppressed crossing-over between itself and pallid (*pa*) (R.J.S. Phillips, 1970b) while increasing crossing-over between *a* and brachypodism (*bp*) (R.J.S. Phillips, 1968), it was not until recently that this was confirmed. Thus E.P. Evans using G-banding has demonstrated the presence of an inversion of chromosome 2, involving about 40% of the chromosome, extending from about the T7Ca breakpoint proximally to the *A*-locus distally (Evans and R.J.S. Phillips, 1978). This raises the possibility that other "alleles" of the agouti series may also represent structural alterations. In this regard R.J.S. Phillips (personal communication) suggests that nonagouti lethal (a^l)

could be a deletion, and agouti umbrous (a^u), which so far has shown no crossing-over with *bp* in 935 animals, may also be structural. a^x too belongs in this category.

[11]A phenotypically similar umbrous factor as A^s, known as a^{6H}, has been described by Searle (1966, 1968a; see also Batchelor et al., 1966). This factor arose in a neutron irradiation experiment and seems to be inseparable from a reciprocal translocation [known as T(2;8)26H] and so may also exemplify a position effect. a^{6H}/a animals are dark umbrous agouti with light pinna hairs. a^{6H}/A mice are normal agouti, and a^{6H} homozygotes are dark agouti with dark pinna hairs.

[12]As emphasized by Wallace (1965) this theory of pseudoallelism does not preclude the idea that different regions of the body may vary in their capacity to respond to some alleles. It also remains to be determined how many "mini-loci" are involved and the exact parts of the pelage they control. Epistasis between some pseudoalleles must be considered too.

[13]Wallace (1965) also points out that whereas the genotypes a^u/a^u and a^e/a^e have dark ears, the alleles a^t and *a*, which stand between them in the series, produce lighter ears, both homozygously and in compounds with a^u and a^e. This, she believes, can best be explained by assuming that there is a locus (say, *E*) which controls ear color, with dominance of the yellow component. Thus a^u and a^e are *e*, and a^t and *a* (and all the other alleles) are *E*.

[14]Dickie (1969a) found that the most common spontaneous mutational events at the agouti locus involve heritable and nonheritable changes from *a* to a^t and from *a* to A^w. Thus at least 31 heritable mutations from *a* to a^t (20 in the C57BL/6J strain) and at least 17 heritable mutations from *a* to A^w (6 in C57BL/6J) occurred at the Jackson Laboratory between July 1, 1961 and January 1, 1968 (see Grüneberg, 1966a).

[15]This situation occurred in a C57BL/6J male. The deviant was phenotypically light-bellied agouti but, when mated to a variety of unrelated nonagouti females, produced three classes of offspring. Some were light-bellied agouti, some were black-and-tan, and some were nonagouti in a ratio of 1:1:2. Extensive testing of all light-bellied agouti, black-and-tan, and some nonagouti offspring indicated that A^w and a^t behaved similarly to other known reoccurrences of these mutations. No animals of the genotype A^w/a^t were detected nor did cytological studies reveal any chromosomal abnormalities. Inasmuch as the three classes of phenotypes were produced in almost a perfect 1:1:2 ratio, it seems most likely that two mutations had occurred at a very early stage of development (Dickie, 1969a).

[16]To overcome the difficulties inherent in both the "complex" and single-locus" theories, R.J.S. Phillips (1966a) has discussed possible models in which the agouti locus would consist of several genes in an operon-type organization. On the one hand, each cistron could relate to one part of the body and all be controlled by an operator. On the other hand, the structural genes could individually be controlled by regulatory genes. In one model she proposes that A^s, which appears to act coordinately on the other alleles of the locus, could be assumed to be an allele of the operator or to be in the first cistron, and to affect transcription through the rest of the complex. According to this scheme a^e might act similarly whereas a^{td} could be an allele in a subsequent cistron and influence only the transcription of a "banding" region. In another model she suggests that A^s and a^x, alleles which are known to show crossing over with the rest, could be considered as mutations of the operator

controlling the cistron, whereas the other alleles would be structural mutants affecting the amount of some substances on which the synthesis of phaeomelanin depends.

[17]It should also be noted in Table 3-1 that this reduction in tyrosine incorporation by A^y was of the same order regardless of the genetic constitution at either the *b*-locus or *c*-locus (i.e., c^{ch}/c^{ch}) (Coleman, 1962).

[18]Prota (1972; see also Prota and Nicolaus, 1967; Fattorusso et al., 1969; Misuraca et al., 1969; Flesch, 1970; Prota and R. Thomson, 1976) has provided evidence from his avian studies that phaeomelanins are amphoteric pigments produced *in vivo* by a deviation of the eumelanin pathway involving a reaction between cysteine and dopaquinone, produced by enzymatic oxidation of tyrosine. He has also shown that this reaction involves 1,6-addition of cysteine to dopaquinone to form two new amino acids, 5-*S*-cysteinyldopa and 2-*S*-cysteinyldopa, and that under physiological conditions these intermediates are converted rapidly into dihydrobenzothiazine derivatives whose further oxidation gives rise to phaeomelanins (see Figure 2-3). Multivesicular bodies have been implicated in the formation of both eumelanosomes and phaeomelanosomes. Thus investigations support the view that the fusion of a Golgi-derived tyrosinase-containing vesicle with a vesicle from the endoplasmic reticulum is involved in the genesis of these granules (see Bagnara et al., 1979). Prota (1970) also contends "that the so-called tricosiderins, the red pigments of human hair, are chemically and biogenetically related to phaeomelanins." It should also be noted that whereas three bands of tyrosinase activity are detected in acrylamide gel electrograms of extracts of eumelanin pigmented skin (see Chapter 3, Section I,I), extracts of $A^y/-$ skin give rise to only a single band, the intensity of which is attenuated when compared to that of other genotypes (Holstein et al., 1967, 1971; see also Geschwind and Huseby, 1972; Geschwind et al., 1972). This indicates that there are "marked quantitative and perhaps qualitative differences between tyrosinases associated with eumelanogenic and phaeomelanogenic melanocytes" (Holstein et al., 1971). Indeed, Holstein and his colleagues contend that this observation does not necessarily support a switch mechanism which is based only on the prevailing titer of cysteine. They believe the observed differences in tyrosinase could "be related to functional properties of the enzyme involved in shunting the melanogenic mechanism either toward the production of eumelanin or of phaeomelanin."

[19]Takeuchi (1970) has also investigated the activity of the agouti locus *in vitro*. He found that when a piece of dorsal *A*/*A* skin from a 2-day-old mouse was cultured, no phaeomelanin was produced. On the other hand, the formation of yellow pigment was observed when skin from a 3-day-old agouti animal was maintained in culture for 4 days. Addition of dopa to the culture medium of 2-day-old *A*/*A* skin resulted in the formation of phaeomelanin. However, phaeomelanin synthesis was prevented in 3-day-old *A*/*A* explants maintained with excess tyrosine. On the basis of these observations Takeuchi suggests that the formation of the agouti pattern may depend upon a regulatory system of feedback type, involving both the *a*- and *c*-loci. His hypothesis is discussed in some detail by Holstein et al. (1971).

[20]According to Geschwind and his colleagues (1972) the few melanocytes which occur in the dermis of the tail and scrotum as well as in the connective tissue of the nipple of $A^y/-$ mice contain yellow granules, i.e., phaeomelanin. This observation, however, remains to be confirmed.

[21]This influence of tract specificity not only holds for black-and-tan and yellow-bellied-agouti genotypes but for nonagouti animals as well. For example, the hairs round the mammae of *a/a* mice contain yellow pigment and if this region is included in $a/a;c^e/c^e$ grafts, all potentially intensely pigmented (*C*/–) melanocytes which invade these follicles produce some phaeomelanin, i.e., they produce the characteristic type of pigmented hair for this region.

[22]Not only do mouse melanoblasts respond to the agouti locus genotype of mouse hair follicles but *rat* melanoblasts can likewise respond to the agouti locus genotype of these follicles. This was demonstrated by transplanting agouti (but albino) BALB/c (*A/A;c/c*) mouse skin from 18- to 20-day-old fetuses to immunologically tolerant newborn (Lewis × BN)F_1 (*a/a;C/c*) rats. About 2 weeks following this procedure some pigmented hairs, displaying the typical agouti pattern, were observed within the borders of the mouse xenografts (Silvers, 1965).

[23]One wonders whether this is responsible for the fact that the number of melanocytes in the ear skin of A^y/a mice appears drastically reduced from the number in *a/a*, *A*/– or a^t/– ears. Moreover, a similar difference has been reported in A^y/A vs *A/A* plantar skin (Quevedo and J. Smith, 1963) and dorsal body skin (Quevedo and McTague, 1963) after exposure to ultraviolet light (UV). Whereas the body skin of lethal yellow ($A^y/A;B/B$) mice tans poorly in response to UV, the skins of black (*a/a;B/B*) and black agouti (*A/A;B/B*) genotypes tan noticeably. Following UV exposure only a few dopa-positive melanocytes occur at the dermoepidermal junction and only a few melanin granules are released to malpighian cells in $A^y/A;B/B$ mice. On the other hand, in the skin of *A/A;B/B* animals there is a marked increase in the number of dopa-reactive epidermal melanocytes and in the number of melanin granules within malpighian cells (Quevedo et al., 1967). The tanning response of viable yellow ($A^{vy}/a;B/B$) mice varies with the phenotype. In general, the skin of "clear yellow" A^{vy}/a mice and the "clear yellow" areas of mosaically pigmented animals present a histological picture similar to that found in similarly exposed lethal yellow ($A^y/A;B/B$) mice, while the skin of "agouti" A^{vy}/a animals and the "agouti"-pigmented areas of mosaics respond like *A/A;B/B* genotypes (Quevedo et al., 1967).

[24]According to Galbraith (1964) at about 6.5 days after plucking the dorsal hairs of adult agouti mice, many follicles begin transition from synthesis of eumelanin to production of phaeomelanin and at about 9 days postplucking the reverse process is most evident. This switch from production of black pigment to yellow pigment is exceedingly rapid.

[25]Galbraith and Arceci (1974) found that individual zigzag hair bulbs contained from 5 to 18 melanocytes and that the average number of these cells in each hair bulb ranged from 10 at day 7 in both black and yellow mice, to 12 in yellow and 13 in black animals at day 13. As noted by these investigators these values are a little higher than Chase's (1951) estimate of 4 to 8 melanocytes per zigzag follicle and are lower than, but within Potten's (1968) observed range of a mean of 7 to 22 melanocytes per bulb in the Strong F ($a/a;b/b;c^{ch}/c^{ch}$; *d/d;s/s*) mouse strain.

[26]This mutation was designated as *So* until its allelism with *e* was established by Searle (1968b).

[27]It appears that the phenotypes of the $A/A;E^{so}/E$ mice described by N. Bateman and the $A/a^e;E^{so}/E$ animals of G. Wolff et al. were slightly different. Whereas the former investigator noted that the flanks of his heterozygotes were "flecked with

yellow hairs," the latter investigators found no evidence of phaeomelanin in their heterozygotes. It also is not clear whether the "yellowed hairs" which Bateman observed on the perineum of E^{so} homozygotes and heterozygotes resulted from the occurrence of phaeomelanin.

[28]Poole and Silvers (unpublished) have produced lethal yellow sombre ($A^y/a;E^{so}/E$) mice and they too are completely black with no yellow pigment and show the same tendency toward obesity as "yellow lethal yellows."

[29]Wolff et al. also observed that zigzags of $A/a^e;E^{so}/E$ mice seemed to be more intensely and uniformly pigmented than those of $A^{vy}/A;E^{so}/E$ animals.

[30]As pointed out by Wolff et al. a similar black and white "banding" pattern occurs, albeit infrequently, among hairs of phenotypically agouti (A/A) animals (Galbraith, 1964).

[31]For details concerning the distribution, morphology, biometrical characteristics, and cytogenetics of this subspecies see Gropp et al. (1969, 1970).

[32]Accordingly its designation was changed from *Tob* to E^{tob}.

[33]According to Searle and Beechey (1970), because young e/e mice are very sooty it is difficult, but not impossible, to discriminate between young e/e and $+/+$ mice on dilute (d/d) or leaden (ln/ln) backgrounds. They also note that the degree of dilution of yellow areas in $A^y/-;d/d$ and $e/e;d/d$ mice of similar sootiness is about the same; that chinchilla (c^{ch}) removes less yellow from e/e than from $A^y/-$, so that it is hard to differentiate between $e/e;c^{ch}/c^{ch}$ and $e/e;C/-$ mice before weaning age; and that the ears of $A^y/-$ mice usually look a little lighter than e/e ears, presumably because of the frequently greater sootiness of the latter.

[34]Although this comparison was made employing hairs from $a/a;B/B;e/e$ and $A^y/a;b/b;E/E$ genotypes, it does not seem likely that the difference at the *b*-locus was responsible for the disparity since E. Russell (1948) obtained nearly identical counts for cortical phaeomelanin granules in $A^y/a;B/B$ and $A^y/a;b/b$ hairs.

[35]There is some evidence that total tyrosinase activity is greater in e/e than in A^y/a skin. This evidence stems from the observation of Geschwind et al. (1972) that the tyrosinase specific activities in 90-day-old e/e skin was about double that of A^y/a skin.

[36]Nevertheless, Pomerantz and Chuang (1970) found that injections of β-MSH not only produced a 45–50% increase in tyrosinase levels when administered to neonatal black or brown mice but the former, but not the latter, animals became darker than untreated controls. Hirobe and Takeuchi (1977a,b) too have shown that if neonatal C57BL/10J (a/a) mice are inoculated subcutaneously with α-MSH or with dibutyryl cyclic AMP, or if the skin of these mice is exposed *in vitro* to these agents, the number of dopa-positive melanocytes in the epidermis increases (see also Hirobe and Takeuchi, 1978). They believe this increased activity requires *de novo* transcription and translation since it can be suppressed by actinomycin-D or cycloheximide.

[37]In some of these studies $A^y/-$ animals from 75–90 days of age were plucked (to stimulate hair growth) 8 days prior to receiving a subcutaneous injection of MSH [0.05 ml of a solution of natural or synthetic α-MSH (3 mg/ml) suspended in beeswax peanut oil]. Twelve to 24 hours following this inoculation skin biopsies indicated a gradual shift from phaeomelanin to eumelanin production in follicular

melanocytes. This production of black pigment seemed to reach its peak approximately 36 hours after the injection and the black granules predominated until some time between the 72nd and 96th hour postinoculation when phaeomelanin synthesis resumed. Geschwind and his associates also demonstrated that the kinetics of the appearance of eumelanin in the hair follicles of these treated mice, as determined by histological sections, was generally consistent with the increase in tyrosinase activity. Thus a 2.5- to 5-fold increase in tyrosinase activity was produced in plucked A^y/– skin within 24 hours of MSH administration, and similar large increases in tyrosinase activity were produced within 24 hours of injecting the hormone into 4-day-old A^y/– mice. In A^y/– mice treated with hormone for a prolonged period an increase in dermal melanocytes as well as a significant increase in the number of melanosomes in the epidermis of the nipple, tail, and scrotum was also observed (Geschwind and Huseby, 1972). Finally it should be noted that while in preliminary experiments Geschwind and Huseby found that both cycloheximide and colchicine prevented darkening of MSH-treated lethal yellow mice, puromycin and actinomycin-D did not (but see note 36).

[38]Since sombre (E^{so}) produces intensely black phenotypes, even in the presence of A^y/–, Geschwind and his associates (1972) determined whether E^{so} normally exerted its effect by increasing MSH levels in sombre mice. As might have been expected from the observation that MSH had no effect on altering the color of *e/e* mice, assays of plasma from *A*/–;E^{so}/– mice revealed no elevation of melanocyte-stimulating activity levels when compared to levels in *A*/–;*E*/*E* animals.

[39]Because such very dark sables appear to have a glossy black back and a yellow belly, they came to be known as "black-and-tans." This designation should not be confused with the genuine black-and-tan caused by the agouti locus allele, a^t.

[40]This umbrous effect can be recognized also in A^y/–;*b*/*b* and *A*/–;*b*/*b* mice (Grüneberg, 1952).

[41]When some wide crosses between strains are made and selection is practiced toward either darkening or making lighter the *A*/– segregants, the results described by Dunn can still be obtained and indicate a considerable potential reservoir of modifiers, i.e., polygenes (Chase, unpublished).

[42]The recombination between *mg* and *a* was 9.8 + 1.2% in males and 13.2 ± 1.0% in females. This sex difference is significant and is in the same direction as that observed by Fisher and Landauer (1953) for this region of the second chromosome (Lane and M.C. Green, 1960).

[43]*md*, by virtue of its association with Robertsonian (whole arm) translocation (16.17)7Bnr (hereafter *Rb*7), has been shown to be located close to the centromere on chromosome 16 (Roderick et al., 1976).

[44]Because *md* was shown to be "0%" from the centromere of chromosome 16, R.J.S. Phillips and G. Fisher (personal communication) crossed mice homozygous for *nc* to animals carrying *Rb*7 and so far have found 0/50 recombinants between *nc* and *Rb*7 in the appropriate backcross. From this observation it appears that *nc* is closely linked to the centromere of either chromosome 16 or 17 and, on the basis of other information, it has been located on 16.

[45]On the other hand, recessive yellow nonagouti curly (*e*/*e*;*nc*/*nc*) mice are yellow with curly whiskers indicating that *e* is epistatic over *nc* with respect to coat color (Beechey and Searle, 1978).

Chapter 3

The *b*-Locus and *c* (Albino) Series of Alleles

I. The *b*-Locus

A. Influence of *B/B* vs *b/b* Substitution on Coat Color

Four mutations have been assigned to the *b*-locus (or to one very closely associated with it) on the fourth chromosome of the mouse. *B*, the wild type allele at this locus, produces black (Plate 1–E) and the most recessive allele, *b*, produces brown (Plate 1–F) eumelanin. Thus *b*, when homozygous, changes the gray color of the wild mouse to a brownish hue, which is generally known as cinnamon agouti or cinnamon. This phenotype results from the occurrence of yellow banded brown (instead of black) hairs.

Inasmuch as *b*-locus alleles have no demonstrable influence on phaeomelanin synthesis, black and brown yellow mice are usually distinguishable only by their eye color (or by whether their extra-follicular melanocyte population is "brown" or "black"). However, because yellow mice often display variable amounts of sootiness as a consequence of the cumulative effect of "umbrous" genes, which promote the synthesis of eumelanin in the hairs of the dorsum (see Chapter 2, Section III), some $A^y/-;B/-$ and $A^y/-;b/b$ animals can be distinguished by coat color, especially if *B* and *b* are known to be segregating.

B. Other Alleles and Dominance Relationships

The three other genes of the *b*-series are light (B^{lt}), cordovan (b^c), and white-based brown (B^w). Whereas B^{lt} and B^w appear to be dominant to *B*, i.e., the phenotypes associated with each of these genes is distinguishable when the allele is heterozygous with *B*, b^c is recessive to black and dominant over brown.[1] These dominance–recessive relationships are not complete however. For example, B^{lt} and B^w yield different phenotypes when homozygous and when heterozygous, with *B* or *b*, and cordovan (b^c/—) mice are indistinguishable from brown (*b*/*b*) animals in the presence of dilute (*d*/*d*) (D.S. Miller and Potas, 1955). In fact, there is evidence that *b* can sometimes express itself when heterozygous with *B*. Thus Dunn and Thigpen (1930) found that the effect of the silver gene, *si* (see Chapter 6, Section II), is greatly intensified in *B*/*b* black mice, and Durham (1911) and Snell (1931) observed that *a*/*a*;*B*/*b*;*p*/*p* mice are a shade lighter than *a*/*a*;*B*/*B*;*p*/*p* animals (but see Little, 1913).

C. Influence of *B*/*B* vs *b*/*b* Substitution on Pigment Granules

The predominant effect of substituting *b*/*b* for *B*/*B* is to change the nature of the eumelanin from a black, fuscous black, or fuscous pigment to a carob brown or mummy brown pigment (E. Russell, 1949b). This change in the color of the pigment granule is accompanied by a change in its size and shape. Brown granules are significantly smaller than black granules and display much less variation in size (E. Russell, 1949a, 1949b).[2] Whereas the granules found in the septules of black-pigmented hairs are usually either long-oval or oval in shape, the granules of brown-pigmented hairs are round (see Figure 4-6) (E. Russell, 1949a). Because of these differences it has been judged that intensely pigmented brown phenotypes have only slightly more than one-third the volume of pigment of intensely pigmented black mice (E. Russell, 1948). Nevertheless this is still about twice as much pigment as occurs in intensely pigmented yellow hairs.

D. Effects of *b*-Locus Alleles on Tyrosinase Activity

The effect of genic substitution at the *b*-locus on tyrosinase activity has been well investigated (L. Russell and W. Russell, 1948; Foster, 1951, 1959; Fitzpatrick and Kukita, 1959; Coleman, 1962) and, contrary to what one might expect, brown mice have at least as much tyrosinase as black mice. Indeed, in most of these studies brown mice had twice the amount of tyrosinase as black animals. The study of Coleman (1962) is particularly pertinent inasmuch as he measured the incorporation of ^{14}C-labeled tyrosine into *a*/*a*;*B*/*B*, *a*/*a*;B^{lt}/B^{lt}, *a*/*a*;b^c/b^c, and *a*/*a*;*b*/*b* skin. Whereas brown skin incorporated about twice as much tyrosine as black (*B*/*B* or *B*/*b*) skin and light skin, cordovan skin was intermediate between black and brown in its uptake (Table 3-1). Moreover, to rule out the possibility that these paradoxical findings were a consequence of the *in vitro* assay employed (see Foster, 1949) Coleman (1962) measured the uptake of tyrosine when injected into 4-day-old *b*/*b* and *B*/*B* mice and obtained similar results. This *in vivo* study

Table 3-1. The Effect of Lethal Yellow (A^y) and Chinchilla (c^{ch}) on the Incorporation of Tyrosine by Different *b*-Locus Alleles[a]

Alleles	Genotype[b]	Tyrosine incorporation[c] (cpm)
Brown	$a/a;b/b;C/C$	2680 ± 51
Cordovan	$a/a;b^c/b^c;C/C$	2290 ± 87
Black	$a/a;B/B;C/C$	1200 ± 36
Light	$a/a;B^{lt}/B^{lt};C/C$	1170 ± 60
Black × brown	$a/a;B/b;C/C$	1180 ± 58
Yellow brown	$A^y/a;b/b;C/C$	1060 ± 49
Yellow black	$A^y/a;B/B;C/C$	460 ± 14
Brown chinchilla	$a/a;b/b;c^{ch}/c^{ch}$	482 ± 17
Black chinchilla	$a/a;B/B;c^{ch}/c^{ch}$	442 ± 15
Yellow brown chinchilla	$A^y/a;b/b;c^{ch}/c^{ch}$	157 ± 5
Yellow black chinchilla	$A^y/a;B/B;c^{ch}/c^{ch}$	140 ± 8

[a]Based on D.L. Coleman (1962).
[b]All mice were nondilute (D/D).
[c]Values represent cpm/mg of dried protein ± standard error of the mean.

also indicated that brown granules develop as rapidly as black granules so that the increased activity of *b/b* skin cannot be attributed to the fact that brown granules develop more slowly and hence retain more active tyrosinase sites on their surface (Fitzpatrick and Kukita, 1959).

E. Site of *b*-Locus Activity

To pinpoint the stage in the development of the melanosome when the *b*-locus is believed to operate requires a brief description of the "normal" development of this organelle. According to Moyer (1961, 1963, 1966) melanosome development is initiated when thin "unit fibers," often contiguous with polysomes, aggregate to form "compound fibers" within a membranous boundary. As these fibers cross-link and become oriented parallel to each other the shape of the melanosome[3] becomes apparent and melanin is deposited at definite sites along the fibers of its matrix.[4] This deposition of melanin continues so that ultimately the details of the matrix of the melanosome are obscured by the electron dense pigment. When melanin synthesis ceases the melanosome represents a typical mature pigment granule.[5]

Observations of *B/–* and *b/b* granules indicate that it is during that stage of melanosome development when melanin is deposited at discrete sites along the fibers of its matrix that the *b*-locus operates. Thus Moyer (1961) observed that when melanin was just beginning to be deposited, the second-order periodicity in black granules was longer (about 200 Å) than the second-order periodicity of brown granules (about 113 Å). If this second-order periodicity represents the active site of tyrosinase activity, the shorter distance between sites in brown granules would allow for a greater number of sites for this activity and, hence, for the greater overall tyrosinase activity of *b*/*b* skin. Thus, according to this scheme, subunits of the protein controlled by the *b*-locus are possibly involved in the formation of the parallel fibers of the melanosome which bind tyrosinase in a certain fixed ratio (Wolfe and Coleman, 1966).[6]

The *b*-locus also influences the size and shape of the pigment granule as well as the final molecular structure of the melanin deposited. When viewed under the electron microscope, the melanin in mature brown granules is flocculent and coarsely granular, whereas that of black granules is very finely granular, almost appearing homogeneous (Moyer, 1966) (see Figure 3-1).[7] It therefore appears that the protein produced by the *b*-locus not only must provide a structural framework for the attachment of tyrosinase but also must somehow affect its activity.

F. Other Ultrastructural Observations

Besides the observations of Moyer (1961, 1963, 1966), Lutzner and Lowrie (1972), and Hearing et al. (1973) (see note 5), the ultrastructure of black and brown melanosomes has also been examined by Rittenhouse (1968a). She observed that sections of black hair bulb melanocytes contain a mixture of round and elongated melanized bodies suggesting a population of granules round in cross-section and oval in longitudinal section. On the other hand, the granules found in intensely pigmented brown hair bulbs were usually nearly spherical. Rittenhouse also noted that maltese dilution (*d*/*d*) (see Chapter 4, Section I) influenced the morphology of *b*/*b* granules; *a*/*a*;*b*/*b*;*d*/*d* granules, like *B*/– granules, were usually oval in longitudinal section and round in cross-section. Moreover, she observed that whereas the unmelanized *B*/– granule consists of one or more rolled membranes, the internal framework of *b*/*b* granules resemble a tangled ball of strands. Although this last observation contrasts with those of Moyer (1963, 1966) who could not find any disruption of the pattern of granule framework in *b*/*b* mice, it should be noted that most of his preparations were derived from the pigmented epithelium of the eye rather than from the hair bulb.

In an attempt to integrate the mode of action of *b*-locus alleles, Foster (1965) has speculated that this locus is the structural gene for a protein the alteration of which can greatly influence the properties of the melanosome.

G. Light (B^{lt})

The B^{lt} mutation, first described by MacDowell (1950) and extensively studied by Quevedo and his associates (Quevedo and Chase, 1958; McGrath and Quevedo, 1965; S. Sweet and Quevedo, 1968) and Pierro (1963a), is ei-

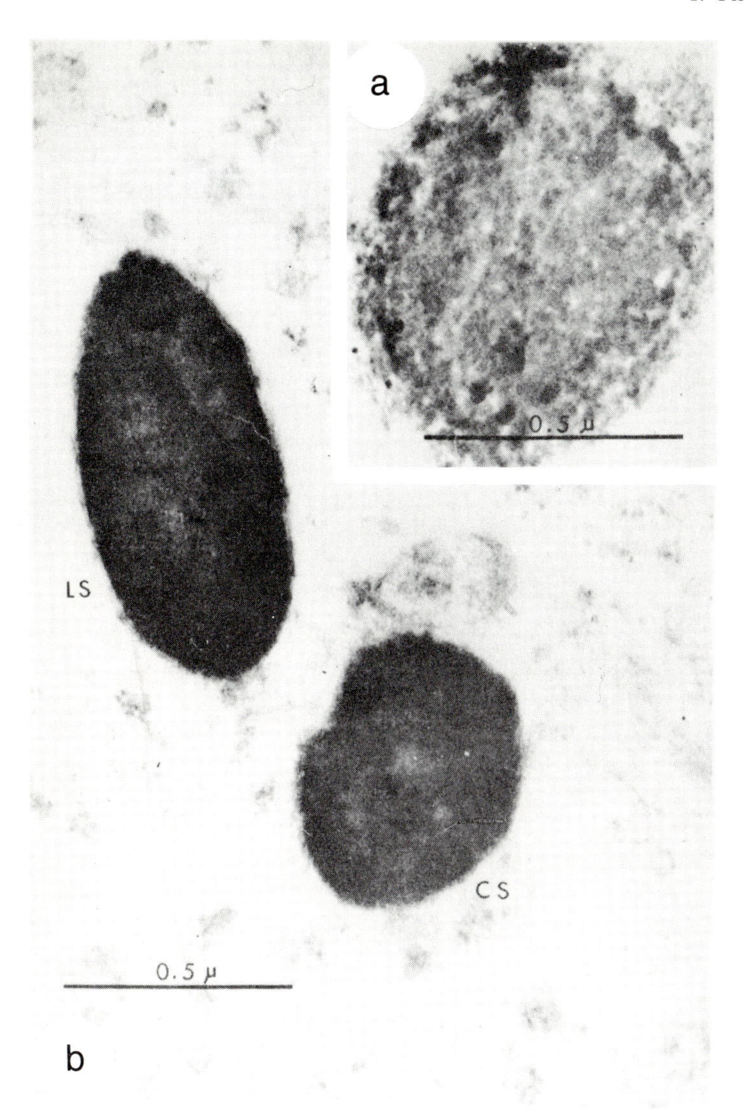

Figure 3-1. (a) A typical brown (*a*/*a*;*b*/*b*) melanin granule at maturity. Note the granularity of the melanin. (b) Mature black melanin granules in both longitudinal section (LS) and cross-section (CS). Note the near homogeneity and extreme apparent density of the melanin. Although black granules are larger than brown ones, these black granules are smaller because they are from a ruby-eye (*ru*/*ru*) animal (see Chapter 4, Section III, B). From F.H. Moyer (1966). Reproduced with permission of the *American Zoologist*.

ther an allele of the *b*-locus, or is closely linked and epistatic to this locus. Homozygous B^{lt} mice (Plate 1–H and Figure 3-2) are phenotypically distinct from heterozygous B^{lt} animals (Figure 3-2). Whereas B^{lt} homozygotes (known as "lights") have almost white fur, except for the hair tips which are

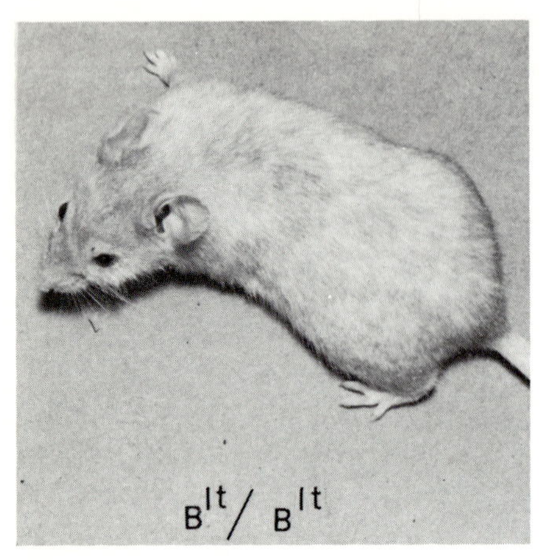

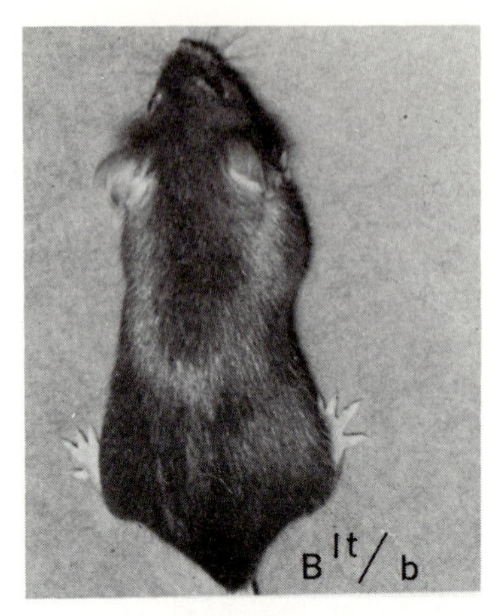

Figure 3-2. Adult *light* ($a/a;B^{lt}/B^{lt}$) and *dark* ($a/a;B^{lt}/b$) mice. From W.C. Quevedo, Jr. and H.B. Chase (1958). Reproduced with permission of the authors and The Wistar Institute Press.

"hair brown," B^{lt} heterozygotes (known as "darks") display darker hair tips (chaetura drab in color) with pigment extending further down the shaft (Grüneberg, 1952; Quevedo and Chase, 1958). The ventral hairs of both homozygotes and heterozygotes are substantially less pigmented than dorsal hairs, and with advancing age the coat color of both lights and darks become progressively lighter. $A^{y}/a:B^{lt}/B$ mice are indistinguishable from non-B^{lt} yellow animals.

In agouti animals, the yellow subterminal band conceals the color of the hair tip in both lights and darks so that the influence of the B^{lt} gene cannot be recognized until the hair is well grown. The distinction between B^{lt} heterozygotes and homozygotes is also less conspicuous in agouti than in nonagouti mice (Grüneberg, 1952).

Although darks of the same age heterozygous for *B* or *b* are essentially identical in appearance (Quevedo and Chase, 1958), the number of uveal melanocytes is lower in B^{lt}/b genotypes than in B^{lt}/B mice (Pierro, 1963a).

One of the most interesting coat-color interactions occurs between $B^{lt}/-$ and P/p. Whereas *P* usually is dominant to *p* (pink-eyed dilute) this is not the case in $B^{lt}/-$ genotypes (nor is it the case in b/b mice—see Chapter 4, Section II, B). The hairs of light and dark mice either homozygous or heterozygous for *p* do not display an absence of pigment in the lower section of their hair shafts (Quevedo and Chase, 1958). Thus the uniform dark sepia coat color of $B^{lt}/-;P/p$ mice is considerably darker than the color of $B^{lt}/-;P/P$ animals. $B^{lt}/-;p/p$ mice are a light metallic grey (McGrath and Quevedo, 1965).

The specific effects of the B^{lt} mutation on pigment deposition have been described in detail by Quevedo and Chase (1958). Cleared hairs of

$a/a;B^{lt}/B^{lt}$ mice display large clumps of pigment granules which occur predominantly in the medulla of the shaft (Figure 3-3a). The degree of pigmentation is variable from hair to hair but almost invariably there is a reduced number of pigment granules as one proceeds from the tip of the hair to its base. Indeed, the bottom half of many hairs are completely devoid of pigment and when it does occur most of it is contained in sporadic clumps, separated from one another by empty septa (Figure 3-3b). The clumps of pigment are highly variable in size; some are large enough to disrupt several septa. Hairs from B^{lt} heterozygotes present a similar picture except that they possess on the average more pigment per hair than B^{lt} homozygotes. The clumps of granules also are fewer and appear smaller in heterozygotes.

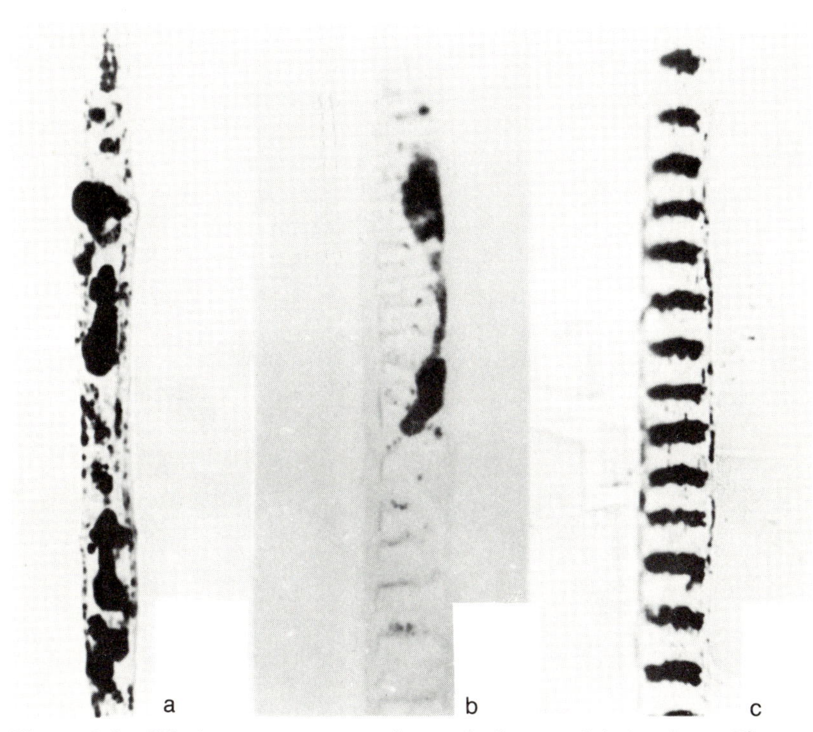

Figure 3-3. Whole mount preparations of pigmented hairs from B^{lt} genotypes. Approx., ×550. (a) Hair from a *light* ($a/a;B^{lt}/B^{lt}$) mouse. Note the large clumps of pigment and their irregular distribution in the medulla of the shaft. (b) Hair from an animal of the same genotype, but from a region deficient in pigment granules. This particular region possesses a single large dendritic melanocyte extending over several septules. (c) Hair from a dark sepia ($a/a;B^{lt}/b;P/p$) mouse. Note that the pigment granules are arranged in a uniform banded pattern in contrast to the clumped condition frequently characteristic of $a/a;B^{lt}/-;P/P$ hairs. Figs. (a) and (c) are from W.C. Quevedo, Jr. and H.B. Chase, 1958 and are reproduced with permission of the authors and The Wistar Institute Press. Fig. (b) is from E.P. McGrath and W.C. Quevedo, Jr., 1965 and is reproduced with permission of the authors and the Australian Academy of Science.

As far as the individual pigment granules are concerned, they are essentially identical in B^{lt}/B^{lt}, B^{lt}/B, and B^{lt}/b genotypes and possess similarities with both B/B and b/b granules. Their color, although lighter than those of B/B mice, is definitely of the black, rather than brown, species. On the other hand, although the size and shape of the granules vary, most of them are round and closer to the size for brown than they are to the size for black (Quevedo and Chase, 1958).

As might be expected from the phenotype, $a/a;B^{lt}/-;P/p$ animals possess pigment throughout the length of their hairs (Figure 3-3c). These animals are also dificient in the large clumps of pigment typical of $B^{lt}/-;P/P$ animals although some small clumps occasionally occur.[8]

The behavior of the melanocytes in the hair bulbs of $B^{lt}/-$ genotypes, as well as the pigment they contain, are a direct reflection of the situation in the hair shaft. Thus the pigment granules within the follicular melanocytes of $a/a;B^{lt}/B^{lt}$,$a/a;B^{lt}/B$, and $a/a;B^{lt}/b$ mice, like those in the hair shaft, resemble black granules in intensity and brown granules in morphology. $B^{lt}/-$ melanocytes appear to contain fewer granules than B/B melanocytes and some of this pigment occurs in clumps, similar in size to those observed in the hair shaft (Figure 3-4). This suggests either that large portions of $B^{lt}/-$ melanocytes or entire melanocytes are incorporated into the growing hair

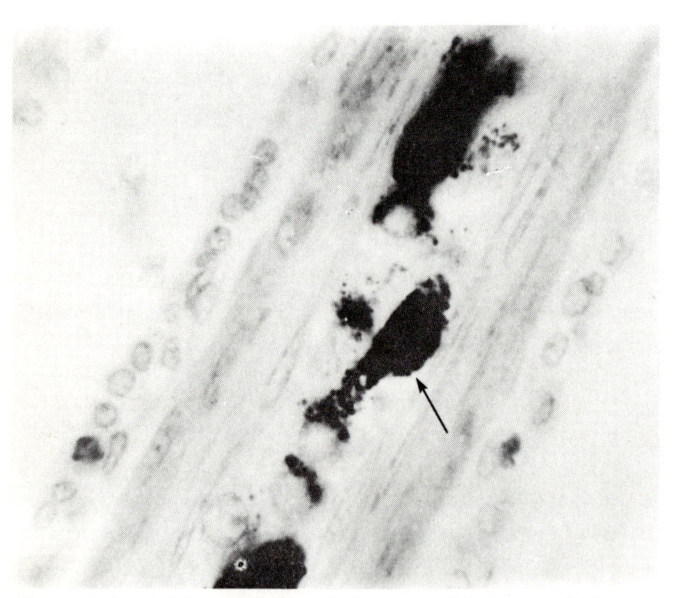

Figure 3-4. Mid-region of a *light* ($a/a;B^{lt}/B^{lt}$) hair follicle (thirteenth day of hair growth cycle). Approx. ×900. A dislodged follicular melanocyte (arrow) continues to discharge pigment granules into a migrating epithelial cell. Note clumped nature of granules within cells. From E.P. McGrath and W.C. Quevedo, Jr. (1965). Reproduced with permission of the authors and the Australian Academy of Science.

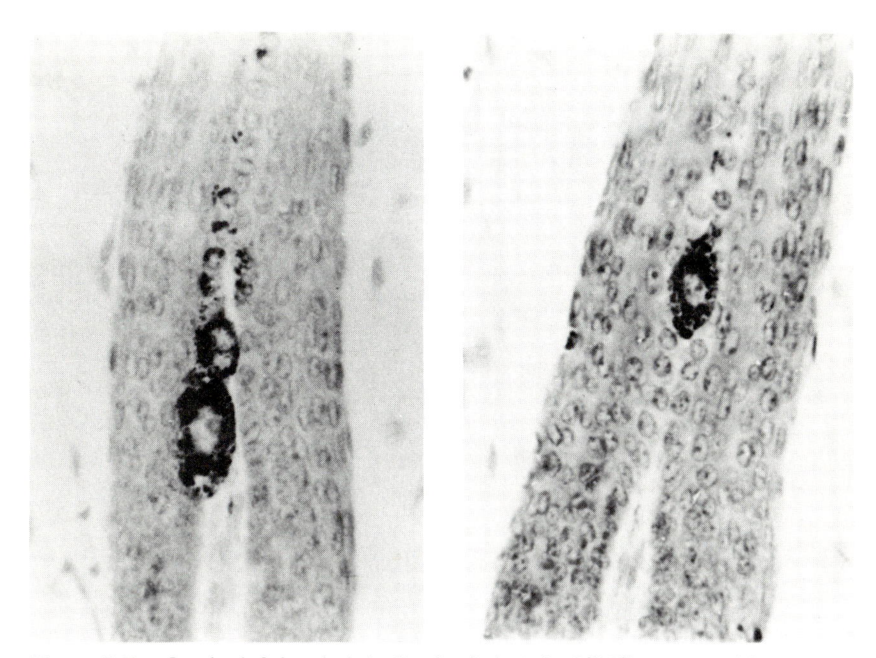

Figure 3-5. On the left is a hair bulb of a *light* ($a/a;B^{lt}/B^{lt}$) mouse (thirteenth day of hair growth cycle). Note that only two large, round melanocytes remain on one side of the papilla and that these cells lack well-developed dendritic processes. On the right is another *light* ($a/a;B^{lt}/B^{lt}$) follicle (tenth day of hair growth cycle). This one has only one large unbranched melanocyte attached to the apex of the attenuated dermal papilla. Approx. ×400. From E.P. McGrath and W.C. Quevedo, Jr. (1965). Reproduced with permission of the authors and the Australian Academy of Science.

(Chase and Mann, 1966). Consistent with this conclusion is the fact that many hair bulbs lack melanocytes completely or possess only one or two of these cells during the latter half of hair growth (Figure 3-5). Indeed, 2 weeks after a new hair growth cycle is induced in light mice by the plucking of resting hairs, very few follicles contain active melanocytes and some of these resemble the amelanotic melanocytes of albinos (Figure 3-6). By the termination of hair growth, many of the hair follicles of light mice resemble those in "white-spotted" areas in that no trace of any melanocytes can be found (Figure 3-7) (Chase, 1958). The situation in B^{lt} heterozygotes is similar except that active melanocytes persist in more hair follicles over a greater course of the cycle (Quevedo and Chase, 1958).

In accord with the situation in the hair shaft of $B^{lt}/-$ mice heterozygous (P/p) or homozygous for p, the melanocytes of the hair bulb persist throughout the growth cycle, apparently releasing pigment granules to epithelial cells at a normal rate. Moreover, the melanocytes of these genotypes remain dendritic and have not been observed to be incorporated into the hair shaft (McGrath and Quevedo, 1965; S. Sweet and Quevedo, 1968).

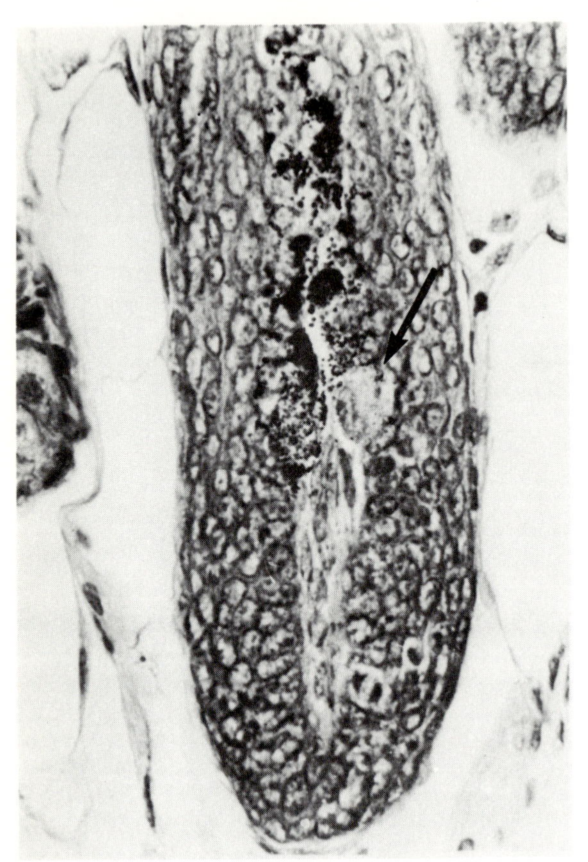

Figure 3-6. Hair bulb of a *dark* ($a/a;B^{lt}/b$) mouse (13th day of hair growth cycle). Approx., ×700. The arrow indicates a cell with chromophobic cytoplasm similar to the follicular amelanotic melanocytes (clear cells) of albino animals (see Fig. 3-12a). From W.C. Quevedo, Jr. and H.B. Chase, 1958. Reproduced with permission of the authors and The Wistar Institute Press.

Although all the melanocytes in the hair bulb of light mice are clearly dendritic during early stages of the hair growth cycle, some rounded, densely melanized cells deficient in dendrites (known as "nucleopetal" cells—see Chapter 4, Section I, D) are subsequently found in the upper part of the follicle (Figure 3-8). The frequency of these cells increases so that the uprooted cells (or portions thereof) which become incorporated into the developing hair shaft also usually lack dendrites, are densely melanized, and may contain pycnotic nuclei (S. Sweet and Quevedo, 1968). Indeed, as will be reiterated when the effects of the *d* and *ln* loci are discussed (Chapter 4, Section I, D), the very fact that nucleopetal melanocytes develop in B^{lt} hair follicles may be responsible for their being uprooted and incorporated into the hair, i.e., a well-developed dendritic system may serve

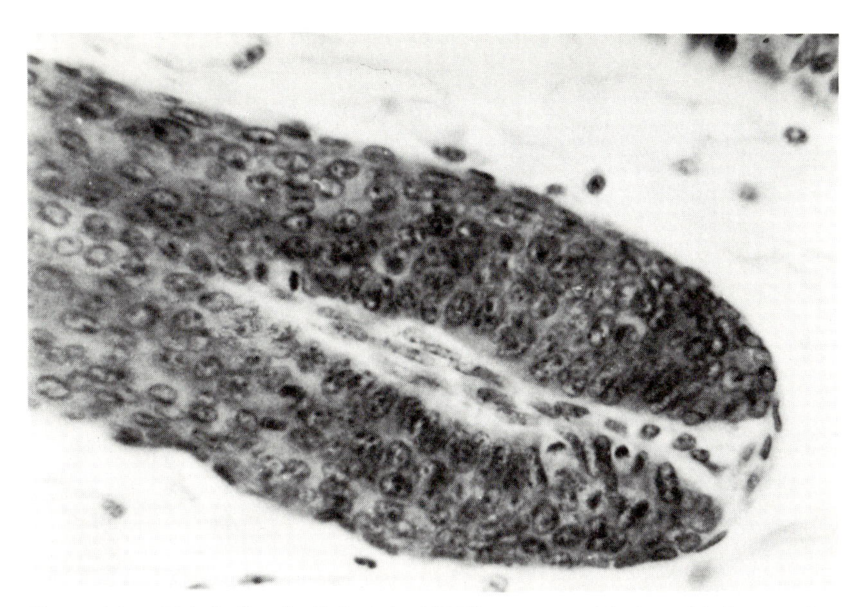

Figure 3-7. Hair bulb of a *light* ($a/a;B^{lt}/B^{lt}$) mouse (10th day of hair growth cycle). Approx., ×550. In this bulb melanogenesis has ceased early, active melanocytes are absent from the hair bulb, and the epithelial cells lining the papilla cavity form a highly uniform border as they do in hair bulbs in white spotted areas (compare with Fig. 3-10 and 3-11). From E.P. McGrath and W.C. Quevedo, Jr., 1965. Reproduced with permission of the authors and the Australian Academy of Science.

as a mechanical anchor. Why these cells develop, however, is not known. One possibility is that the B^{lt} gene causes follicular melanocytes to produce pigment granules at a faster rate than they can deliver them to keratinocytes and as a consequence the melanocytes enlarge, become detached from their normal matrix positions, and are swept into the hair shaft (Quevedo and Chase, 1958; S. Sweet and Quevedo, 1968).[9] If this mechanism is correct one would expect the rate of melanin synthesis to be reduced in B^{lt} genotypes heterozygous for *p* (since these animals do *not* display either nucleopetal melanocytes in their follicles or large clumps of pigment in their hairs). Unfortunately, concrete evidence that this is the case is lacking. Nevertheless, $B^{lt}/-;p/p$ as well as yellow-B^{lt} mice almost certainly have a slower rate of pigment synthesis than $B^{lt}/-;P/P$ mice and these genotypes also display melanocytes of normal morphology.

1. Interaction With Leaden and Dilute

Since the recessive mutations leaden (*ln*) and dilute (*d*) (see Chapter 4, Section I) also affect melanocyte morphology by transforming dendritic (nucleofugal) cells into cells with a poorly developed dendritic system (nucleopetal), some mention of how these genes interact with B^{lt}, as well as a comparison of the effect each of these genes has when combined with other

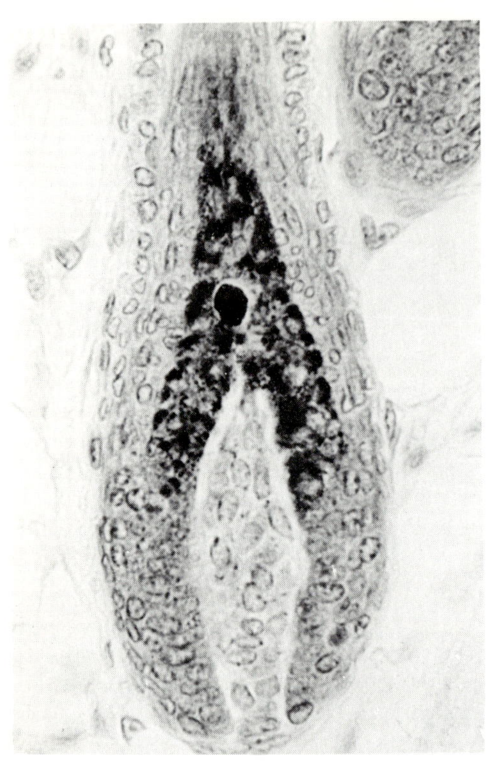

Figure 3-8. Hair bulb of a *light* ($a/a;B^{lt}/B^{lt}$) mouse (6th day of hair growth cycle). Note that among the nucleofugal melanocytes of the upper bulb is a rounded, densely pigmented nucleopetal cell. Approx., ×550. From S.E. Sweet and W.C. Quevedo, Jr., 1968. Reproduced with permission of the authors and The Wistar Institute Press.

coat-color determinants, is germane. Indeed, the results of these interactions demonstrate clearly that the influence B^{lt} has on melanocyte morphology is quite distinct from that of *d* or *ln* (McGrath and Quevedo, 1965; S. Sweet and Quevedo, 1968).

Inasmuch as *d*/*d* hair bulb melanocytes are always nucleopetal, one might anticipate that when *d*/*d* is combined with $B^{lt}/-$ these nucleopetal cells would aggravate the "diluting" action of B^{lt} by accelerating the rate at which melanocytes become incorporated into the hair shaft. This, however, is *not* the case. Although in $a/a;d/d;B^{lt}/-$ mice the morphology of the follicular melanocytes are consistent with their *d*/*d* constitution, they do not appear to be lost from the hair bulbs more frequently than those of nondilute light or dark nonagouti mice. Indeed the dynamics of melanocyte behavior in dilute-B^{lt} mice seems to be more comparable to those of B^{lt} heterozygotes than homozygotes (McGrath and Quevedo, 1965). This observation is important because it indicates that factors other than melanocyte shape are undoubtedly involved in determining whether melanocytes are dislodged from the hair bulb and incorporated into the septules of the hair (McGrath and Quevedo, 1965).

Among the observations which indicate that B^{lt} differs from *ln* and *d* in the manner in which it influences melanocyte morphology are the following: (1) Whereas B^{lt} has no influence on the morphology of yellow melanocytes, both *d/d* and *ln/ln* yellow mice display nucleopetal cells (Poole and Silvers, unpublished); (2) *p*, even when heterozygous, can influence the morphology and behavior of $B^{lt}/-$ melanocytes, whereas this is not the case even when *p* is homozygous in combination with either *ln/ln* (S. Sweet and Quevedo, 1968) or *d/d* (Poole and Silvers, unpublished); and (3) while the hairs of B^{lt} genotypes, particularly homozygotes, become unpigmented prematurely as a consequence of the fact that the melanocytes which are incorporated into the hair are not replaced, this is not the case in *ln/ln* and *d/d* animals. Thus, although the melanocytes of leaden and dilute mice are also incorporated periodically into the hair, they appear to be replaced so that a population of follicular melanocytes persists to pigment the entire shaft.

Finally it should be noted that in contrast to the situation in *ln/ln* and *d/d* genotypes and, indeed, in contradistinction to the situation in B^{lt} hair bulbs, the extrafollicular melanocytes of light mice are usually nucleofugal in morphology. Thus, light melanocytes in the dermis and epidermis of the sole of the foot, ear pinna, scrotum, and tail (S. Sweet and Quevedo, 1968; Quevedo, 1969b), as well as those in the harderian gland (Markert and Silvers, 1956), all display well-developed dendrites (Figure 3-9). Why this is

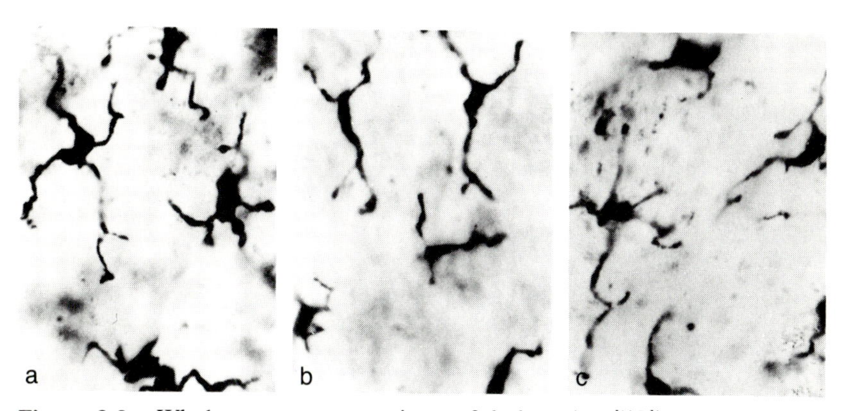

Figure 3-9. Whole mount preparations of *light* ($a/a;B^{lt}/B^{lt}$) melanocytes in the epidermis of the (a) ear pinna, (b) sole of foot and (c) tail. Dopa-reagent. Approx., ×450. Note the well-developed dendrites of these cells (compare with Fig. 4-3c and d). From S.E. Sweet and W.C. Quevedo, Jr., 1968. Reproduced with permission of the authors and The Wistar Institute Press.

the case is not known but one possibility, which will be discussed further when we concern ourselves with the action of genes at the *d* and *ln* loci (Chapter 4, Section I, E), is that light melanocytes have an intrinsically greater capacity to develop well-melanized dendrites than either leaden or dilute melanocytes (Quevedo, 1969b). In the compact environment of a hair bulb light, leaden and dilute melanocytes are unable to extend their dendrites whereas in less compact extrafollicular locations, light melanocytes, but not dilute or leaden cells, develop normally.

2. Can a Melanocyte Pigment More Than One Hair?

The observation that by the end of a given hair cycle, but when the hair is still growing, the hair bulbs of lights and darks appear to be devoid of melanocytes, and are morphologically similar to the nonpigmented bulbs of white-spotted mice which are known to lack pigment cells (Chase and Rauch, 1950; Silvers, 1956), is important because it relates to the question of whether melanocytes (or their mitotic descendents) can pigment more than one hair. Thus if the disappearance of mature melanocytes from the hair follicles of B^{lt} mice is due to *all* of them being incorporated into the shaft, the melanocytes for the next hair generation would have to be derived either from some unpigmented stem cells in the epithelium, or from cells in the dermal papilla (Chase, 1951, 1958) as is the case for feathers (Foulks, 1943). On the other hand, it is also possible that at least some melanocytes revert to a dormant form which is indistinguishable from the epithelial cells of the follicle and are reactivated, perhaps producing mitotic descendents, when the next hair cycle is initiated (Chase, 1958; Quevedo and Chase, 1958). Although according to Chase (1958) the bulk of evidence supports the contention that there are stem cells at definite sites of the hair follicle which give rise to the mature and expendable melanocytes (see also discussion following Chase and Mann, 1960), and the situation in B^{lt} follicles is consistent with this, there is also evidence (see Chapter 11, note 17) which indicates strongly that melanocytes (or their mitotic decendents) can, in fact, pigment more than one hair. Perhaps both possibilities occur, i.e., while some melanocytes may be able to pigment more than one hair, a stem cell source of new cells may also operate as an insurance mechanism.

3. Influence on the Pigmentation of the Eye

Although B^{lt} heterozygotes and homozygotes have black eyes, the substitution of B^{lt} for B does result in a decrease in the amount of eye pigment, a decrease which is more pronounced in homozygotes than in heterozygotes (Pierro, 1963a). While this decrease is caused predominantly by a reduction in the number of melanocytes in the uveal tract, the number of pigment cells is reduced in the choroid of both lights and darks, and probably in the iris region of lights, as well. Nevertheless, the amount of pigment in the eyes of B^{lt} homozygotes and heterozygotes is somewhat greater than in b/b eyes (Pierro, 1963a).

H. White-Based Brown (B^{w})

The last member of the *b*-series, white based brown (B^{w}), occurred following the exposure of spermatogonia to 600 r low-dose rate γ-irradiation (Hunsicker, 1969). This mutation resembles B^{lt} in some of its effects and, like B^{lt}, the possibility that it is closely linked to *B* and *b* has not yet been ruled out completely. As in the case of B^{lt}, B^{w} mice display a reduction in the amount of pigment at the base of the hair which is more pronounced in homozygotes than in heterozygotes. However, in contrast to the situation in B^{lt} genotypes (where animals heterozygous for *B* or *b* are phenotypically identical), B^{w}/B mice are clearly different from B^{w}/b animals. In B^{w}/B animals only the extreme base of the hair is light, the remainder being just barely distinguishable from *B*/*B* dorsally, though clearly lighter ventrally. On the other hand, one-quarter to one-third of B^{w}/b dorsal hairs are near-white proximally, and the portion which is pigmented is phenotypically brown. Moreover, whereas the eyes of B^{w}/B mice are full colored, those of B^{w}/b animals appear brown. B^{w} homozygotes resemble B^{w}/b heterozygotes except that the proximal, near-white portion of the hair is about twice as wide (Hunsicker, 1969).

I. Influence of *b*-Locus on Multiple Forms of Tyrosinase

The tyrosinase of mice, like that in other organisms, has been shown to occur in a number of forms separable by acrylamide-gel electrophoresis (Wolfe and Coleman, 1966; Burnett and Seiler, 1966; Burnett et al., 1967, 1969; Holstein et al., 1967, 1971; Burnett, 1971; Quevedo, 1971). Although *B*, B^{lt}, and b^{c} have no significant effect on the typical tyrosinase pattern consisting of three uniformly darkened bands (T_1, T_2, and T_3) of dopa melanin, this is not the case for the *b* allele. *a*/*a*;*b*/*b* animals display normal darkening of T_1, but considerably less dopa melanin is deposited in the T_2 and T_3 bands (Holstein et al., 1967; Quevedo, 1971). The significance of this difference is not known.

II. The *c* (Albino) Series of Alleles

Before considering the phenotypic effects of the so-called albino or *c*-series of alleles (chromosome 7) it should be emphasized that although albinism is epistatic to all other coat-color determinants, i.e., all mice, regardless of genotype, lack pigment in the presence of *c*/*c* (Plate 2–A), albino mice nevertheless possess a full complement of pigment cells. Thus, the inability of albino animals to produce pigment stems not from an absence of melanocytes, as is the case for white spotting, but from a deficiency and/or alteration of the structure of tyrosinase in melanocytes which are otherwise normal.

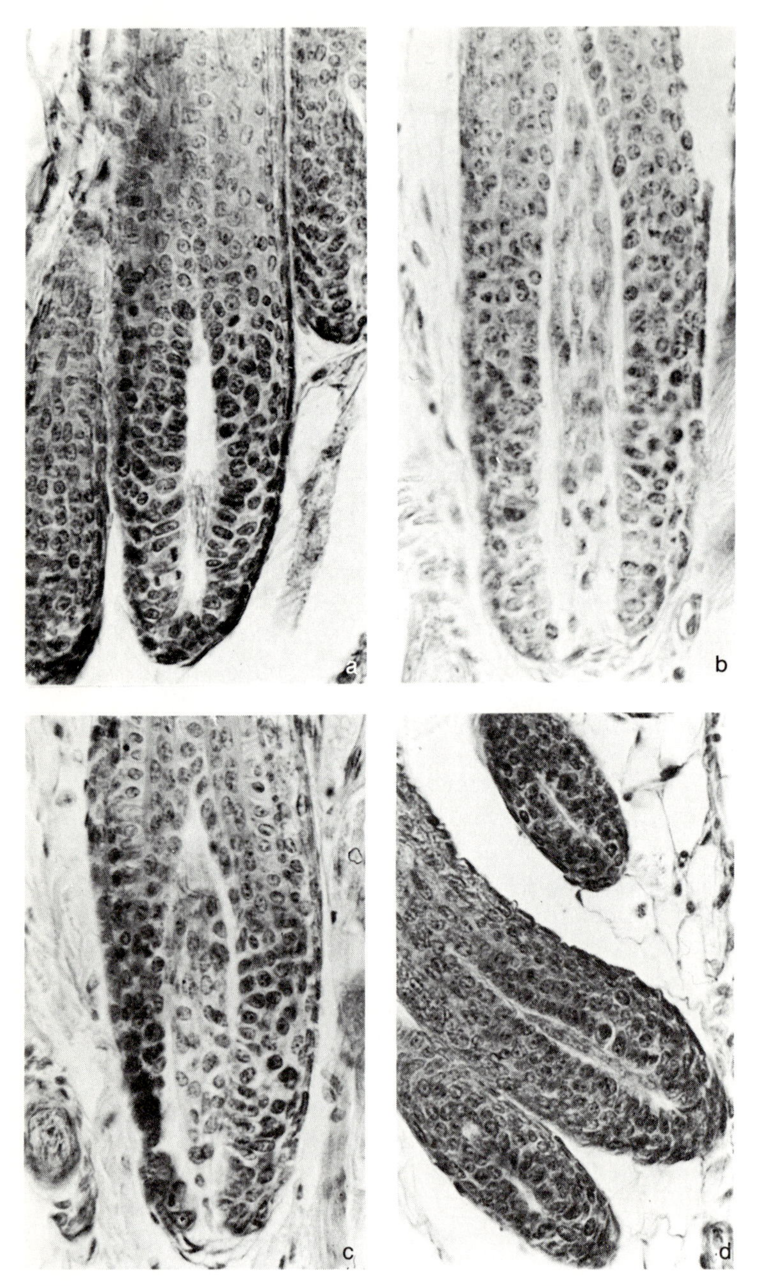

Figure 3-10. Hair bulbs (7th day of hair growth) from the white spotted areas of (a) varitint-waddler (*Va*/+); (b) belted (*bt*/*bt*); (c) splotch (*Sp*/+); and (d) piebald (*s*/*s*) mice. Note the regular size and even arrangement of all the cells of the matrix. Approx., ×450. From W.K. Silvers, 1956. Reproduced with permission of The Wistar Institute Press.

A. Evidence for the Occurrence of Amelanotic Melanocytes

Evidence that albino animals possess a nonfunctioning population of melanocytes—appropriately known as "amelanotic melanocytes"—stems from a variety of observations, most of which were made in the mouse.

When the hair bulbs of albino mice are examined histologically and compared with those originating from white-spotted areas, they are strikingly different. Whereas the hair bulbs of white-spotted areas are characterized by matrices consisting of regularly arranged cells of equal size, (Figures 3-10 and 3-11), albino hair bulbs contain, in addition, many large "clear" cells in their upper bulb region (Chase and Rauch, 1950; Silvers, 1956) (Figure 3-12a). Since these large cells with an apparently hyaline cytoplasm are similar in morphology and location to the pigment-containing cells found in lightly pigmented phenotypes (Figure 3-12b), they are considered to be amelanotic melanocytes (Silvers, 1956).

Further evidence for this conclusion stems from the observation that the

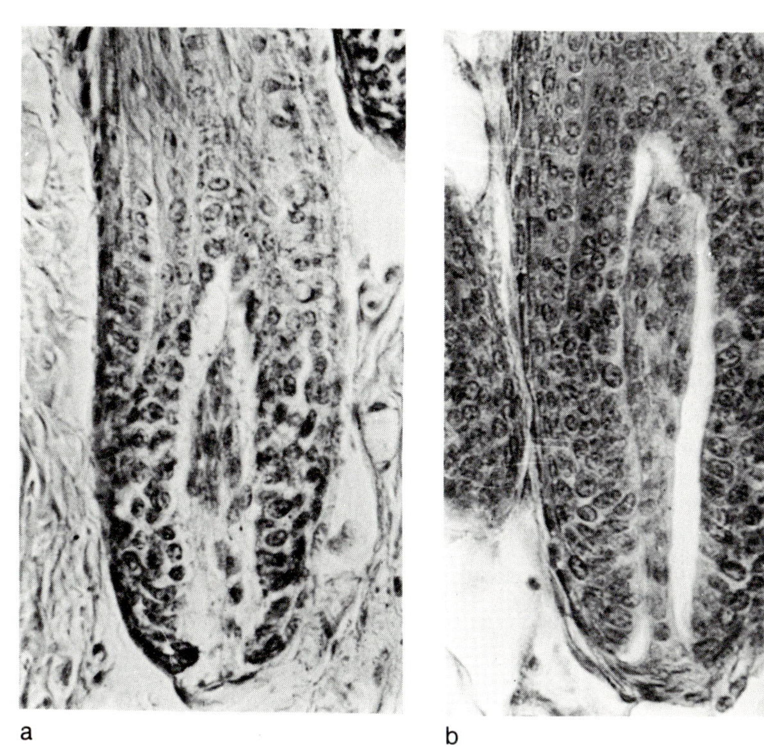

Figure 3-11. Hair bulbs (seventh day of hair growth) from (a) white (Mi^{wh}/Mi^{wh}) and (b) viable dominant spotting (W^v/W^v) mice. Both of these genotypes are "one big spot" lacking melanocytes throughout their integument. Approx., ×500. From W.K. Silvers (1956). Reproduced with permission of The Wistar Institute Press.

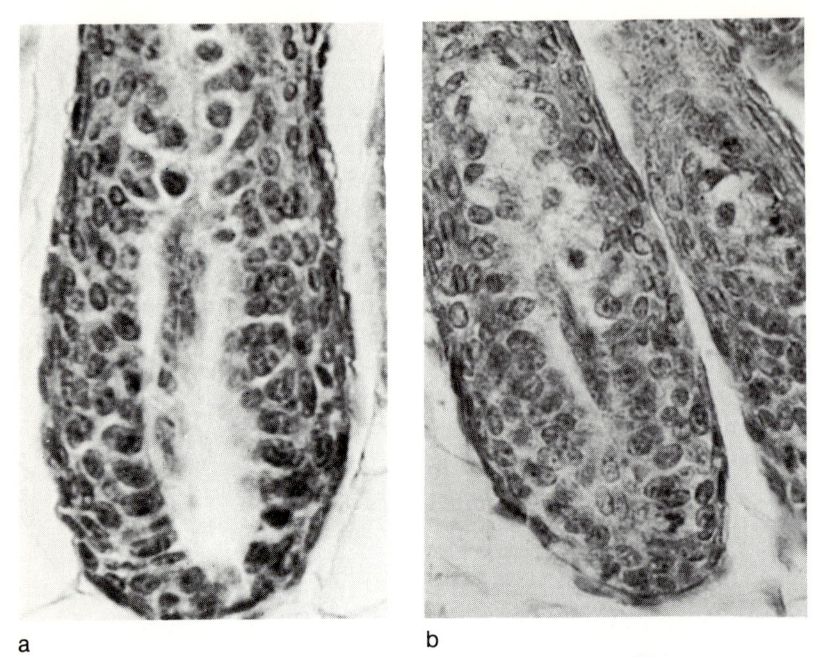

a b

Figure 3-12. Hair bulbs from (a) an albino mouse (7th day of hair growth). Note the large clear cells in the upper portion of the matrix; (b) a pink-eyed chinchilla ($A^w/A^w;c^{ch}/c^{ch};p/p$) mouse (7th day of hair growth). This bulb displays the same large clear cells but with some pigment granules in the cytoplasm. Approx., ×600. From W.K. Silvers, 1956. Reproduced with permission of The Wistar Institute Press.

experimentally depigmented melanocytes of black and yellow mice maintained on a biotin-deficient diet are indistinguishable from the clear cells of albinos, though they retain their dopa-positive character (Quevedo, 1956). Moreover, both clear cells and melanocytes exhibit similar sensitivities to X-rays. Thus, when the skin of albino mice in the resting stage of hair growth is exposed to 1200 r of irradiation, a dose known to destroy almost completely the melanocyte population in resting hairs of pigmented animals (Chase, 1949; Chase and Rauch, 1950), there is a marked destruction of follicular clear cells (Quevedo, 1957). This similar radiosensitivity of clear cells and melanocytes, added to the morphological evidence noted above, indicate further that clear cells are in fact amelanotic melanocytes. This conclusion was substantiated again when it was demonstrated that clear cells, like melanocytes, are derived from the neural crest.

The neural crest originates embryologically between the junction of the neural tube and its overlying ectoderm and is initially continuous from head to tail. As development proceeds, however, its constituent cells migrate ventrolaterally on either side of the spinal cord and at the same time become segmentally clustered (see Chapter 1, note 3). In the mouse this anterior to posterior and mediolateral migration of neural crest cells, from their place of origin to their definitive positions, takes place between the eighth and twelfth day of embryonic development (the gestation of the mouse is about 20 days), as demonstrated in the classic experiments of Rawles (1940, 1947, 1953). Thus, by transplanting tissues derived from various regions of C57BL/6 mouse embryos of different ages to the coelom of the chick embryo, Rawles was able to demonstrate that only those explants which included cells of neural crest origin produced melanocytes. She found neural crest cells to be confined to the region of the neural tube in 8.5- to 9-day-old embryos and only when this region was included in grafts of this age did melanocytes develop. By approximately 11 days of age, however, she found that cells of neural crest origin had made their way into almost all regions of the body so that skin ectoderm and adhering mesoderm removed from almost any level of the trunk (but not from the limb buds) produced pigmented hairs when transplanted to the chick coelom. Limb buds receive migrating melanoblasts between the eleventh and twelfth days of gestation and only at this time did limb-bud ectoderm and adhering mesoderm give rise to pigmented hairs.

Once this "timetable" for the migratory pathway of neural crest cells was established, and it was substantiated that the melanocytes of pigmented animals were derived from these cells, it was easy to demonstrate that they likewise differentiated into the clear cell or putative amelanotic melanocyte population of albino animals. This was accomplished by showing that hair bulbs in the skin of grafts which differentiated from albino embryo explants possessed clear cells only when the explant was known to contain cells of neural crest origin (Silvers, 1958c) (Figure 3-13a and d). Indeed, the fact that the hair bulbs of skin known to be deprived of its neural crest component were indistinguishable from those normally originating in white-spotted areas (Figure 3-13b and c) provided the strongest evidence that white spotting resulted from an absence of melanocytes, pigmented or otherwise (Silvers, 1958c).[10]

B. *c*-Series Alleles

1. Extreme Dilution (c^e)

When one considers that the albino mutation was already known and maintained in mice in Greek and Roman times, and, in fact, was the first mammalian trait to be analyzed following the rediscovery of Mendel's principles (Cuénot, 1902; Castle and Allen, 1903), it is somewhat surprising that descriptions of other *c*-locus mutations are of much more recent vintage. The second mutation to be recorded at this locus, c^e (for extreme dilution), was first described in 1921 by Detlefsen.[11] The mutant animal was caught in a corn crib in Illinois and on a first and cursory examination gave the appearance of being a slightly stained or dirty black-eyed white. The animal

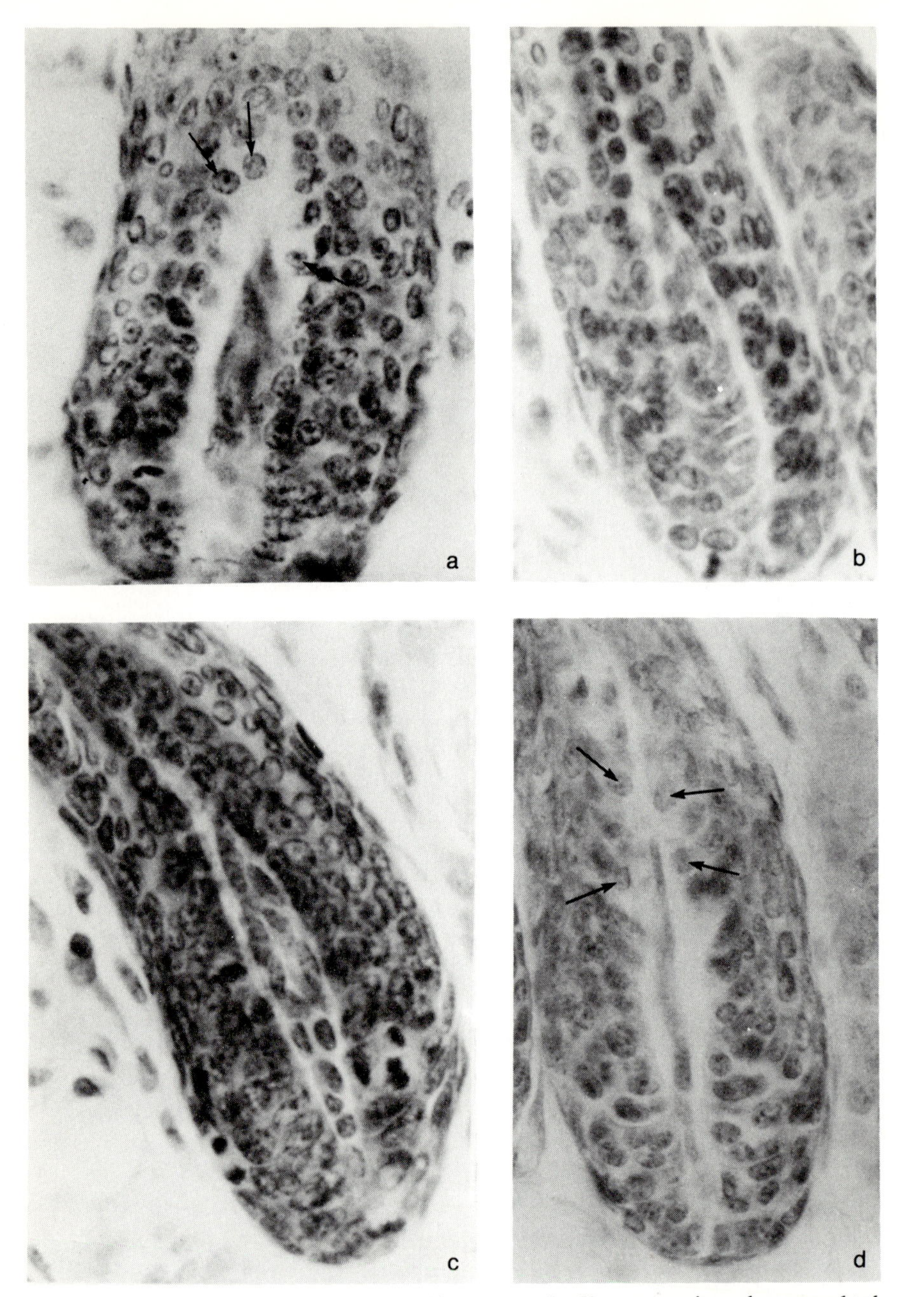

Figure 3-13. Hair bulbs that developed in (a) a graft of intact somites plus neural tube taken from a 10½-day-old albino (*c/c*) embryo. This graft possessed neural crest cells, and clear cells (arrows) are present. (b) A graft from the limb bud of a 10 to 10½-day-old albino (*c/c*) embryo. This graft did not include neural crest cells, and note the absence of clear cells and the similarity to bulbs shown in Figs. 3-10 and 3-11. (c) A neural crest-free limb bud graft from a 10½-day-old potentially pigmented (*a/a;b/b;d/d*) embryo. Note similarity to (b). (d) A graft from the limb bud of a 12½-day-old albino (*c/c*) embryo. This graft included neural crest, and as in (a) clear cells are present (arrows). Approx., ×600. From W.K. Silvers, 1958c. Reproduced with permission of The Wistar Institute Press.

darkened with age however and eventually acquired a brownish shade, "a little lighter than an ordinary pink-eyed brown with a slight dull yellowish cast" (Detlefsen, 1921). Examination of the hair revealed no clear evidence of an agouti pattern, the base of the hair being light and the apical portion being pigmented. The ventral surface of the mutant was noticeably lighter than the dorsum and dark pigment was quite pronounced in the skin of the ears and scrotum. Detlefsen also observed that the eyes of mutant animals at birth were somewhat less heavily pigmented than the wild type and that this difference persisted for some time. The eyes of adult animals were, however, very similar in color to those of intensely pigmented mice.

Detlefsen demonstrated quite clearly that this mutation was a c-locus allele and that it expressed itself when heterozygous with albinism. c^e/c mice (on a nonagouti background) are roughly intermediate in color between c^e and c homozygotes and, although the amount of pigment in their eyes is reduced from normal, they nevertheless have enough pigment to make them appear black-eyed on superficial inspection (Grüneberg, 1952).

The c^e allele when combined with lethal yellow (A^y) completely removes all phaeomelanin and consequently $A^y/-;c^e/c^e$ and $A^y/-;c^e/c$ mice are black-eyed white. $a/a;c^e/c^e$ mice can be described as pale brown and $A/-;c^e/c^e$ animals, the genotype described by Detlefsen, are significantly lighter as a consequence of the fact that the yellow band of the agouti hair is completely diluted out.

2. Chinchilla (c^{ch})

Almost immediately after c^e was found and described, Feldman (1922) recorded another c-locus allele which he designated c^r (for ruby-eyed) but which has come to be known as chinchilla (c^{ch}). This mutation, which is higher on the scale than c^e, was procured from a fancier. It too has a more drastic effect on reducing the intensity of phaeomelanin than eumelanin. Thus, as described by Feldman (1922) "the black agouti type of the homozygous mutant possesses black pigment which is reduced to a very dark dull slate-color, while yellow is greatly reduced and appears about intermediate between white and the normal yellow of the wild type." $a/a;B/B;c^{ch}/c^{ch}$ animals are readily distinguishable from phenotypically gray $A/A;B/B;c^{ch}/c^{ch}$ mice (Plate 2–B) and are best described as a medium shade of sepia. $A^y/-;B/-;c^{ch}/c^{ch}$ mice are a faint cream or ivory-color.

When the black agouti type of the mutant is heterozygous for albinism ($A/-;B/-;c^{ch}/c$), the eumelanin pigment is further reduced to a brownish shade while the yellow band is reduced almost to white. Thus $a/a;B/-;c^{ch}/c$ mice are a dull brown color, a little lighter than the typical chocolate ($a/a;b/b$) phenotype.

One of the most interesting features of the c^{ch} mutation, and one to which we will return (Section D), is that while as noted above it drastically reduces phaeomelanin production, and significantly reduces the expression of black pigment, it does not influence the deposition of brown pigment [a dilution of brown (b/b) pigment starts only with c^{ch}/c and is progressively more severe in c^e/c^e and c^e/c genotypes, respectively (Grüneberg, 1952)]. Thus $a/a;b/b;c^{ch}/c^{ch}$ and $a/a;b/b;c^{ch}/c^e$ mice are virtually indistinguishable from $a/a;b/b;C/C$ animals. Indeed the only difference is that the hairs which occur outside and around the ears of nonagouti brown mice are yellowish in $a/a;b/b;C/C$ animals but much lighter in the corresponding c^{ch}/c^{ch} and c^{ch}/c^e genotypes (Grüneberg, 1952).[12]

3. Intense Chinchilla (c^i)

A fifth allele in the albino series was also described by Feldman (1935) and was known as intense chinchilla (c^i). As far as I am aware, this mutation is no longer available. The mutation, which represented the most intense of the series below C, was first noted in a chinchilla stock. When homozygous this allele had such a very slight affect on the synthesis of black pigment that it was not possible to distinguish $a/a;B/-;c^i/c^i$ mice from $a/a;B/-;C/-$ animals with certainty. On the other hand, in accord with the other alleles at the locus, it had a more pronounced influence on phaeomelanin diluting the intense yellow pigment characteristic of C/C genotypes to a pale ocherous yellow color, bordering on a lemon tint (Feldman, 1935). Because c^i/c^i had very little, if any, influence on black pigment and a less drastic effect on yellow than the lower alleles of the series, $A/-;B/-;c^i/c^i$ animals appeared to be not very different from the wild type. Intense chinchilla like the other members of the series was phenotypically completely recessive to full color, at least on a nonagouti background. It also displayed incomplete dominance over the lower alleles of the series.

4. Summary of Phenotypic Effects of C, c^i, c^{ch}, c^e, and c

The effects which the c-series of alleles considered above have in the presence of nonagouti (a/a), lethal yellow ($A^y/-$), and pink-eyed dilution (p/p) are summarized in Table 3-2. Inasmuch as p/p has a drastic effect on eumelanin synthesis, but little influence on phaeomelanin production (see Chapter 4, Section II, A), while these c-series alleles reduce yellow pigment more than black, it is not surprising that $A^y/-;c^i/c^i$ and $A/A;c^i/c^i;p/p$ mice are significantly lighter than $a/a;c^i/c^i$ animals and that $A^y/-;c^{ch}/c^{ch}$ and $A/A;c^{ch}/c^{ch};p/p$ genotypes are lighter than $a/a;c^{ch}/c^{ch}$ mice. Indeed, these interactions are especially conspicuous on a black-and-tan ($a^t/-$) background where one can compare the effects of these genes on eumelanin and phaeomelanin production in the same host.

5. Platinum (c^p)

In 1964 a mouse slightly different from an albino was observed in an AKR ($a/a;c/c$) × DBA/2J ($a/a;b/b;d/d$) litter. At maturity this animal had pink eyes and a coat a shade darker than albino with a luster or sheen. Breeding

Table 3-2. Interaction of Albino Alleles with a/a, A^y, and p/p[a,b]

	Nonagouti (a/a)	Yellow (A^y)	Pink-eyed (p/p)
Full color (C/C)	Black	Orange	Tawny
Intense chinchilla (c^i/c^i)	Black	Cream	Buff
Chinchilla (c^{ch}/c^{ch})	Sepia	Ivory	Cream
Extreme dilute (c^e/c^e)	Pale brown	White	White
Albinism (c/c)	White	White	White

[a]Feldman (1935).
[b]From H. Grüneberg (1952).

tests revealed that it carried a mutant allele at the *c*-locus which was named "platinum" and designated c^p. $a/a;B/-;c^p/c^p$ animals are lighter than the corresponding c^e/c^e genotype and have pink eyes. $a/a;c^p/c$ heterozygotes have a phenotype intermediate between c^p/c^p and albinism (Dickie, 1966a). The most interesting feature of this mutation is its expression in lethal yellow mice. Inasmuch as c^p/c^p has a more drastic effect on eumelanin than c^e/c^e and the latter produces black-eyed whites in combination with $A^y/-$, one might expect $A^y/a;c^p/c^p$ mice likewise to be white but with pink-eyes, i.e., indistinguishable from true albinos. However, this is *not* the case for while such animals have pink-eyes they nevertheless are pigmented, resembling $a/a;c^p/c^p$ but with a definite yellowish tinge (Plate 2-C) (Poole and Silvers, unpublished).[13] This mutation is therefore enigmatic in that it has a greater influence than c^e/c^e in inhibiting eumelanin synthesis but a less drastic effect than c^e/c^e on phaeomelanin production.

6. Himalayan (c^h)

Inasmuch as mutant genes which produce light or white body fur and dark extremities are common in mammals, e.g., the Himalayan rabbit, Siamese cat, "albino" guinea pig, and "partial albino" hamster, and the mutation responsible for this phenotype appears in all cases to be one of an allelic series similar to the *c*-series in mice, it is surprising that it has been only within the last two decades that a "himalayan" mutation has been recorded in mice.[14] The deviant animal occurred also in an AKR/J ($a/a;c/c$) × DBA/2J ($a/a;b/b;d/d$) litter and breeding tests confirmed that a mutation at the *c*-locus, appropriately designated "himalayan" (c^h), was responsible for its phenotype (M.C. Green, 1961).

According to M.C. Green (1961) $a/a;c^h/c^h$ mice (Plate 2–D) "are indistinguishable from albinos at birth. At about one week of age they are slightly darker than albinos, particularly on the tail. The juvenile coat, when fully grown, is pale tan with little evidence of darkening of the extremities except on the tail. The body color may be uniform or may be slightly lighter across the shoulders and darker toward the tail. At the first molt the nose, ears, tail and scrotum, but not the feet, become considerably darker, and the rest of the body becomes lighter. The dark hair on the nose extends back to about the anterior border of the eyes. The body fur, both dorsal and ventral, is lightest on the anterior half of the body and darkens gradually from the middle of the body back to the tail. There is often a particularly dark ring of body fur next to the tail and the skin of the scrotum may be very dark. The ears are darkest on their anterior border. The feet never become dark as they do in the Himalayan rabbit. At subsequent molts the body may become somewhat lighter and the extremities darker, but there is considerable variation." Green also notes that "the eyes of c^h/c^h mice are not pigmented at birth but become darker with age, and are ruby colored at weaning."

Of course the most interesting feature of the himalayan mouse is that, like the Himalayan rabbit (Schultz, 1915) and Siamese cat (N. Iljin and V. Iljin, 1930), its ability to produce melanin is promoted by low temperatures (M.C. Green, 1961; Coleman, 1962; Moyer, 1966).

When heterozygous with albinism, himalayan (c^h/c) mice are indistinguishable at birth from either c/c or c^h/c^h homozygotes (they all lack pigment). Within about a week, however, they develop a very pale buff phenotype which is roughly intermediate between the color of c^h/c^h and albino mice. Moreover, this intermediate level of pigmentation persists since the extremities of c^h/c mice never become as dark as in c^h/c^h animals.

The expression of an intermediate level of pigmentation also occurs when c^h is heterozygous with c^e and c^{ch}. On an agouti background, c^h/c^e mice are dark eyed from birth with a pale coat which subsequently darkens at its extremities. Consequently the extremities of the adult heterozygote are darker than the extremities of the corresponding c^e/c^e homozygote, but lighter than the extremities of c^h/c^h mice. On the other hand, the rest of the body displays a level of pigmentation lighter than $A/-;c^e/c^e$ but darker than $A/-;c^h/c^h$ adult mice (M.C. Green, 1961). c^{ch}/c^h heterozygotes are significantly lighter than c^{ch}/c^{ch} animals. Whereas, as noted above, the $a/a;c^{ch}/c^{ch}$ mouse is a dull black or sepia color, the $a/a;c^{ch}/c^h$ adult is a light golden brown. Furthermore, although the nose and tail of this animal are darker than the rest of the body, they too are not as dark as in the c^{ch}/c^{ch} mouse (M.C. Green, 1961). Thus the himalayan allele like the other lower alleles of the c-locus produces intermediate phenotypes when heterozygous. However, unlike the other alleles, c^h appears to occupy two positions in the c-series hierarchy. While in depth of color produced it is below c^{ch} in the series, it is above c^e in color of the extremities but below this allele insofar as its effect on the color of the trunk is concerned.

7. Chinchilla-mottled (c^m)

The final c-locus allele chinchilla-mottled (c^m)—originally known as c^{22H}—was found in the progeny of a neutron-irradiated male (R.J.S. Phillips, 1966b). c^m/c^m and c^m/c^{ch} animals on a nonagouti, black ($a/a;B/-$)

background are mottled displaying patches of normal c^{ch}/c^{ch} fur and patches of a lighter fur fairly similar to c^{ch}/c in intensity (R.J.S. Phillips, 1966b, 1970a). The amount of each color varies considerably. c^{m}/c^{m} mice can be distinguished from c^{m}/c^{ch} heterozygotes by their whiter belly color (R.J.S. Phillips, 1970a). While there is no evidence that this variegation is caused by a translocation, the c^{m} phenotype is influenced by a dominant modifying gene, probably independent of *c*, which has provisionally been called "modifier" and designated $M(c^{m})$. $c^{m}/c^{m};M(c^{m})/M(c^{m})$ mice are almost white; they also appear to be smaller than normal (R.J.S. Phillips, 1970a) (see Chapter 7, Section VI).

8. Other Alleles

In addition to the specific alleles noted above, L. Russell (1979) has reported the occurrence of 16 mosaic, or fractional, *c*-locus mutants which were characterized by area(s) of lighter fur or mottling.[15] Although these deviants arose in the course of radiation experiments, they probably were not radiation induced. If they were, this would indicate that the bulk of spontaneous *c*-locus mutations are fractionals (L. Russell, 1979). Since in this group of mutants about one-half of the germinal tissue carried the mutation, it appears that they were derived from an overall blastomere population that was one-half mutant. As pointed out by Russell, "such a population could result from mutation in one strand of the gamete DNA; in a daughter chromosome derived from pronuclear DNA synthesis of the zygote; or in one of the first two blastomeres prior to replication."

A considerable number of *lethal* *c*-locus deviants have also been recovered from radiation experiments (see Russell et al., 1979). Many of these deviants, in addition to preventing melanin synthesis (see Rittenhouse, 1970), have other very different effects. Indeed, because they are associated with such a diverse, and seemingly unrelated, series of abnormal conditions they undoubtedly represent chromosomal deletions for genetic material other than just the albino locus (Erickson et al., 1968, 1974a; Gluecksohn-Waelsch et al., 1974; L. Russell et al., 1979; L. Russell and Raymer, 1979).[16]

C. Dominance of C

Before leaving the subject of the phenotypic effects of *c*-locus alleles, it should be noted that there are some situations in which *C* is not completely dominant over the other alleles of the series. This occurs in the presence of p/p (pink-eyed dilute); the genotypes $a/a;B/B;p/p;C/c$ and $a/a;b/b;p/p;C/c$ are clearly lighter in color than the corresponding C/C types (Snell, 1941). It likewise occurs in lethal yellow, black-and-tan (a^{t}), and agouti mice (and presumably in all phaeomelanin containing regions of agouti-locus genotypes). In A^{y} (and A) mice C/c, C/c^{h}, and C/c^{e} genotypes are noticeably lighter, especially on the ventrum, than the corresponding *C*-homozygote. The influence in agouti animals is limited to the yellow portion of the hair and in a^t/a^t animals to the ventrum.[17] Contrary to what might be expected, C/c^e genotypes appear slightly lighter than C/c mice (Poole and Silvers, unpublished). That this affect is a consequence of a direct interaction between the *c*- and *a*-loci, and not due to a general influence of the *c*-locus on phaeomelanin synthesis, is provided by the fact that it does not occur in recessive yellow mice. Thus $a/a;e/e;C/C$ and $a/a;e/e;C/c$ animals are indistinguishable (Poole and Silvers, unpublished).[18]

D. Effect on Pigment Granules

Dunn and Einsele (1938) compared the amount of pigment present in the hairs of both black and brown agouti animals bearing different *c*-locus alleles and found that, in general, the reduction in intensity of hair color was accompanied by a parallel graded reduction in the amount of melanin as measured by weight. Moreover, since they found that in combinations with black (but not brown) the chief tangible factor accompanying this decrease was the size of the pigment granule, they concluded that it was via this influence on granule size that the *c*-locus controlled the quantity of melanin produced in black animals. Although this effect on granule size was confirmed by E. Russell (1946, 1948, 1949a), her more detailed investigations revealed that in some cases changes in the level of pigmentation produced by the albino series also involved the number, shape, color intensity, and distribution of the granules in the hair.

Some estimates of the total eumelanin present in B/B and b/b animals of different *c*-locus genotypes, as reported in the studies of Dunn and Einsele (1938) and E. Russell (1948), are presented in Table 3-3. It is clear from

Table 3-3. The Total Eumelanin in B/B and b/b as Affected by the Albino Series[a,b]

Genotype	Black (B/B)	Brown (b/b)	Black (B/B)	Brown (b/b)
$C/-$	7.2 (20)	3.4 (14)	1332	480
c^{ch}/c^{ch}	5.0 (18)	3.4 (4)	695	431
c^{ch}/c^e	2.8 (19)	3.4 (3)		
c^{ch}/c	2.9 (10)	2.7 (12)		
c^e/c^e	1.8 (12)	1.7 (7)	147	159
c^e/c	1.4 (7)	1.3 (5)		
c/c	1.2 (14)		0	0

[a]On the left, the percentage weight of melanin + impurities (number of estimations in paretheses) after Dunn and Einsele (1938). On the right proportional values based on histological methods, from E.S. Russell (1948).

[b]From H. Grüneberg (1952).

Table 3-4. Effect of the Albino Series on Pigment Granule Characteristics[a,b]

Genotype		Granules per meduleary cell	Mean granule diameter (μm)	Granule shape	Granule Color
$a/a;B/B;$	C/C	90 ± 2	1.44 ± 0.013	Long oval	Intense black
	c^{ch}/c^{ch}	86 ± 2	1.05 ± 0.010	Oval	Fuscous black
	c^{e}/c^{e}	39 ± 2	0.94 ± 0.014	Round	Fuscous
$a/a;b/b;$	C/C	90 ± 2	0.77 ± 0.007	Round	Carob-brown
	c^{ch}/c^{ch}	85 ± 2	0.79 ± 0.008	Round	Carob-brown
	c^{e}/c^{e}	43 ± 2	0.77 ± 0.012	Round	Carob-brown
$A^{y}/a;B/B;$	C/C	44 ± 2	0.83 ± 0.013	Round	Aniline yellow
	c^{ch}/c^{ch}	30 ± 1	0.77 ± 0.009	Round	Aniline yellow
	c^{e}/c^{e}	0			
$A^{y}/a;b/b;$	C/C	40 ± 2	0.82 ± 0.010	Round	Aniline yellow
	c^{ch}/c^{ch}	24 ± 1	0.76 ± 0.010	Round	Aniline yellow
	c^{e}/c^{e}	0			

[a]Based on data of E.S. Russell (1946b).
[b]From H. Grüneberg (1952).

these data that *c*-locus substitutions reduce pigment much less in brown than in black mice. Indeed as already noted, on a nonagouti brown background the genotypes $C/-$, c^{ch}/c^{ch}, and c^{ch}/c^{e} all possess the same amount of pigment. This is especially conspicuous in brown-and-tan ($a^{t}/-;b/b$) mice where these *c*-locus genotypes all have an intensely pigmented (chocolate) dorsum but can readily be distinguished by the color of their phaeomelanin-containing ventral hairs.

More detailed data on the effect of three albino series alleles (C, c^{ch}, and c^{e}) on a number of pigment granule attributes in black, brown, and yellow mice, as derived from the data of E. Russell (1946) by Grüneberg (1952), are given in Table 3-4. Here it can be noted that in $a/a;B/B$ genotypes the number of pigment granules is not reduced significantly from step C/C to c^{ch}/c^{ch} (see Figure 4-6) but is very significantly reduced in c^{e} homozygotes. On the other hand, there is a progressive diminution of granule size from C/C to c^{e}/c^{e} accompanied by, and believed to be responsible for (E. Russell, 1949a), changes in granule shape and color intensity. Hence, whereas the significant reduction in color intensity displayed by $a/a;B/B;c^{e}/c^{e}$ mice results predominantly from a reduction in the number of granules, the much less conspicuous phenotypic difference between $a/a;B/B;C/C$ and $a/a;B/B;c^{ch}/c^{ch}$ animals is primarily a consequence of granule size (see Figure 4-6).

In brown animals the albino series does not influence granule size, though, as note in Table 3-4, it reduces granule number. In yellow mice, however, as well as in the yellow-pigmented regions of other agouti-locus genotypes, *c*-locus substitutions have a slight influence on granule size although here, too, their predominant effect is on granule number.

It therefore appears that in black mice there are a maximum number of pigment granules that can be produced and since this level is attained in c^{ch}/c^{ch} animals, any further increase in the quantity of pigment must stem from an increase in granule size. In brown mice, on the other hand, the influence of *c*-locus alleles seems to be limited both by the number of granules and their size, i.e., there seems to be a size beyond which b/b granules cannot grow, and since *both* of these attributes reach a maximum in c^{ch}/c^{ch} (actually c^{ch}/c^{e}) mice, this allele has no diluting influence on the chocolate phenotype (E. Russell, 1949b).

E. Effect on Tyrosinase Activity

Since all the evidence indicates that the action of the *c*-series of alleles is a general one, affecting the level of the whole pigmentation reaction rather than influencing the type of pigment produced, it should come as no surprise that these alleles appear to produce their effect by controlling the activity of tyrosinase.

Evidence for this was initially obtained in mice by L. Russell and W. Russell (1948) who incubated lightly fixed, 30-μm sections of skin, from 6- to 7-day-old animals, of different *c*-locus genotypes, in either a buffered dopa or "control" solution. The change in the intensity of the pigment in the hair follicles resulting from this treatment was then compared to a series of "standards" and graded. Although this method demonstrated that the intensity of the pigment formed corresponded to what would be expected from visual examination of the genotypes, i.e., $C/C > c^{ch}/c^{ch} > c^{e}/c^{e} > c/c$, the reactions much more closely reflected the influence of these genotypes on phaeomelanin than on eumelanin synthesis, regardless of whether the piece of skin possessed yellow pigment or not (see also W. Russell et al., 1948).

More recently Coleman (1962) has determined the amount of [2-^{14}C] tyrosine that is incorporated into slices of skin from infant mice of different *c*-locus genotypes and although his results (Table 3-5) are similar to those of the Russells', they nevertheless provide some new and significant information. Probably the most significant is that while $a/a;B/B;C/c$ and $a/a;B/B;C/c^{h}$ animals are indistinguishable phenotypically from the corresponding *C*-homozygote, they incorporate only about 50% as much tyrosine. This suggests that at this biochemical level there is no dominance of *C* over other *c*-series alleles. It may also explain why in the presence of those *a*-alleles which promote the synthesis of phaeomelanin, C/c heterozygotes produce less of this pigment than *C*-homozygotes. Coleman's results, as noted in the table, also indicate that $a/a;c^{ch}/c^{ch}$ mice incorporate less tyrosine than either $a/a;C/C$ or $a/a;C/c$ animals, a result consistent with their slightly lighter phenotype, and that the level of incorporation is still fur-

Table 3-5. Effect of the *C*-Locus on the *in Vitro* Incorporation of DL-[2-^{14}C]Tyrosine into Pigment[a]

Origin of skin	Genotype[b]	Tyrosine incorporation[c] (cpm)
Wild	$a/a;C/C$	1200 ± 36
Wild × albino	$a/a;C/c$	617 ± 33
Chinchilla	$a/a;c^{ch}/c^{ch}$	442 ± 15
Chinchilla (pink-eyed)	$a/a;c^{ch}/c^{ch};p/p$	350 ± 5
Chinchilla × albino (pink-eyed)	$a/a;c^{ch}/c;p/p$	151 ± 17
Himalayan	$a/a;c^h/c^h$	225 ± 15
Himalayan × wild	$a/a;C/c^h$	676 ± 47
Extreme dilution	$A^w/A^w;c^e/c^e$	98 ± 11
Albino	$a/a;c/c$	47 ± 5
Boiled skin	$a/a;C/C$	13 ± 8

[a]Based on D.L. Coleman (1962).
[b]Only alleles deviating from wild type are listed.
[c]Values represent cpm/mg of dried protein ± standard error of the mean.

ther reduced in the even more lightly pigmented $a/a;c^{ch}/c^{ch};p/p$ and $a/a;c^{ch}/c;p/p$ genotypes. The values for animals of these last two genotypes are included to emphasize that although this situation is similar to $a/a;C/C$ and $a/a;C/c$ in that the amount of tyrosine incorporated by the heterozygote is about half that of the homozygote, in this instance homozygotes (c^{ch}/c^{ch}) and heterozygotes (c^{ch}/c) are easily distinguished visually.[19]

The results with himalayan mice are particularly interesting inasmuch as, already noted, their phenotype is temperature dependent; if raised in a cold environment their pigmentation becomes significantly darker and, conversely, at warm temperatures, lighter.[20] This indicates that in these animals tyrosinase is heat labile and in accord with this is the fact that when skin slices from 5-day-old C/C and c^h/c^h animals are incubated at 55°C for periods up to 1 hour, the incorporation of tyrosine is decreased by 10% in the C/C skin but by about 70% in the himalayan skin. This thermolability of tyrosinase under the influence of the c^h allele is important because it suggests that perhaps all alleles at the *c*-locus control the protein *structure* and not the quantity of the enzyme (Coleman, 1962; see also Foster, 1967; Foster et al., 1972).[21]

F. Influence on Forms of Tyrosinase

As already noted a number of investigations have disclosed that there are multiple forms of tyrosinase in mice (Holstein et al., 1967, 1971; Burnett et al., 1969) and some of these are influenced by the *c*-locus (Wolfe and

Coleman, 1966). Thus an electrophoretic study of tyrosinases isolated from the skin of different *c*-locus genotypes has revealed not only quantitative but qualitative differences (Wolfe and Coleman, 1966). Tyrosinase from *C*/*C* mice produce two distinct tyrosinase bands on electrophoresis whereas similar preparations from c^{ch}/c^{ch} animals produce only a single, albeit faster moving, one. Himalayan tyrosinase also displays two tyrosinases, a fast moving one which corresponds to the fastest-moving wild type form and one which moves much slower than its counterpart in wild type mice (Wolfe and Coleman, 1966).

If the major influence of the *c*-series of alleles is on the structure of tyrosinase, it is difficult to explain how such a structural change in the protein could influence both the size and number of pigment granules. One possibility, however, is that the abnormal protein is not produced at a normal rate, or, more likely, that it is unable to conjugate properly with the other proteins destined to form the melanin matrix (Coleman, 1962). On the other hand, there is some evidence which suggests that the albino locus might be more correctly regarded as a regulator rather than a structural locus for tyrosinase (Hearing, 1973; see also Chian and Wilgram, 1967; Pomerantz and Li, 1971).

G. Ultrastructural Effects

The influence of the *c*-locus on the ultrastructure of the melanin granule has received considerable attention. Moyer (1966) examined the granules of chinchilla, himalayan, extreme dilute, and albino mice and concluded that the development of the granule in albino genotypes parallels that in nonalbinos. Indeed, the only difference he noted was that the granules of albino mice did not possess any melanin and, therefore, did not display any second-order periodicity (Figure 3-14). In contrast all pigmented *c*-locus genotypes were characterized by a second-order periodicity which was determined by their *b*-locus constitution, an observation consistent with the fact that the *c*-locus affects the amount, rather than the quality, of melanin. Moyer also observed that whereas there was no *obvious* difference in the size, shape, and number of c^{ch}/c^{ch},c^{h}/c^{h} and *C*/– retinal granules, their number, as well as their size, was reduced in c^{e}/c^{e} and *c*/*c* mice. This coincides with the effect these alleles have on the granules of the hair.

Hearing and his colleagues (Hearing et al., 1973) also found that the premelanosomes formed in the retinal and choroid of albino mice were similar to those of pigmented animals except for the absence of melanin. They also noticed in both of these locations that the number of premelanosomes decreased after several weeks postpartum.

Rittenhouse (1968b), on the other hand, has reported that the fine structure of albino and pigmented granules is significantly different. As previously noted (see Section I, F) she found that there was a distinct dif-

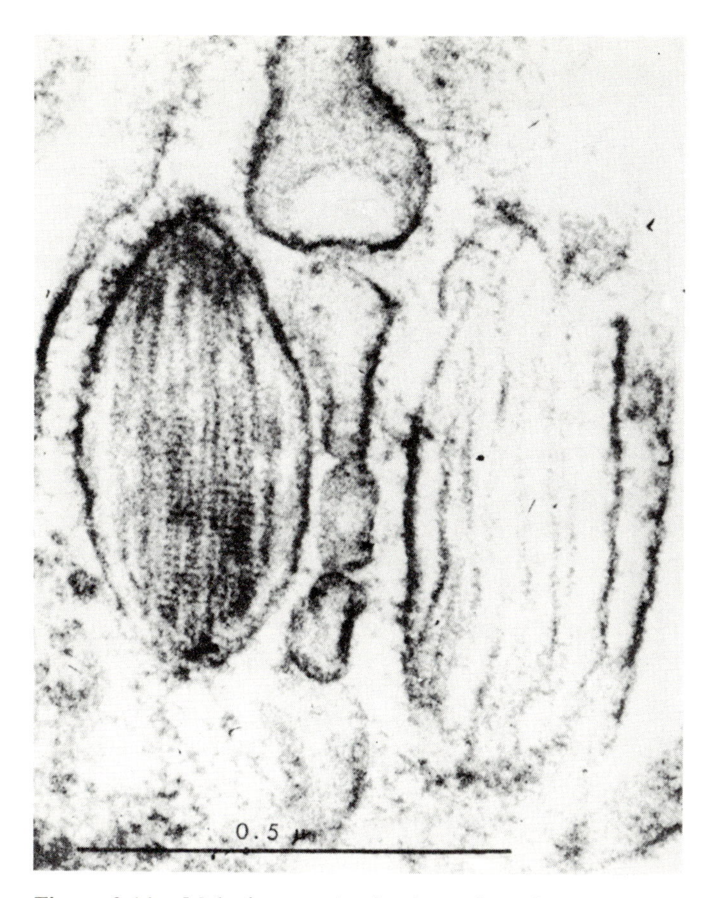

Figure 3-14. Melanin granules in the retina of a 15-day-old albino mouse. The granule on the left is mature and is "normal" in all respects except that no pigment has been deposited on the thickened compound fibers. It is not possible to determine whether the granule on the right is a developmental or a degenerative stage. From F.H. Moyer (1966). Reproduced with permission of the *American Zoologist*.

ference in the ultrastructure of *a*/*a*;*B*/*B* and *a*/*a*;*b*/*b* hair bulb granules and consequently one might have anticipated that *B*/*B* and *b*/*b* albino granules would likewise be different. This, however, was not the case. According to Rittenhouse the failure of albino granules to melanize is accompanied by changes in granule structure such that there appears to be no distinct difference between the hair bulb granules of brown-albino and black-albino mice. Animals of *both* genotypes contain some granules which resemble those of the "black" type (ovals with longitudinal strands, or small circles with spiral or circular patterns) and others which are more typically those of the "brown" variety (large circles with a complex disorderly internal structure). Since this absence of melanization occurs along with a shift toward the "black" pattern in *b/b;c/c* melanocytes, these two changes could be related. Accordingly Rittenhouse suggests that the disorganized framework of *b/b* granules may tend to acquire the more orderly organization of black granules if not stabilized almost immediately by at least a light melanization. Whether this is the case and/or whether at least some of the "brown" type granules represent granules which, because they are not melanized, are disintegrating (Rittenhouse, 1968b), remains to be resolved. Regardless, Rittenhouse (1968b) observed that this shift toward a pattern of longitudinal strands and membranes characteristic of *B/B* granules also occurs when *p/p* and *b/b* are incorporated into the same genome (see Chapter 4, Section II, E).

H. Influence on Central Visual Pathways

In addition to an absence of melanin, the central visual pathways of albino mice [and of other albino mammals (Lund, 1965; Guillery et al., 1971; Creel, 1971; Witkop et al., 1976)] are abnormal. Parts of the retina that normally give rise to uncrossed retinofugal axons send axons across the midline. Although the manner in which the albino gene acts upon the retinal cells to affect the chiasmic growth of their axons is not known, Guillery and his associates (1973) have shown that its influence is *extracellular*. This was demonstrated by taking advantage of Cattanach's *flecked* translocation (Cattanach, 1961, 1974; Ohno and Cattanach, 1962; for excellent review see Eicher, 1970b). This translocation involves the transfer of the normal allele at the albino locus to the X-chromosome where, in females, it is subject to "Lyonization" (see Chapter 8, Section I). Thus female mice which are homozygous for *c* and heterozygous for this translocation display the characteristic albino-variegated or flecked coat, and in the pigment layer of the retina one sees small patches of pigmented cells intermingled with patches of albino cells (Deol and Whitten, 1972). Gullery and his colleagues reasoned that if, in the production of the abnormal chiasmatic pathway, each ganglion cell acted independently and the specification of a cell as ipsilateral or contralateral depended upon the enzymes produced within that cell, then a flecked mouse should have an abnormal pathway approximately half as large as that of an albino mouse. This seemed especially reasonable inasmuch as the effect which *c* has on pigmentation is intracellular. On the other hand, if the mechanism responsible for the abnormality was intercellular, and involved materials which diffused between cells, or involved mechanical interactions between cells, then flecked mice could display an abnormality equal to that of albino mice, a defect somewhere between albino and normal animals, or no abnormality at all. This latter situation apparently prevails as in contrast to the albino mouse where the ipsilateral component is, as expected, markedly smaller than in pigmented animals, in flecked mice this component is, surprisingly, somewhat larger than in pigmented mice.

The precise relationship between the influence of the albino locus on pigment formation and on the normal development of the eye remains to be determined. On the one hand the lack of tyrosinase may affect the concentration of some other substance which can diffuse between cells and influence the course taken by some of the ganglion axons; on the other hand, *c* could affect a process that is independent of tyrosinase synthesis, and this process could act upon the ganglion cells (Guillery et al., 1973).

I. Influence on Behavior

Before concluding this section there is one investigation which should be noted because it provides good evidence that albinism (or the absence of this condition) can influence behavior. Taking advantage of a mutation from *C* to *c* in the C57BL/6 strain, Fuller (1967) demonstrated that the performance of mice in certain test situations was affected solely on the basis of their *c*-locus genotype. *c*/*c* mice escaped more slowly from water, were less active in an open field, and made more errors on a black-and-white discrimination task than *C*/– animals. Although the cause of these differences in performance is not known, it raises the possibility that other coat color determinants also may influence this kind of behavior.[22]

Notes

[1]Cordovan received its name from the fact that it produces a phenotype which resembles the rich brown of cordovan leather. The mutation was recognized in an F_1 litter from a cross of a C57BL/10 male by a DBA/1 female. Hence the mutation must have originated in the C57BL/10 stock. The same male which sired the litter containing the mutant sired six other F_1 litters (five by the same female), in none of which cordovan appeared (D.S. Miller and Potas, 1955).

[2]According to Moyer (1966) in melanocytes of neural crest origin brown granules average about 0.5×0.8 μm whereas black granules average about 0.5×1.5 μm.

[3]There have been some discrepancies in the utilization of the terms "premelanosome," "melanosome," and "melanin granule." As originally proposed (Seiji et al., 1961) the term *premelanosome* was defined as a distinctive particulate protein matrix upon which melanin was usually deposited with consequent formation of the *melanosome*. Thus the *melanosome* represented a premelanosome after the onset, but prior to the completion, of melanin synthesis which characteristically possessed an active tyrosinase system. The *melanin granule*, on the other hand, represented a melanin-containing organelle in which melanization was complete and no tyrosinase activity could be detected. However, because certain difficulties arose in the utilization of these terms (see Fitzpatrick et al., 1967) in 1967 Fitzpatrick and his associates proposed new definitions. They requested that the term *melanosome* be employed to designate the fully pigmented melanin-containing organelle only. The *premelanosome* would then be used to designate all stages in the genesis of melanosomes that precede the fully developed state. Finally, they suggest that the

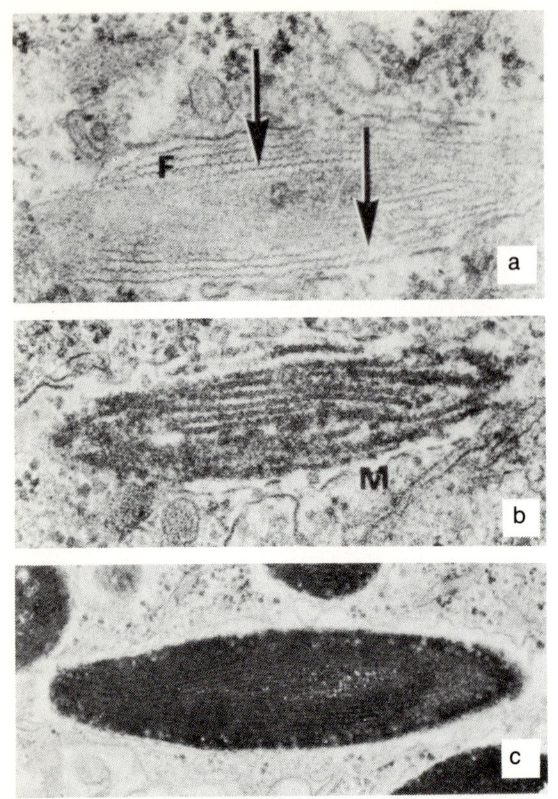

Figure 3-15. Three stages in the development of black retinal granules prepared from 20-day-old embryos. In (a) the fibriller nature of the premelanosome is seen. The filaments (F) are spaced about 200 Å from each other. The pitch of the helices seen in these filaments (arrows) is about 150 Å. Approx. ×40,000. In (b) a partial melanization of the filaments has occurred and the limiting membrane (M) of the organelle can be seen. Approx. ×46,000. In (c) melanin deposition is almost complete but filaments can still be resolved, particularly near the center of the granule. Circular electron lucent areas, with a diameter of 200–300 Å, are evident near the periphery of the granule. Approx. ×25,000. From V.J. Hearing et al. (1973). Reproduced with permission of the authors and Academic Press, Inc.

term *melanin granule* be used to include all melanin-containing particulates that can be observed with light microscopy.

[4]According to Moyer (1966), "inspection of the fibers in melanin granules at high magnifications reveals further structural details. The unit fiber is approximately 35Å in diameter and generally appears as a loosely coiled helix with a variable pitch of several hundred Ångström units. The compound fibers are approximately 90 Å in diameter. In longitudinal section they appear to be hollow cylinders with a lumen approximately 35 Å in diameter and walls approximately 35 Å thick. In cross section they do not appear to be hollow. The cross-links that form between compound fibers during their aggregation into matrix sheets are fairly uniform in size. They are approximately 100 Å long and 30 to 40 Å wide. Thus, the final matrix on which melanin is deposited appears to be a lattice of parallel fibers spaced approximately 100 Å apart and cross-linked by fibers approximately 35 Å in diameter. Cross sec-

tions of melanin granules show that this flat lattice-like sheet is rolled up like a rug to give the mature granule its three dimensional form."

[5]The ontogeny of *a*/*a*,*B*/*B* pigment granules has also been described by Lutzner and Lowrie (1972) and by Hearing et al. (1973). Hearing and his colleagues observed that premelanosomes in the retina and choroid of *a*/*a*;*B*/*B* mice are fibrillar in appearance (Figure 3-15a), measure about 35–45 Å in diameter, and sometimes possess a helical or zigzag structure. "The axial diameter of the filaments is about 60 Å, and the pitch of the helices is between 150 and 200 Å." Occasionally a striated periodicity is visible, oriented perpendicular to the length of the filament, and is in the order of 95–100 Å. Moreover, sometimes "the filaments are arranged in phase so that the striated periodicity observed imparts a periodicity on the entire granule." Melanin is deposited rather uniformly on the filaments as the granule matures (Figure 3-15b), and as this deposition builds up the filaments fuse (figure 3-15c). Frequently in mature granules, "circular electron lucent areas are observed, especially at the periphery of the granule; these measure 200–300 Å in diameter" (Figure 3-15c). Lutzner and Lowrie's observations, which also were confined to granules of the eye (choroid, retina, and iris), are similar to those of Hearing *et al.* Their paper includes a schematic diagram of how they believe melanin formation proceeds and this diagram is reproduced here (Figure 3-16) as a means of providing a concise summary of their

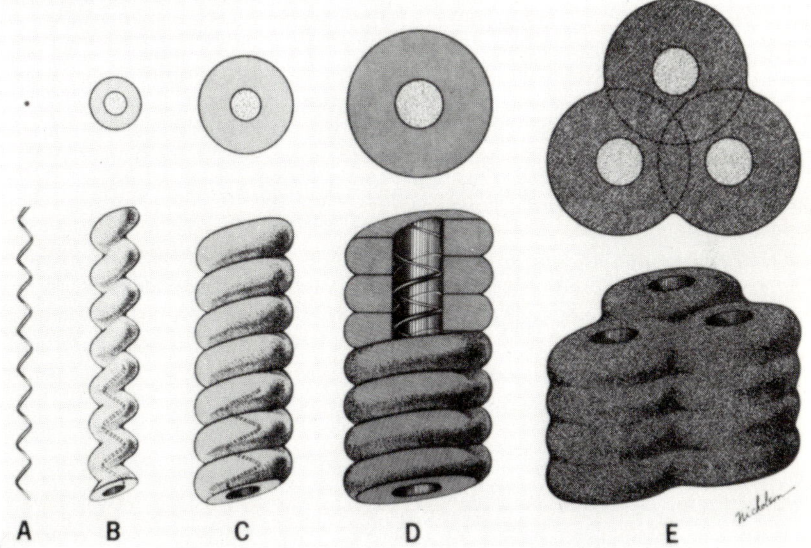

Figure 3-16. Schematic diagram of speculative sequence of melanin formation as viewed by M.A. Lutzner and C.T. Lowrie (1972). The melanofilament (A) is 30 Å thick, coiled into a helix with an axial diameter of 90 Å and a pitch of 150 Å. This melanofilament thickens (B) and becomes more electron dense. It thickens even more, approaching a tubule in contour (C). The early melanotubule (C) at this stage has an axial diameter of 400 Å, which widens eventually to 600 Å in diameter (D). The walls of this mature melanotubule (D) are 200 Å thick and the lumen is about 200 Å in diameter. This mature melanotubule represents a 20-fold increase over the diameter of the melanofilament. Crossbars within the lumen of the melanotubule may be a helix with the same approximate dimensions as the melanofilament. Mature melanotubules may partially fuse as shown in (E). Reproduced with permission of the authors and Plenum Publishing Co.

views. In contrast to Moyer (see note 4; and it should be noted that Moyer's unit and compound fibers compare with Lutzner and Lowrie's melanofilaments and youngest melanotubules, respectively), these investigators suggest that "rather than cross-linking to form matrix sheets, young melanin tubules widen to 600 Å in diameter and then partially fuse." Moreover, they prefer the notion that melanin is deposited around each melanofilament and that the melanin itself forms the walls of the melanotubule, rather than being deposited on a matrix sheet of aggregated compound fibers.

[6]Since, as noted (Table 3-1), tyrosinase activity in b^c/b^c skin is intermediate between that of $B/-$ and b/b mice this hypothesis would be strengthened if it was found that the second-order periodicity of b^c/b^c melanosomes was likewise intermediate between those of black and brown melanosomes. As far as I am aware this information is not available.

[7]Moyer (1966) also points out that as a result of the fact that black granules are finely granular and brown granules flocculent (see also Hearing et al., 1973), black granules appear much denser than brown granules; they also tend to shatter more readily during sectioning than brown ones.

[8]The fact that clumping of eumelanotic pigment granules is significantly reduced in mice heterozygous for both B^{lt} and p is surprising. Since some clumping is observed in pink-eyed non-B^{lt} mice (see Chapter 4, Section II, D), more clumping might be anticipated in darks ($B^{lt}/-$) heterozygous for p than in P/P darks (Quevedo and Chase, 1958).

[9]On the other hand, the possibility should not be overlooked that the primary affect of B^{lt} is on the capacity of the melanocyte to extend dendrites. If such an effect influenced the rate at which these cells delivered granules to keratinocytes, this, in turn, might influence their stability in the bulb (S. Sweet and Quevedo, 1968).

[10]Mayer and Oddis (1977) have isolated amelanotic melanocytes from cultures of embryonic albino epidermis. The isolation procedure was based upon the fact that while melanoblasts differentiate *in vitro* only when in contact with epidermal cells, once differentiated these cells persist even though the epidermal cells do not survive in culture for more than 11 days. When 14-day-old pigmented ($a/a;b/b$) and albino (c/c) embryonic epidermis, respectively, were cultured *in vitro* the pure populations of cells which persisted were identical in every respect save that the former were pigmented while the latter were not. On the other hand, no such cells were found when similar procedures were carried out on epidermal explants from white-spotted (W/W and S_l/S_l) mutant embryos (see Chapter 7, Section VII).

[11]Detlefsen referred to this mutation as c^d.

[12]In 1967 A.J. Bateman described an $a/a;c^{ch}/c$ mouse which possessed two differently colored areas on each side of the dorsum just posterior to the midline of the trunk. One of these patches was a medium shade of sepia, i.e., it resembled the $a/a;c^{ch}/c^{ch}$ phenotype; the other was white. To explain this mosaicism Bateman suggests that it resulted from somatic crossing-over between the *c*-locus and the centromere in a primordial melanoblast of the neural crest, and that the complementary daughter cells (c^{ch}/c^{ch} and c/c) lined up on opposite sides of the mid-dorum (see Chapter 7 and Chapter 8, Section I). An alternative explanation for the twin spots would be nondisjunction of the c^{ch}-bearing chromosome, producing complementary clones which would be trisomic ($c^{ch}c^{ch}/c$) and monosomic (c). However, because

chromosome 7 is quite large and important, Bateman does not believe that this explanation is as likely since the viability of the monosomic line would be doubtful. Other possible examples of somatic crossing-over in the mouse include phenotypes described by Keeler (1931), Dunn (1934), and Carter (1952; see Chapter 10, note 7). For a detailed review on the case for somatic crossing-over in mice and other mammals see Grüneberg (1966a). For other mechanisms by which mosaicism may be produced see Robinson (1957), L. Russell (1964), and L. Russell and Woodiel (1966).

[13]$A^y/a;c^p/c^p$ mice bear a striking resemblance to $A^w/A^w;c^{ch}/c;p/p$ animals, i.e., c^{ch}/c heterozygotes of strain 129.

[14]In 1939 Mohr described a mouse caught on the German island of Spiekerog, which resembled the Himalayan rabbit in its fur pigmentation. Although no breeding tests were carried out with this animal it is quite likely that its phenotype also resulted from a *c*-locus mutation to "himalayan." An independent mutation to c^h has also been reported by Wallace (1972). It occurred in a $c^{ch}/c^{ch} \times c^{ch}/c^{ch}$ mating. According to Wallace $a/a;c^h/c^h$, unlike $A/A;c^h/c^h$, "are darker in the nest than as adults."

[15]All but one of these fractional mutants were fully viable when homozygous; a situation which contrasts with those for the *d*-locus where only 3 of 15 tested spontaneous mutations proved viable, most of the remainder being of the d^{op} type (see Chapter 4, note 8) (L. Russell, 1971).

[16]These lethal albino "alleles," which have been studied most extensively by Gluecksohn-Waelsch and her associates (Erickson et al., 1968; Thorndike et al., 1973; Trigg and Gluecksohn-Waelsch, 1973; Erickson et al., 1974a,b; Gluecksohn-Waelsch et al., 1974, 1975; D.A. Miller et al., 1974; Garland et al., 1976; S. Lewis et al., 1976), include the following: c^{6H} and c^{25H}, which are early embryonic lethals and c^{14CoS}, c^{112K}, c^{65K}, and c^{3H} which are lethal within a few hours after birth [nevertheless c^{3H}/c^{6H} compounds are viable but stunted and sterile, usually dying at approximately 2 months of age (Searle, 1961)]. Studies of these latter four "alleles" have revealed ultrastructural abnormalities of the endoplasmic reticulum and Golgi apparatus in the liver and kidney of newborn homozygotes. These abnormalities are associated with deficiencies of at least three enzymes (glucose-6-phosphatase, tyrosine aminotransferase, and serine dehydratase) which in normal mice increase in the liver and kidney shortly after parturition. In addition, some serum proteins have been found to be significantly lower in these neonatal albino deviants. It has been suggested that the mutational effects observed in these animals result from deletions involving regulatory rather than structural genes at or near the *c*-locus (Garland et al., 1976). Twenty four other lethal albino deviants known to represent deficiencies have been studied by L. Russell and Raymer (1979). These include 13 which are lethal before implantation. 10 shortly after implantation, and 1 neonatally.

[17]Microscopic examination of cleared hairs from $A^y/a;C/c$ as well as from the ventrum of $a^t/a^t;C/c$ mice has confirmed the visual reduction in phaeomelanin. In these hairs yellow granules are located in the distal portions of the shaft with virtually no granules in the proximal regions. This reduction in granule number is most striking in ventral hairs where, in some instances, they are completely absent (Poole and Silvers, unpublished).

[18]C/C and C/c genotypes have also been reported to be distinguishable by means of X-rays. Following low doses of irradiation Hance (1928) reported that the regenera-

ting hairs of $A/A;C/c$ mice were mostly white whereas most hairs of $A/A;C/C$ animals tended to be even darker than those of untreated controls, with only a slight admixture of white hairs. This observation, however, remains to be confirmed (Chase, 1949). In fact, extensive tests of C/C and C/c in combination with agouti and nonagouti have failed to indicate a difference in this "greying" response from X-irradiation (Chase, unpublished) and the only phenotype that does respond differently is silver (Chase and Rauch, 1950) (see Chapter 6, Section II, G).

[19]It should also be noted that although $a/a;b/b;c^{ch}/c^{ch}$ mice resemble $a/a;b/b;C/C$ animals they incorporate only about one-fifth as much tyrosine. Moreover, the increased incorporation of tyrosine displayed by intensely pigmented brown over intensely pigmented black mice is not evident in the presence of c^{ch}/c^{ch} (see Table 3-1).

[20]Quevedo et al. (1967) have studied the influence of environmental temperature on UV-induced tanning of himalayan (c^h/c^h) and intensely pigmented black (C57BL/St) mice maintained at 10, 25, and 34°C. They found that whereas in himalayan animals there was a significant activation of epidermal melanocytes in UV-irradiated plantar skin at 10 and 25°C but not at 34°C, no UV activation of epidermal melanocytes occurred in the depilated hairy skin at any of these temperatures. On the other hand, at all temperatures the irradiated plantar and hairy skin of C57BL/St mice displayed a striking increase in melanogenically active epidermal melanocytes as well as in heavily melanized malpighian cells. Quevedo and his associates have also shown that c^h/c^h and C/C skin grafts transplanted to (C57 × himalayan)F_1 histocompatible hosts (as well as grafts exchanged between histocompatible $A^y/A;B/B$ and $A/A;B/B$ littermates) respond to UV as *in situ*. Thus the response of epidermal melanocytes to UV appears to be determined primarily by the genetic and physiologic constitution of the irradiated skin and not by more remote systemic factors (Quevedo et al., 1967).

[21]The effects of various *c*-series alleles on tyrosine hydroxylase and melanin production has also been investigated at the University of Edinburgh (Bulfield, 1974). In these studies c^{ch}/c^{ch} mice were found to produce an enzyme with lower V_{max} and higher K_m than C/C animals. c/c had less than 10% of C/C activity. It was not established whether this was "residual," unspecific, activity or allele-specific activity. Some of the other alleles investigated (c^e,c^h,c^p) as well as c^{ch} also had enzymes differing in heat stability. Heterozygotes of $c/-$ with all other alleles displayed intermediate values of activities compared to homozygotes. These findings are important because they further support the view that the *c*-locus is the structural one for the enzyme.

[22]Studies by Winston and Lindzey (1964) and by Thiessen et al. (1970) have also demonstrated that albino mice differ from pigmented animals in a number of behavioral situations. Moreover, employing coisogenic subjects King and Rush (1976) have shown that albino mice consume significantly less alcohol and sleep (following injection of ethyl alcohol) significantly longer than either heterozygous (C/c) or homozygous (C/C) pigmented animals. These investigators also found that pigmented mice heterozygous for albinism (C/c) tended to consume less alcohol and to sleep longer than those which were homozygous for the normal allele (C/C). In this case, however, there was some disparity in the results suggesting a maternal effect and environmental influences.

Chapter 4

Dilute and Leaden, the *p*-Locus, Ruby-Eye, and Ruby-Eye-2

I. Dilute (*d*) and Leaden (*ln*)

Recessive mutations at two loci of the mouse produce coat-color deviants which are phenotypically very similar. One of these mutations, described initially by Murray (1931, 1933), is known as leaden (*ln*; chromosome 1) because when homozygous it transforms the intensely pigmented nonagouti coat color to bluish-grey;[1] the other is dilute (*d*; chromosome 9), originally known, because of its influence on black, as the Maltese or blue dilution gene of the mouse fancy. These determinants produce a similar dilution of brown (Plate 2–E) and yellow pigment, giving the coat an overall "washed-out look" (Searle, 1968a).

A. *d*-Locus Alleles

Mutations at the *d* locus are particularly interesting because in addition to their diluting effect on hair pigment, they may also affect the nervous system.[2] Thus the d^l (for dilute-lethal) mutation[3]—a mutation recessive to both *D* and *d*—is phenotypically indistinguishable from *d*/*d* but, unlike *d*, produces a severe neuromuscular disorder characterized by convulsions and opisthotonus (arching upward of the head and tail). Myelin degeneration occurs in the CNS of d^l homozygotes (Kelton and Rauch, 1962) and they usually die at about 3 weeks of age (Searle, 1952).[4] Although it has been reported that phenylalanine metabolism is seriously disturbed in these mice (Coleman, 1960; Rauch and Yost, 1963) this remains to be confirmed (Zannoni et al., 1966; Mauer and Sideman, 1967).[5]

Two other mutations, d^s (slight dilution) and d^{15} (dilute-15), have also been described[6] which affect the nervous system. d^s/d^s homozygotes have a coat which is darker than d/d (Dickie, 1965a) and develop a peculiar behavior at about 4 months of age that is more pronounced in females than in males. The animals sway, lean, and lose their balance. The feet seem unable to support the body properly and slide out sideways, and the back arches in what seems to be an opisthotonic spasm. This characteristic, however, does not interfere with reproduction or significantly shorten life span (Dickie, 1967b). The color of d^s/d heterozygotes is intermediate between d^s/d^s and d/d (Dickie, 1965a). The d^{15} mutation when homozygous produces a slightly diluted pelage, i.e., darker than d/d but lighter than $D/-$, as well as behavior abnormalities similar to, but less severe than, d^l/d^l.[7] d/d^{15} heterozygotes are similar in color to d^{15} homozygotes but behave normally (R.J.S. Phillips, 1962).[8]

B. Influence on Hair Pigment

In spite of the fact that mutations at the *d* locus have a dilution effect when introduced into genotypes which otherwise provide for intensely pigmented eumelanotic and phaeomelanotic hairs, this effect is not due to a reduction in the amount of pigment in the hair. To the contrary, mutations at the *d*-locus produce phenotypes which probably have on the average more hair pigment than the corresponding nondilute animals. This is certainly the case for d/d animals irrespective of whether the mutation is on a lethal yellow or nonagouti background (Brauch and W. Russell, 1946; E. Russell, 1948). The d/d phenotype is brought about by the fact that between one-third and two-thirds of the pigment is deposited into a few very large, conspicuous clumps with clear-cut edges (Figure 4-1a, b, and c; see Figure 4-6) (E. Russell, 1949b) and these clumps, like the contracted melanophores in amphibian skin, have little effect on light absorption (E. Russell, 1948; Grüneberg, 1952). This clumping of granules in the septules of the hair is accompanied usually by a reduction of cortical granules in proportion to medullary number, by the irregular arrangement of nonclumped granules, and by some degree of pigmentation lag, i.e., the tips of the hairs often possess little pigment (E. Russell, 1949a), and these factors also undoubtedly contribute to the phenotype. Indeed, Onslow (1915) has suggested that melanin in the hair cortex maintains color intensity by preventing light from reflecting off the air spaces in the medulla. As might be expected the most dramatic clumps in d/d hairs occur in the medulla where they sometimes extend over several septules.

C. Dilute and Leaden: Similarities and Differences

The situation in leaden mice is essentially the same as in dilute (Figure 4-1d) except that some leaden genotypes, e.g., chocolate leaden animals

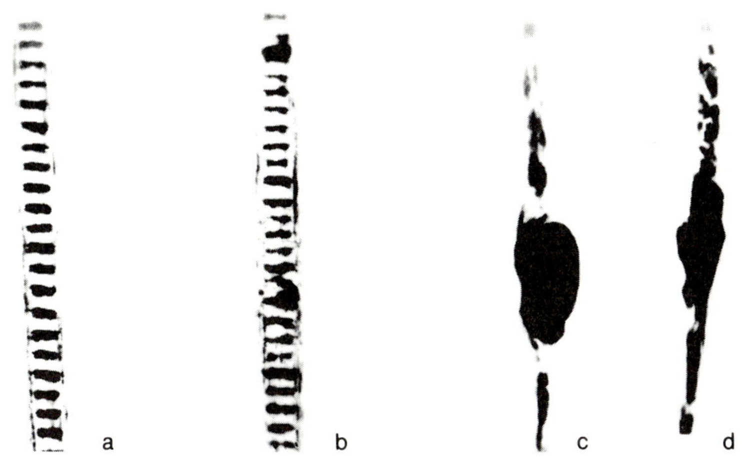

Figure 4-1. Whole mount preparations of pigmented hairs. Approx., ×300. (a) Hair from an intense black (*a*/*a*) mouse. Note the regular banded arrangement of melanin granules within the medulla. (b) Hair from a dilute black (*a*/*a*;*d*/*d*) mouse. Several small clumps of melanin granules disrupt the otherwise normal pattern of pigment distribution. (c) Another hair from a dilute black animal but in this case showing a region containing an unusually large clump of melanin. This clump of pigment is only partially surrounded by the hair shaft but remains anchored to it. (d) Hair from a pink-eyed leaden (*a*/*a*;*p*/*p*;*ln*/*ln*) mouse. A large medullary clump of melanin is evident as a swelling in the terminal portion of the hair shaft. From E.P. McGrath and W.C. Quevedo, Jr., 1965. Reproduced with permission of the authors and the Australian Academy of Science.

(*a*/*a*;*b*/*b*;*D*/*D*;*ln*/*ln*), are a little lighter in color than the corresponding dilute type (*a*/*a*;*b*/*b*;*d*/*d*;*Ln*/*Ln*). This appears to be due to a more pronounced pigment lag in *ln*/*ln* hairs rather than to any noticeable differences in pigment clumping (Poole and Silvers, unpublished). When genes for dilute and leaden occur together (*d*/*d*;*ln*/*ln*) their effect on pigment is no different than that observed when leaden occurs alone. Nevertheless, there are some important differences in the behavior of these mutants. These include the fact that (1) leaden, but not dilute (Poole and Silvers, unpublished) is epistatic to recessive yellow (*e*/*e*;*ln*/*ln* animals are indistinguishable from *E*/*E*;*ln*/*ln* mice) (Hauschka *et al.*, 1968); (2) whereas A^y/*a*;*D*/*d* mice are lighter than A^y/*a*;*D*/*D* animals, A^y/*a*;*Ln*/*Ln* and A^y/*a*;*Ln*/*ln* mice are indistinguishable (Poole and Silvers, unpublished)[9]; and (3) leaden black skin is more active than dilute black in tyrosinase and dopa oxidase activity, as well as in the ability to undergo *in vitro* darkening (Foster and L. Thomson, 1958). It should also be noted that while choroidal pigment granules tend to clump in both *ln*/*ln* and *d*/*d* mice (Hearing et al., 1973), according to Moyer (1966) retinal granules are clumped in *d*/*d* but *not* in *ln*/*ln* animals (Hearing et al. report no clumping in the retina of either genotype). Finally, there is no evidence that leaden mice display any reduction in phenylalanine hydroxylase activity (Wolfe and Coleman, 1966). These differences are important because they indicate that while these mutations often have similar phenotypic effects, their primary action(s) may be quite different.

D. Influence on Melanocyte Morphology and Behavior

Although the primary effects of these genes remain to be determined, histological examination of both *d/d* and *ln/ln* skin reveals a difference in the morphology of their melanocytes as compared with the melanocytes of the wild type (*D/D;Ln/Ln*) which can account for how these loci produce their phenotypic effect. Whereas the melanocytes of wild type animals are characterized by the possession of long, relatively thick dendritic processes which contain a substantial portion of the cell substance, in *d/d* and *ln/ln* mice the pigment cells have fewer and thinner dendritic processes (Figure 4-2). Because of this altered morphology the melanin granules are largely clumped around the nucleus in the body of the cell (Markert and Silvers, 1956; Moyer, 1966; Rittenhouse, 1968a) [such a melanocyte has been described as *nucleopetal* as contrasted with the "normal" *nucleofugal* type (Markert and Silvers, 1956)] and this crowding, in conjunction with the inadequate development of dendrites, undoubtedly results in an uneven release of granules from the melanocyte to the epidermal cells of the hair bulb (Rittenhouse, 1968a). This is especially the case since there is evidence that pigment granules may be released into hair cells from the cell body as well as from the tips of dendrites (Straile, 1964). The fact that *d/d* and ln/ln melanocytes have few dendrites, which are often stubby, may also prevent them from establishing significant contact with those epithelial cells destined to form the hair cortex and, if this is the case, it would account for the reduction in cortical pigment in dilute and leaden hairs (Straile, 1964; McGrath and Quevedo, 1965).

The nucleopetal morphology of *d/d* and *ln/ln* melanocytes could also explain why some of these cells appear to be incorporated into the hair shaft, especially at the termination of the hair growth cycle. Melanocytes with short, stubby dendrites may not be anchored as well as nucleofugal cells in the hair bulb and, as a consequence, become dislodged and swept into the hair shaft in the face of flowing epithelial cells (Quevedo and Chase, 1958; McGrath and Quevedo, 1965; Straile, 1964; S. Sweet and Quevedo, 1968). Indeed, this incorporation of nucleopetal melanocytes into the hair is undoubtedly responsible for the two or three enormous clumps which are sometimes found at the base of the hair causing local enlargement of the hair diameter (Figure 4-1c and d) (E. Russell, 1949b; McGrath and Quevedo, 1965).

Inasmuch as active dendritic melanocytes are not limited to the hair follicles of mice but occur in the dermis and epidermis of certain areas of the body (ears, soles of the feet, tail, scrotum, muzzle, and genital papilla), as well as throughout the connective tissue that encapsulates and subdivides the harderian gland,[10] it is scarcely surprising that in *d/d* and *ln/ln* genotypes the

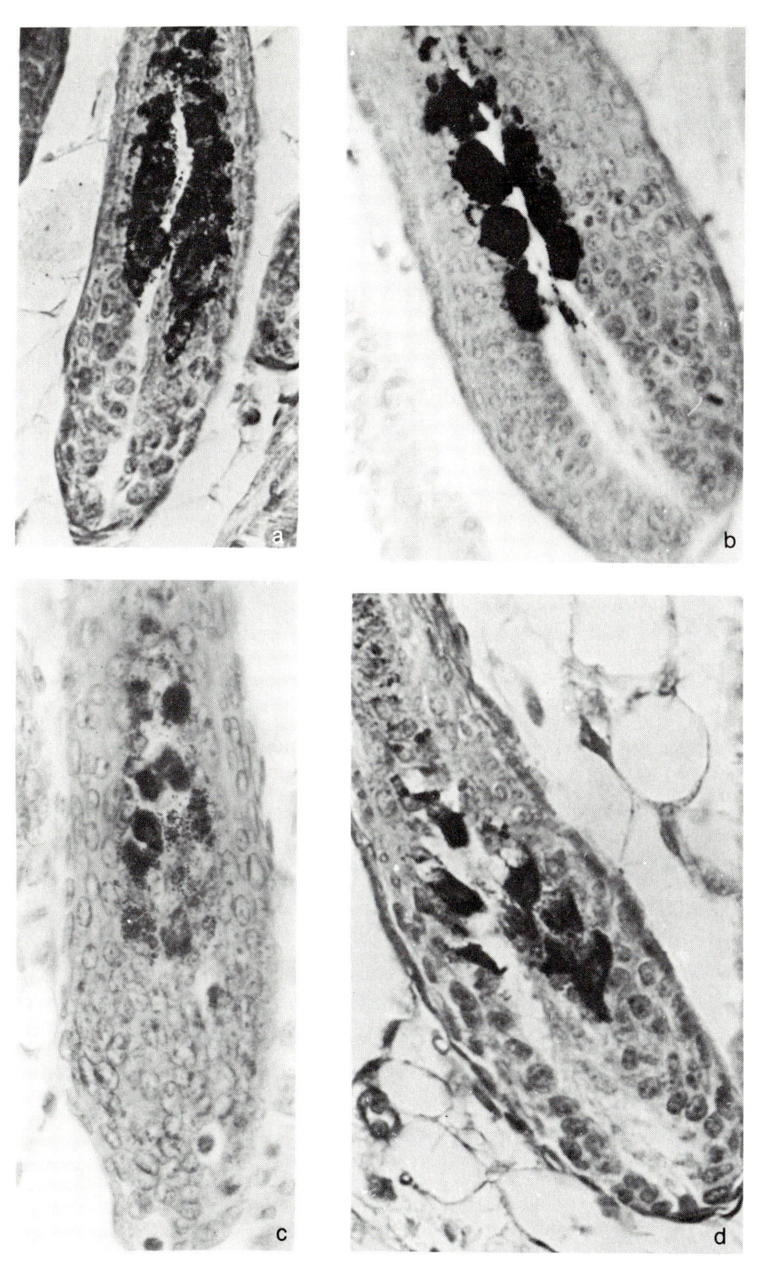

Figure 4-2. (a) Hair bulb from a 5-day-old intense black (*a*/*a*) mouse. Numerous dendritic melanocytes border the papilla cavity in the upper bulb. (b) Hair bulb from an adult leaden (*a*/*a*;*ln*/*ln*) mouse (12th day of hair growth cycle). (c) Hair follicle from a 5-day-old dilute (*a*/*a*;*b*/*b*;*d*/*d*) mouse. (d) Hair bulb from another 5-day-old dilute (*a*/*a*;*d*/*d*;*p*/*p*) mouse. In (b), (c), and (d) note clumping of granules within melanocytes and absence of dendrites. Approx., ×400. Fig (b) is from E.P. McGrath and W.C. Quevedo, Jr., 1965 and is reproduced with permission of the authors and the Australian Academy of Science.

melanocytes in these regions are likewise predominantly nucleopetal (Markert and Silvers, 1956) (Figures 4-3 and 4-4). Thus while Gerson and Szabó (1968) observed no difference in the *number* of melanocytes in either the dermis or epidermis of ear, tail, palm, or scrotal skin of *d/d* vs *D/D* mice, they found that the melanocytes in these regions of *d/d* skin could always be distinguished from those in *D/D* skin on the basis of differences in cell size, congestion of perikarya, or number and shape of dendrites. Moreover, these investigators also observed significantly less melanin in the malpighian cells of *d/d* mice than in *D/D* animals, a difference manifested both by a decrease in the number of pigmented malpighian cells as well as by a smaller number of pigment granules in those cells which were melanized. This decrease in pigmentation, which is likewise exemplified by the observation that dilute black (*d/d;B/*–) mice tan less well than intense black (*D/D;B/*–) animals when exposed to UV (Quevedo, 1965), undoubtedly results from the fact that because of fewer dendrites, nucleopetal melanocytes come into contact with fewer epidermal cells (Gerson and Szabó, 1968), a situation which could also interfere with their cytocrine activity.[11]

E. Effect of Cellular Environment on Melanocyte Morphology

To determine whether the genes at the leaden and dilute loci govern melanocyte morphology through the cells of the tissue environment or whether these genes act primarily within the melanoblasts themselves, Markert and Silvers (1959) transplanted embryonic tissue containing melanoblasts from normal, leaden, and dilute animals into the anterior chambers of the eyes of adult albino or pink eyed mice having the same or different *Ln* and *D* constitution as the graft. In all instances melanocytes of nucleofugal genotype (*D/D;Ln/Ln*) displayed a nucleofugal morphology in the anterior chamber of the host, regardless of its genotype (Figure 4-5a). On the other hand, donor melanocytes of nucleopetal genotype (*d/d* or *ln/ln*) always assumed shapes which varied all the way from typical nucleopetal to typical nucleofugal (Figure 4-5b and c). These results are consistent with the hypothesis that although genes at both the dilute and leaden loci exert their activity from within the developing melanoblast, the number and size of dendritic extensions of a melanocyte is probably a function of the environment in which the cell resides. Melanocytes of *d/d* and *ln/ln* mice have an innately weak capacity for extending dendrites, as indicated by their altered morphology in the rather compact tissue environments in which they normally occur. However, in less restrictive environments, such as in the anterior chamber of the eye, these melanocytes do extend more and longer dendritic processes and in many instances are nucleofugal (Markert and Silvers, 1959).[12]

Further evidence that the *d* locus acts within the melanoblast is provided by the work of Reed (1938). He found that when skin from newborn albino

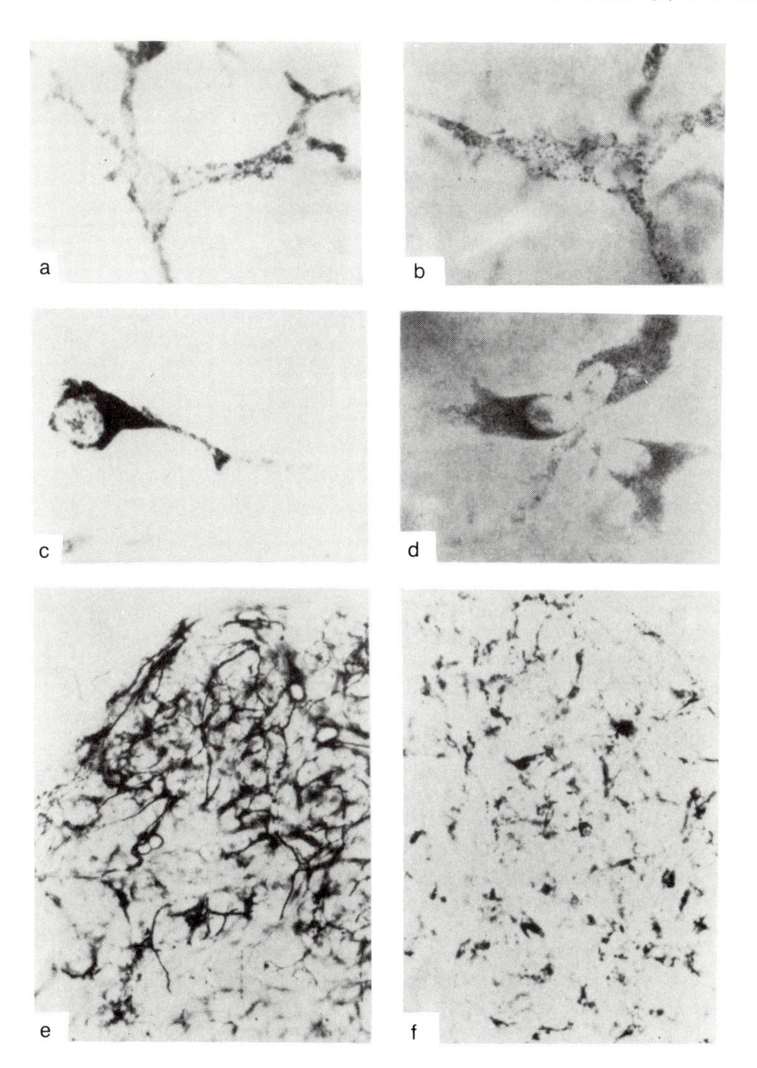

Figure 4-3. (a) Melanocyte from the harderian gland of a 10-day-old *a*/*a* mouse. Note the distribution of melanin in thick dendritic processes. Approx. ×650. (b) Melanocyte from the harderian gland of a 10-day-old *a*/*a*;*b*/*b* mouse. Approx. ×650. (c) Melanocyte from the harderian gland of a 4-day-old *a*/*a*;*d*/*d* mouse. Note the reduction in dendritic processes as compared to (a) and (b). The granules are clustered around the nucleus (nucleopetal type). Approx. ×650. (d) Three melanocytes from the harderian gland of a 6-day-old *a*/*a*;*b*/*b*;*d*/*d* mouse. Note the nucleopetal morphology. Aprox. ×650. (e) Nucleofugal melanocytes imbedded in the connective tissue capsule of the harderian gland of a 10-day-old *a*/*a* mouse. Approx. ×130. (f) Nucleopetal melanocytes imbedded in the connective tissue capsule of the harderian gland of a 4-day-old A^y/*a*;*d*/*d* mouse. Note the dilution effect produced by the concentration of melanin around the nuclei of the melanocytes. Compare with (e). Approx. ×130. From C.L. Markert and W.K. Silvers (1956). Reproduced with permission of *Genetics*.

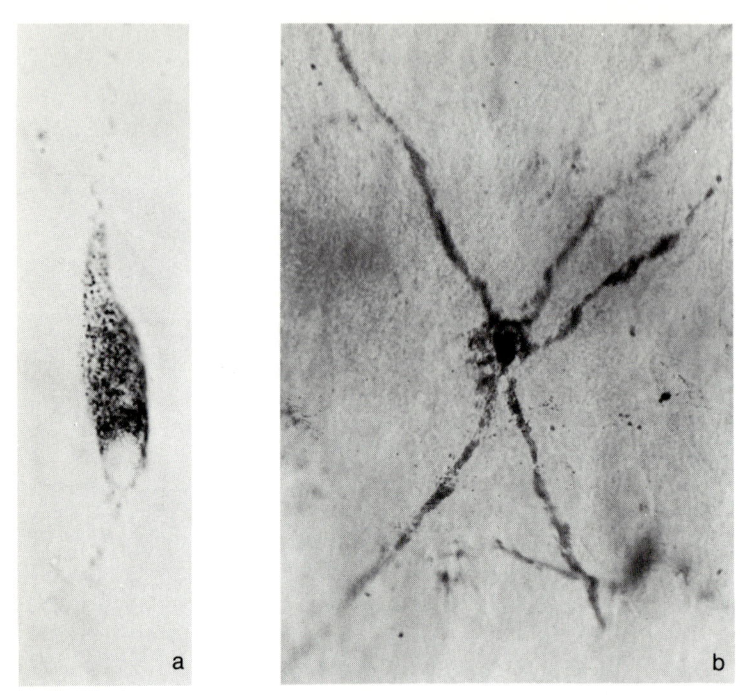

Figure 4-4. (a) A typical nucleopetal melanocyte from the harderian gland of a leaden (*a*/*a*;*b*/*b*;*ln*/*ln*) mouse. Note the two thin, sparsely pigmented dendrites and the aggregation of most of the pigment in the cell in an area near the nucleus. Approx., ×1,100. (b) A typical nucleofugal melanocyte from the harderian gland of a nonagouti (*a*/*a*) mouse. Note the dispersion of a large fraction of the melanin granules of the cell into the five large dendrites. Approx., ×450. From C.L. Markert and W.K. Silvers, 1959. Reproduced with permission of Academic Press Inc.

(nondilute) mice was transplanted to both neonatal dilute and nondilute hosts, the phenotype of the pigmented hairs originating within the border of the graft was always determined by the *d*-locus genotype of the invading cell.

II. The *p*-Locus (Pink-Eyed Dilution)

The *p*-locus on chromosome 7[13] is characterized by 13 alleles. The wild type allele, *P*, produces an intense pigmentation of both the hair and eyes, whereas the oldest and most common mutant at this locus, *p* (for pink-eyed dilution), a mutation carried in many varieties of the mouse fancy, reduces the pigmentation of both the eyes and the coat. The eyes of *p*/*p* mice resemble those of albinos, possessing a beautiful pink tint. However, in contrast to albino eyes, *p*/*p* eyes are not entirely free of pigment; small amounts of melanin are found in the iris and retina (Durham, 1908, 1911; Little, 1913) and a few melanocytes occur in the choroid as well (Markert and Silvers, 1956).

A. Influence on Coat Color

Insofar as the pigmentation of the hair is concerned *p* drastically reduces black and brown pigments, but has only a slight influence on phaeomelanin synthesis. Thus, except for the color of the eyes, $A^y/a;p/p$ and $A^y/a;P/-$ genotypes look much alike. Agouti pink-eyed animals also superficially resemble orange or yellow animals although they possess dull or slate grey hair bases which are recognized most readily when the hair is blown. The genotype *a/a;B/B;p/p* (Plate 2–F) is known as "blue lilac" and resembles

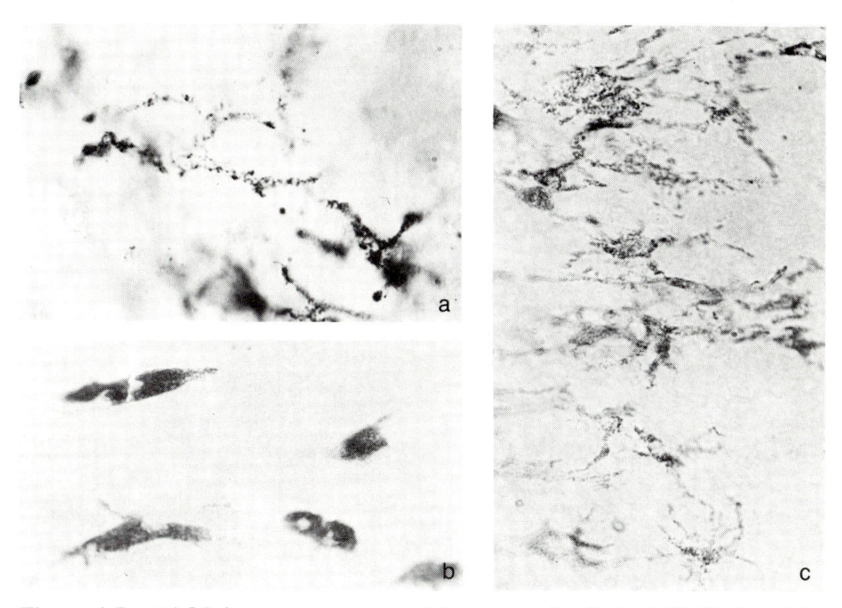

Figure 4-5. (a) Melanocytes recovered from a graft of *a/a;D/D* 12-day-old embryonic skin to the eye of an *a/a;b/b;d/d;p/p* host. The melanocytes have a nucleofugal morphology in accord with their own genotype but in contrast to the genotype of the host. Approx. ×970. (b) Nucleopetal melanocytes recovered from a graft of *a/a;b/b;ln/ln* 10-day-old embryonic dorsal tissue to the eye of a nonleaden albino host. The shape of the melanocytes is in accord with their own genotype but in contrast to that of the host. Approx. ×450. (c) Melanocytes recovered from a graft of *a/a;b/b;d/d* 12-day-old embryonic dorsal tissue to the eye of a nondilute albino host. These pigment cells exhibit a typical nucleofugal morphology although their own genotype is nucleopetal. The behavior of these cells is in striking contrast to those shown in (b), and is presumably due to the peculiar physical properties of the graft site. From C.L. Markert and W.K. Silvers (1959). Reproduced with permission of Academic Press, Inc.

"maltese blue" (*a*/*a*;*B*/*B*;*d*/*d*), but is lighter (Grüneberg, 1952). Similarly, *a*/*a*;*b*/*b*;*p*/*p* mice, which are known as "champagne" or "café au lait," are essentially a lighter edition of dilute chocolate (*a*/*a*;*b*/*b*;*d*/*d*) (Grüneberg, 1952).

B. Interaction with *B*, B^{lt}, and *b*

We have noted already that *p*/*p* is a dominance modifier of B (on some genetic backgrounds *a*/*a*;*B*/*b*;*p*/*p* mice are a little lighter than *a*/*a*;*B*/*B*;*p*/*p* animals) and that B^{lt} is a dominance modifier of *P* (*a*/*a*;B^{lt}/–;*P*/*P* mice have a different phenotype from *a*/*a*;B^{lt}/–;*P*/*p* genotypes). There is, however, another dominance modifier of *P*, namely, *b*/*b*. Thus Wallace (1953) reported that brown nonagouti mice heterozygous for pink-eyed (*a*/*a*;*b*/*b*;*P*/*p*) are distinguishable from brown animals homozygous for the wild type allele (*a*/*a*;*b*/*b*;*P*/*P*). Indeed, as a result of the interaction of *B* and *b* with *P* and *p*, six different phenotypes may occur (Wallace, 1953). These are presented below:

<table>
<tr><th></th><th>PP</th><th>Pp</th><th>pp</th></tr>
<tr><td>BB</td><td colspan="2" rowspan="2">1</td><td>2</td></tr>
<tr><td>Bb</td><td>3</td></tr>
<tr><td>bb</td><td>6</td><td>5</td><td>4</td></tr>
</table>

Although, depending upon the genetic background, there may be various degrees of overlap among the phenotypes expressed by some of these different genotypes, e.g., genotypes 2 and 3 may overlap as may 5 and 6, it is evident that on at least some genetic backgrounds the dominance of *P* over *p* (and *B* over *b*) is incomplete (see Chapter 12, Section I, C).

C. Other Alleles

Other reported alleles at the *p*-locus include Japanese ruby (p^r); dark pink-eye (p^d); *p*-sterile (p^s); *p*-black-eyed sterile (p^{bs}); *p*-cleft palate (p^{cp}); pink-eyed mottled-1 (p^{m1}); pink-eyed mottled-2 (p^{m2}); pink-eyed unstable (p^{un}); *p*-extra dark (p^x); *p*-darkening (p^{dn}); p^{6H} and p^{25H}.

1. Japanese Ruby (p^r)

Japanese ruby (p^r), which unfortunately may now be extinct (M.C. Green, 1966a), was first discovered in a stock of Japanese waltzing mice (Sô and

Imai, 1926). It is also recessive to wild type and has no effect on yellow (Searle, 1968a). The eyes of p^r/p^r mice are very variable in color, ranging from almost pink to almost black. Moreover, the two eyes of a single individual may differ (heterochromia iridis). Because of this variability the classification of p^r/p^r mice must usually be based on the color of the coat which (on a nonagouti background) is intermediate between that of blue lilac ($a/a;p/p$) and black. This allele seems to be completely dominant over p (Grüneberg, 1952).

2. Dark Pink-Eye (p^d)

Dark pink-eye (p^d), a mutation which was probably X-ray induced, when homozygous slightly dilutes the color of the coat. The eyes of these homozygotes are only slightly pigmented at birth but darken within the next few days. The coat of p^d/p mice is lighter than p^d/p^d but darker than p/p. Their eyes are colorless at birth but darken during the first 2 weeks (Carter, 1958).[14]

3. Pink-Eyed Sterile (p^s)

Pink-eyed sterile (p^s) is evidently another X-ray induced mutant (Hollander et al., 1960a,b). The interesting feature of this allele, which is recessive to p, is its pleiotropic effects. In addition to its influence on pigmentation, which is the same as p, this allele has a number of other effects some of which seem unrelated. p^s/p^s mice not only weigh less at birth than their normal littermates but grow poorly, partly because their incisors wear abnormally and they have difficulty in chewing, and are noticeably smaller as adults (Hollander et al., 1960a,b). The locomotion of these mice is also somewhat uncoordinated and at times almost waltzing. Moreover, males are almost completely sterile as a consequence of both a poor mating response and a very high incidence of abnormal sperm. The sperm defect has been traced to an abnormality of the acrosome cap which occurs at the supermatid stage resulting in sperm heads which are very variable in shape and structure, although sperm motility is not lost. The Golgi body seems implicated in the acrosome defect. Since p^s/p heterozygotes have normal sperm it is apparent that sperm morphology depends on the paternal genotype (Searle, 1968a). p^s/p^s females, although more fertile than males, likewise often fail to breed and, when fertile, produce few litters and are poor mothers (Hollander et al., 1960a,b). p^s/p^s mice also display extensive lesions of the pars nervosa and median eminence. The pituitary gland is reduced in size [in fact, D. Hunt and D. Johnson (1971) suggest that the syndrome is caused by a pituitary dysfunction] and changes are also apparent in their adrenals and pancreas (D. Johnson and D. Hunt, 1972). Even though p^s has a similar linkage relationship to the albino locus (c) as p (Hollander et al., 1960a) the many, seemingly unrelated, abnormalities associated with this allele suggest that it could represent a small deletion rather than a point mutation (Melvold, 1974).

4. p^{6H} and p^{25H}

p^{6H} and p^{25H} are two other alleged *p*-locus alleles which affect coat color the same as *p*. Like p^s, these alleles were radiation induced and cause abnormal reproduction and behavior. D. Hunt and D. Johnson (1971) concluded from electron microscope studies that the abnormal sperm head shape in males homozygous for either of these alleles results from abnormal proacrosome formation, and multiple tail number results from a failure in cytokinesis. Until recently, a good case could be made for arguing that p^{6H} and p^{25H} represented the same mutation. This is no longer tenable; while both, when homozygous, produce grossly abnormal sperm, the frequency of morphological types is different. Moreover, unlike the situation in p^{6H} or p^{25H} homozygotes, where copulation is never observed, somewhat fewer than 50% of p^{6H}/p^{25H} males copulate with normal females and form vaginal plugs (Wolfe et al., 1977).

5. *p*-Black-Eyed Sterile (p^{bs})

p-black-eyed sterile (p^{bs}), originally known as p^{24H} (R.J.S. Phillips, 1965), was found in the progeny of an irradiated male. $A/A;B/B;p^{bs}/p^{bs}$ animals have dark eyes at birth, are lighter than $A/A;B/B;p^d/p^d$ animals, and bear a closer resemblance to $A/A;B/B;p/p$ than to $A/A;b/b;P/P$ mice (R.J.S. Phillips, 1977). Like p^s/p^s animals, p^{bs} homozygotes have a behavioral disorder as well as a high incidence of sterility accompanied, at least in males, by reduced pituitary gonadotropin levels (Melvold, 1974). The sterility of homozygous p^{bs} males is also caused by a sperm abnormality but one distinguishable from the abnormality in p^{25H}/p^{25H} mice (D. Johnson and D. Hunt, 1972; Wolfe *et al.*, 1977). Both abnormalities are expressed in p^{bs}/p^{25H} males (D. Johnson and D. Hunt, 1972). The coat color of p^{bs}/p heterozygotes is intermediate between p/p and p^{bs}/p^{bs} and, like the latter, they have dark eyes at birth but behave normally (R.J.S. Phillips, 1977) and have normal spermiogenesis (D. Johnson and D. Hunt, 1972).

6. *p*-Cleft Palate (p^{cp})

p-cleft palate (p^{cp}), formerly p^{11} (R.J.S. Phillips, 1973), was found at Harwell in the progeny of a newborn irradiated male. p^{cp}/p mice are indistinguishable from p/p animals, but most p^{cp} homozygotes, while also of the same color as p/p, have cleft-palate and die soon after birth. Although occasionally p^{cp} homozygotes survive either because they are unaffected or only slightly affected, they nevertheless produce offspring with the same high incidence of cleft palate (R.J.S. Phillips, 1973).

7. *p*-Extra Dark (p^x) and *p*-Darkening (p^{dn})

p-extra dark (p^x) and *p*-darkening (p^{dn}) are two *p*-alleles which have only recently been reported (R.J.S. Phillips, 1977) although they occurred some time ago. p^x occurred as a spontaneous mutation in C3H and p^{dn} was induced with ethyl methanesulfonate (EMS). p^x/p^x homozygotes have dark eyes and are slightly lighter than wild type. p^{dn} homozygotes have pink eyes at birth which turn dark by weaning. Their coats are slightly darker than p/p and they have light ears.

The mutations p^x, p^d, p^{bs}, p^{dn}, p^{6H}, p^{25H}, and p^{cp} are all maintained at the MRC Radiobiological Unit at Harwell, England by R.J.S. Phillips who has prepared a tabular summary of their phenotypic effects (see Table 4-1). Dr. Phillips also notes that when these alleles are heterozygous, "dark on the whole appears to be dominant to light but there are some intermediate effects—for instance eye color in p^d/p, ear color in p^d/p or p^d/p^{bs}, and coat color in p^{bs}/p. It may be that most or all of the effects are intermediate but that the differences are not observable by eye." Moreover she adds "p^x and p^d fall into a darker, less *p*-looking category than p^{bs} and p^{dn} and this effect is phenotypically dominant" (R.J.S. Phillips, 1977).

8. Pink-Eyed Mottled-1 (p^{m1}) and Pink-Eyed Mottled-2 (p^{m2})

Pink-eyed mottled-1 (p^{m1}) and pink-eyed mottled-2 (p^{m2}) are two *p*-locus alleles which arose at Oak Ridge in different radiation experiments, but which behave similarly (L. Russell, 1964, 1965). When heterozygous with *p*, each of these mutations produces a coat in which wild-type areas and areas of typical pink-eyed dilute phenotype are freely intermingled (L. Russell, 1964, 1965). While, on the average, about 50% of the coat expresses the wild-type coloration and 50% that of p/p mice, there may be a great deal of variation among animals. Inasmuch as at least 50% of the more than 1000 offspring produced from mating p^m/p animals with p/p mice were mottled, the remainder being pink-eyed dilute (L. Russell, 1964, 1965), it appears that this condition cannot be attributed to a somatically, highly mutable "wild-type" allele.[15] It is therefore apparent that p^m occurs in all cells but for some unknown reason produces intense pigment in some and typical pink-eyed dilute pigment in others. Although there is no evidence that p^m is the result of a position effect, small rearrangements or other types of very closely linked variable suppressors of *P* have not been completely ruled out (L. Russell, 1964, 1965).

There is some similarity between the effect of the p^m alleles and mottled agouti (a^m), an allele at the *a*-locus (see Chapter 2, Section I, A, 12) which often produces agouti and nonagouti patches of fur freely intermingled at their edges (L. Russell, 1964, 1965). This similarity is especially striking since a^m/a animals also produce about 50% mottled progeny when mated to a/a mice (L. Russell, 1964, 1965). Nonetheless, it should be borne in mind that although the mottling of p^m/p and a^m/a may seem similar they actually are quite different. This stems from the fact that whereas the *a*-locus acts via the hair bulb environment (see Chapter 2, Section I, E and Chapter 7, Section VI), the *p*-locus acts autonomously within the melanocyte (Silvers, 1958b).

Table 4-1. Description of the Phenotypes of p-Alleles (Other Than p) Held at Harwell[a]

	Eyes at		On agouti. black			
Genotype	Birth	Weaning	Coat color	Ear color	Behavior[b]	Probable origin
p^x/p^x	Dark	Dark	Only slightly lighter than wild-type		Normal	Spontaneous (in C3H)
p^d/p^d	Light	Dark	Similar in depth of color to b/b but less "orange"		Normal	Radiation induced
p^{bs}/p^{bs}	Dark	Dark	Lighter than p^d/p^d and in p rather than b-color range (darker than p^{dn})		Slightly jerker	Radiation induced
p^{dn}/p^{dn}	Pink	Dark	Slightly darker than p/p	Light	Normal	EMS induced
p^{6H}/p^{6H}, p^{25H}/p^{25H}	Pink	Pink	Indistinguishable from p/p		Slightly jerker	Radiation induced
p^{cp}/p^{cp}	Pink	Pink	Indistinguishable from p/p		Normal[c]	Radiation induced
p^x/p, p^{bs}/p^d	Dark	Dark	Slightly lighter than p^d/p^d	Slightly darker than p^d/p^d	Normal	
p^d/p	Very light	Dark	Slightly lighter than p^x/p and p^{bs}/p^d	Light	Normal	
p^{bs}/p	Dark	Dark	Intermediate between p/p and p^{bs}/p^{bs}		Normal	
p^{bs}/p^{25H}, p^{bs}/p^{6H}	Dark	Dark	Intermediate between p/p and p^{bs}/p^{bs}		Slightly jerker	

[a]Prepared from R.J.S. Phillips (1977).

[b]The slightly jerker phenotypes tend to be smaller than normal and are generally male-sterile. (One of seven p^{bs}/p^{bs} males tested by matings to three females had one litter of two mice, and one of five p^{bs}/p^{25H} males tested had one litter of four mice. Three p^{6H}/p^{6H}, three p^{6H}/p^{25H}, and 19 p^{bs}/p^{6H} were sterile.) The females are poor breeders.

[c]Most die at birth from cleft palate but a few have lived, proved to resemble p/p, and to be fertile.

9. *p*-Unstable (p^{un})

p-unstable (p^{un}) [formerly p' or p^m (Wolfe, 1963)] is the final allele at the p-locus. It also produces a phenotype consisting of areas of dilute and intense pigmentation. However, unlike p^{m1} and p^{m2}, this bicolored pattern seems to be due to a spontaneous somatic reversion to wildtype at a relatively high frequency at all stages of development. Data from experimental matings have shown that the frequency of somatic reversion of p^{un} to + occurs more often in p^{un}/p^{un} progeny of heterozygous parents, p/p^{un} or $+/p^{un}$ mated *inter se*, than from homozygous p^{un}/p^{un} parents mated *inter se* (Melvold, 1971). It is also higher in p^{un}/p^{un} progeny from somatically mosaic parents, even when there is no evidence for germinal mosaicism (Melvold, 1971). The proportion of the coat affected appears to follow a bimodal distribution, i.e., it is usually either very small or quite large with few intermediate cases (the majority of mosaic animals fall into the former category).[16] Melvold has calculated that the spontaneous somatic reversion rates (at day 10 of embryonic life) in p^{un}/p^{un} progeny from matings in which the parents are mosaic or heterozygous, are generally two to three times that of p^{un}/p^{un} progeny from solid colored (nonmottled) p^{un}/p^{un} *inter se* matings.[17] While the exact basis for the expression of this allele is not known, Melvold suggests "that certain heterozygous combinations allow the formation, by recombination, of new alleles of varying stability and that the choice of mosaic mice for experimental matings may, in fact, result in selection for unstable alleles."[18]

When p^{un}/p^{un} mice are bicolored (see Figure 7-5), their phenotypes correspond to those of allophenic mice produced from merging embryos known to bear different alleles for coat-color determinants which act via the melanocyte (Mintz, 1967, 1969a). Indeed, the similarities between the pigment pattern of p^{un}/p^{un} and allophenic mice strongly supports Mintz's contention that all the melanocytes of the adult coat are *clonally* derived from a finite number of primordial melanoblasts originating along the mid-dorsum of the embryo (Mintz, 1967) (see Chapter 7, Section VI).[19]

D. Influence of Pink-Eyed Dilution on Pigment Granules

The influence that substituting p/p for $+/+$ has on the attributes of the pigment granules, and their distribution in the hair of different genotypes, has been carefully examined by E. S. Russell (1946, 1948, 1949a, 1949b). The major effect of this substitution is to alter the size of the pigment granules of all nonagouti genotypes, changing them to a very characteristic irregular shred shape, with flocculent clumping (Figure 4-6). Tiny flecks of pigment are also present, and the edges of the granules never appear as clear or as sharp as in $+/+$ mice (E. Russell, 1949b). As granule size is one of the key pigmentation characteristics affecting shape and color intensity, it is possible that the unique shape of p/p granules results from their small size (E. Russell, 1949b). The color quality of the pigment itself is not altered, but the

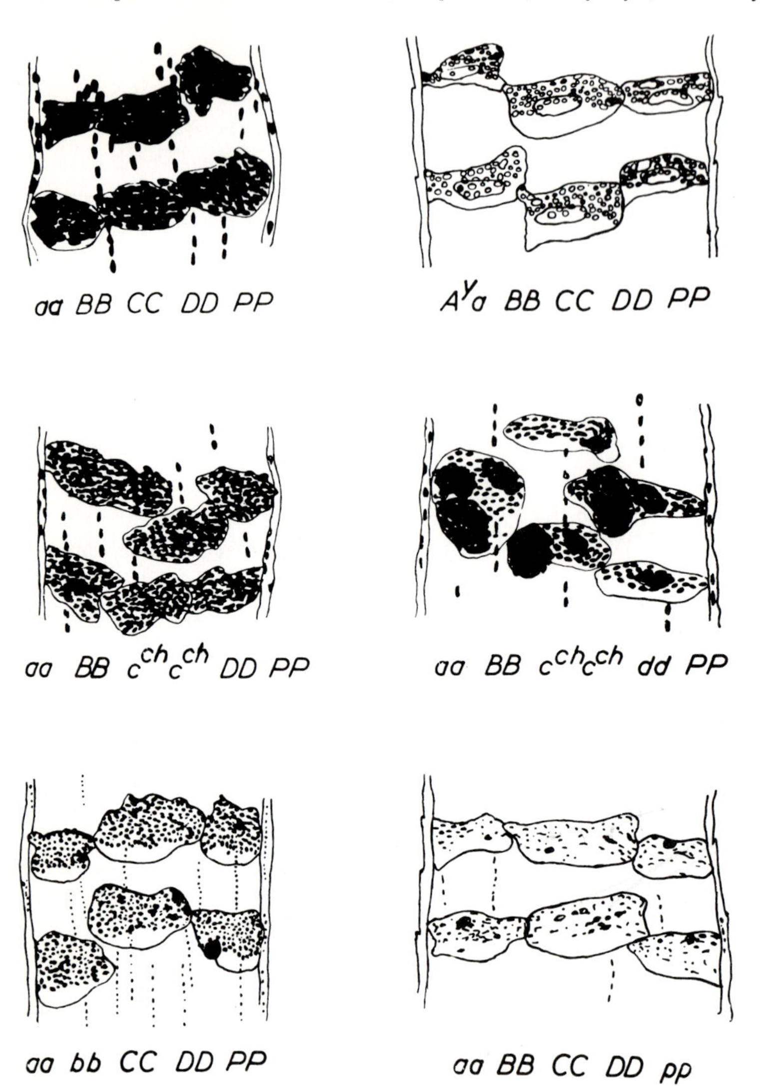

Figure 4-6. Pigment granules in awl hairs of six different genotypes. The large masses are medullary cells while the streaks of granules are from the overlying cortex. In this drawing, the yellow phaeomelanin granules of the A^y mouse are symbolized by open circles; no distinction is made between black and brown eumelanin. In the medullary cells of yellow mice the nucleus is usually visible. Note that whereas the granules in intensely pigmented black ($a/a;B/B$) hairs are usually either long-oval or oval, those in brown ($a/a;b/b$) hairs are round. Note also that the substitution of c^{ch}/c^{ch} for C/C (on an $a/a;B/B$ background) results in a diminution of granule size and not number. Finally note the clumping of granules in the d/d hair and the shred-shaped granules in the p/p shaft. These drawings were made by Dr. E.S. Russell and appeared in H. Grüneberg (1952). They are reproduced with the permission of Drs. Russell and Grüneberg and of Martinus Nijhoff B.V.

degree of pigmentation in both nonagoutis and yellows is decreased as a consequence of fewer cortical granules and an increase in pigmentation lag. Although *a*/*a*;*p*/*p* granules are unique, their attributes may nevertheless be affected, albeit in a minor way, by certain other genic substitutions. Thus, pink-eyed dilute brown (*a*/*a*;*b*/*b*;*p*/*p*) granules are, on the average, shorter and more rounded than those of pink-eyed dilute black (*a*/*a*;*B*/*B*;*p*/*p*) mice (E. Russell, 1949b).

E. Ultrastructural Observations and Possible Gene Action

The influence that pink-eye dilution has on the ultrastructure of the nonagouti melanin granule has received attention from Moyer (1961, 1963, 1966), Rittenhouse (1968b), and Hearing et al. (1973). Unfortunately there is no unanimity in their observations. Moyer observed that in the very early development of the +/+ melanosome thin single fibers aggregate to form compound fibers which cross-link and become oriented parallel to each other, however, in *p*/*p* mice the fibers do not become arranged in the same orderly parallel fashion and there is much less cross-linking between them (Figure 4-7). As a consequence the mature granule never achieves the size of a "normal" granule and its shape is altered. Although this lack of orientation of the fiber matrix in *p*/*p* mice makes it difficult to assess the granularity of the melanin deposited, Moyer did find occasional granules in which at least a portion of the matrix was fairly well oriented. A study of the melanin in these areas indicated that the *p*-locus did not affect the granularity of the melanin nor, as might be expected, was the second-order periodicity altered. Thus, according to Moyer, the *p*-locus appears somehow to be involved with the early structure of the melanosome matrix possibly by controlling the protein which provides the cross-linkages. This cross-linkage is basic to the structure of the final granule and could influence tyrosinase activity in numerous ways (Wolfe and Coleman, 1966).

In contrast to these observations, however, are those of Rittenhouse (1968b). She reported little difference between the structure of *p*/*p* and +/+ granules. Indeed, according to her observations the only striking characteristic of both pink-eyed black and pink-eyed brown melanocyte granules is that, in contrast to the wild-type (nonagouti), a high proportion of them are lightly melanized or unmelanized. *p*/*p* granules also are noticeably smaller than +/+ granules, apparently because they fail to enlarge after melanization is initiated. Rittenhouse's comparison of the granules of pink-eyed black and pink-eyed brown melanocytes with each other, and with intense black and brown granules, is particularly interesting because they indicate that *p*/*p* can shift the disorderly internal framework she found characteristic of *a*/*a*;*b*/*b*;*P*/*P* granules—an internal framework resembling a tangled ball of strands (Rittenhouse, 1968a)—toward the much more organized pattern of

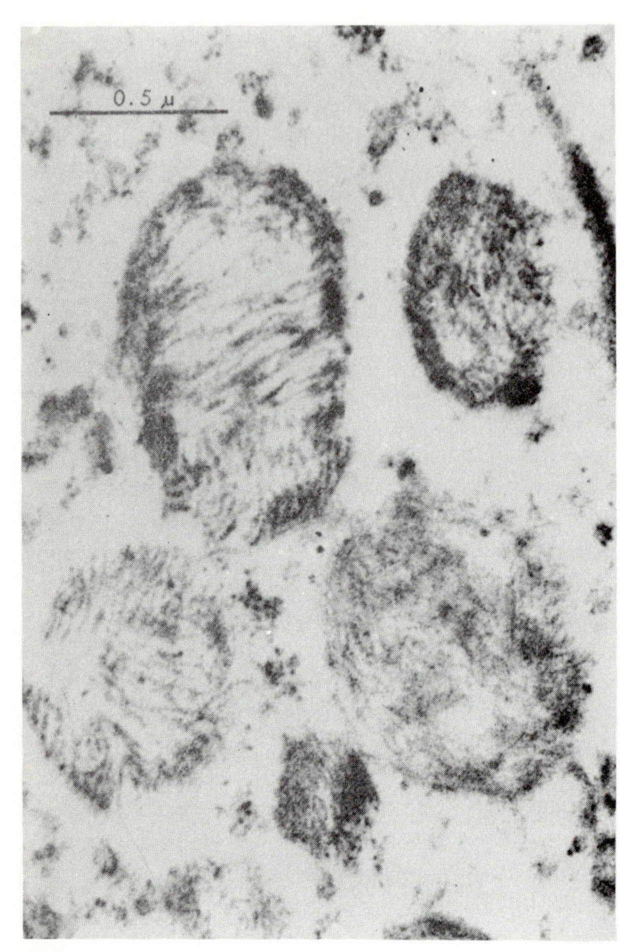

Figure 4-7 Stages in the development of *a*/*a*;*p*/*p* melanin granules. Note that the fibers run in several planes in one section instead of in the orderly parallel fashion seen in *a*/*a*;+/+ genotypes. Note also the reduction in cross-linking. From F.H. Moyer (1966). Reproduced with permission of the *American Zoologist*.

longitudinal strands and membranes characteristic of *B*/*B* granules (see Chapter 3, Sections I, F and II, G).

While one might attribute the different observations of Moyer and Rittenhouse to the fact that the latter examined only the granules of hair bulb melanocytes whereas the former paid particular attention to those of the retina, this does not seem to be completely responsible for the discord. Thus, more recently Hearing and his colleagues have examined pigment granules from retinal epithelial cells and from choroidal melanocytes of *p*/*p* mice and report no decisive differences. Indeed, their observations are similar to those of Rittenhouse since they found that the only significant difference between *p*/*p* and +/+ granules was that the former were much less melanized (Hearing et al., 1973).

The elegant study of Sidman and Pearlstein (1965) deserves special consideration not only because it too includes some ultrastructural observations on pink-eyed granules, but because it focuses attention on another possible mode of action of *p*-series alleles. These investigators cultured eye tissue from postnatal *p*/*p*, p^{un}/p, and p^{un}/p^{un} mice *in vitro* and observed that when tyrosine was present in the culture medium the amount of pigment was greatly increased over that found *in vivo*. This increase in melanin occurred both in melanocytes and pigment epithelial cells and was unrelated to the structure of the granule. Thus, in agreement with Moyer's observations, Sidman and Pearlstein also found numerous melanosomes in their pink-eyed cultures with internal membranes in disarray. Indeed, the fact that these abnormal melanosomes appeared to make pigment as effectively *in vitro* as normal melanosomes while maintaining their abnormal morphology indicates that, contrary to the suggestion of Moyer (1963), their structure is not responsible for the reduced amount of pigment they normally display.

It therefore appears that the effect of the *p*-series of alleles on pigment formation may result from a failure to use an existing melanin-synthesizing enzyme system at a rate adequate to achieve normal pigmentation (Sidman and Pearlstein, 1965). This possibility is in accord with other biochemical studies (Foster, 1963, 1965; Coleman, 1962, 1963) and, if confirmed, might help explain some of the pleiotropic effects of *p*-locus alleles. If for some reason *p*/*p* melanocytes cannot use tyrosine, or a further metabolite in the melanin pathway, at a normal rate because of some intracellular competition which favors some alternate use of available substrate, this diversion of tyrosine could be involved in producing some of the other effects attributed to the locus (Sidman and Pearlstein, 1965). On the other hand, at least some of the pleiotropic effects attributed to this locus could very well be a consequence of chromosomal deletions involving adjacent loci concerned with other functions (Melvold, 1974). Indeed, this possibility is especially attractive since those "*p* alleles" with the most severe effects were all radiation induced.

III. Ruby-Eye (*ru*)

Ruby-eye (*ru*; chromosome 19) was found by Dunn (1945) in a silver piebald stock. The mutation is easily recognized at birth, the eyes of the newborn lacking the iris pigment ring of the wild type and resembling the eyes of the mutants pink-eyed dilute, pallid, and albino. The eyes darken during the first week after birth and when opened are a dark ruby color.

A. Influence on Coat Color

This mutation also influences the pigmentation of the coat. Ruby-eyed, black (*a*/*a*;*B*/*B*;*ru*/*ru*) mice are a dull dark sepia and ruby-eyed black agouti (*A*/*A*;*B*/*B*;*ru*/*ru*) mice superficially resemble brown agouti animals since the black base of the fur is reduced to dark slate, while the yellow tip is only slightly reduced in intensity (Dunn, 1945). Ruby-eye does not have as drastic an effect on reducing the intensity of black pigment as does pink-eyed dilution (*p*/*p*) but has a somewhat greater effect than *p*/*p* in reducing the intensity of phaeomelanin. When *p*/*p* and *ru*/*ru* are incorporated into the same stock, the phenotype more closely resembles that of *p*/*p* (Dunn, 1945).

B. Pigment Granules

The most distinctive feature of the pigment granules of the normal retina is their great diversity of size and shape (Figure 4-8). Their shapes vary from spherical, through all gradations, to thin elongated rods with the largest granules being at least 500 times the volume of the smallest granules (Markert and Silvers, 1956). While the retinal granules of *ru*/*ru* mice also vary in size, their average size, as well as their range of diversity in morphology, is significantly reduced and there are not as many of them. Moreover, their color in animals of black genotype (*a*/*a*;*B*/*B*;*ru*/*ru*) is only slightly darker than typical brown granules (Markert and Silvers, 1956).

The granules found in the melanocytes of the harderian gland of ruby-eyed black mice are also dark brown and their morphology is altered here too. Thus although the sizes and shapes of these melanocyte granules, unlike those of the retina, are more or less uniform, the spheroidal granules of *a*/*a*;*B*/*B*;*ru*/*ru* melanocytes resemble those of brown melanocytes more than they resemble the oval or long-oval-like granules of "normal" black pigment cells (Markert and Silvers, 1956).

Ultrastructural studies of *ru*/*ru* pigment granules have disclosed that in spite of the fact that those of *a*/*a*;*B*/*B*;*ru*/*ru* mice resemble the granules of *a*/*a*;*b*/*b* animals, the quality of pigment deposited on the granule nevertheless is determined by the *b*-locus (Moyer, 1966 (see Figure 3-1)). In fact, according to Moyer the only unique feature of *ru*/*ru* retinal melanosomes is that their development is delayed and they do not become melanized until after birth.[20]

C. Possible Mode of Action

With the limited amount of evidence available, it is difficult to pinpoint the specific action of the *ru* gene. Since it does not appear to alter significantly the ontogeny or fine structure of the melanosome, its only major influence seems to be in delaying the onset of granule synthesis (Moyer, 1966). It is

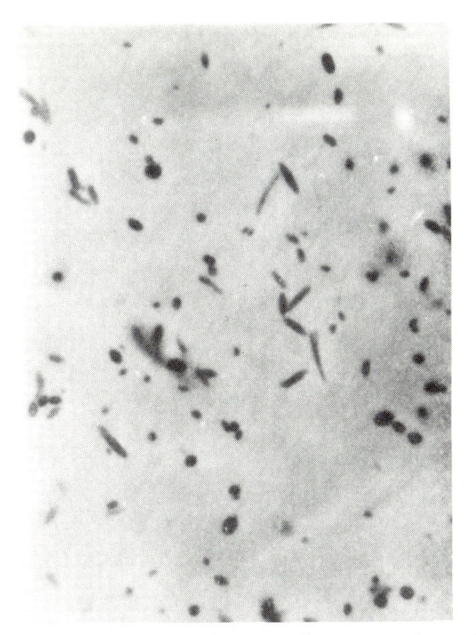

Figure 4-8. Fresh, squash preparation of the retina and choroid from a 14-day-old nonagouti (*a*/*a*) mouse. Note the great diversity in size and shape of the melanin granules, a diversity due entirely to those of retinal origin. Approx. ×1500. From C.L. Markert and W.K. Silvers (1956). Reproduced with permission of *Genetics*.

conceivable that during the course of this delay, the concentrations of substrates or cofactors required for protein synthesis are altered in such a way as to influence the number and the size of the granules produced (Moyer, 1966). It is known that newborn *ru*/*ru* skin incorporates about 14% less tyrosine than +/+ skin (Coleman, 1962), but this reduction in activity could well be due to the reduced number of granules in ruby-eyed animals rather than to any change in the structure of matrix protein (Moyer, 1966). Indeed, as Moyer suggests perhaps the major effect of the *ru*-locus is to regulate the activation of loci that direct the synthesis of matrix protein[21] (see also Chapter 6, note 11).

IV. Ruby-Eyed-2 (*ru-2*)

A. Origin and Description

In 1966 Lilly reported a recessive coat-color mutation which, because it was phenotypically identical to ruby-eye, he designated *ru-2*. The deviant occurred in a substrain of C57BL/6 and was shown to be located near pink-eyed on chromosome 7. Eicher (1970a), in a more detailed linkage analysis, reported *ru-2* to reside 3 cM from *p* and 17 cM from *c* with the order being *c-p-ru-2*. Like *ru*, *ru-2* produces a grayish-brown color in the eumelanin region of the agouti hair and dark red eyes.[22] Phaeomelanin is only slightly affected. The influence of *ru-2* at the cellular level has unfortunately not been investigated (Eicher, 1970a).

B. Other Alleles (Maroon and Haze)

Recently it has been discovered that maroon (*mr*) and haze (mr^{hz}), two recessive coat-color determinants which were believed to be inherited independently from *ru-2*, are actually allelic with *ru-2* (Eicher and Fox, 1977). Hence these two mutants are now designated $ru\text{-}2^{mr}$ (maroon) and $ru\text{-}2^{hz}$ (haze).

Maroon arose spontaneously in a "lactation" stock (N. Bateman, 1957). The eyes of $ru\text{-}2^{mr}/ru\text{-}2^{mr}$ mice are colorless at birth, but darken to a rich maroon. The coat color of maroon mice varies from almost normal to extreme pallid, even among sibs, and darkens with age. The underfur and isolated hairs of this mutant are often silvered (N. Bateman, 1957). According to Roderick (1973, 1975), nonagouti maroon ($a/a;ru\text{-}2^{mr}/ru\text{-}2^{mr}$) animals look very much like nonagouti buff (*a/a;bf/bf*) mice (see Chapter 6, Section IV, B). Nevertheless, these mutants can be readily distinguished from each other by the fact that whereas $a/a;ru\text{-}2^{mr}/ru\text{-}2^{mr}$ mice have light ears, a solid belly, and translucent eyes, *a/a;bf/bf* animals have darker ears, a belly lighter than the rest of the coat, and opaque eyes (Davisson, 1977). According to Davisson "the eyes must be observed by examining the mouse in profile against light." The double homozygote $a/a;ru\text{-}2^{mr}/ru\text{-}2^{mr};bf/bf$ has a distinctly lighter coat color than either homozygote alone (Roderick, 1975; Davisson, 1977). Maroon has been reported to interact with silver (*si*) in the F_1 (N. Bateman, 1957) (see also Chapter 6, note 11).

Haze was originally described by Dickie (1965b) after it arose spontaneously in the DBA/2J strain. This mutation is characterized at birth by light colored eyes. In the adult the hair color is brownish and the ventrum is white. In 1974 E. S. Russell demonstrated that haze was allelic to maroon.

Notes

[1]*ln* occurred spontaneously in a highly inbred chocolate (*a/a;b/b*) stock.

[2]Because Castle and his associates (1936) found that *d/d* mice were heavier than their nondilute littermates, Castle (1940) concluded that *d* also affects body size. In the light of a more carefully controlled analysis by L. Butler (1954), however, it seems more likely that this influence is due to linkage, i.e., close to the *d*-locus there are genes which affect body size.

[3]d^l occurred as a spontaneous mutation in the C57BL/Gr strain (Searle, 1952).

[4]Kelton and Rauch (1962) compared the development of myelination in the brain and rostral portion of the spinal cord in *D/D*, D/d^l and d^l/d^l animals, beginning at 4 days of age, and observed no differences. However, when the brains and spinal cords of these same genotypes were examined for the presence of degenerating myelin, as revealed by the Marchi method, the vestibulospinal, spinocerebellar, and tectospinal systems of the d^l/d^l animals, but not of the others, displayed signs of degeneration. With few exceptions, this degeneration occurred within 1 or 2 days of the onset of myelinization.

[5]Coleman (1960) reported that the enzyme, phenylalanine hydroxylase, which converts phenylalanine to tyrosine, had only 50% of normal activity in DBA/1J (*a/a;b/b;d/d*) mice and only 14% of normal activity in mice homozygous for d^l. Coleman did not attribute this decreased activity to a direct effect of the *d*-locus on the production of the enzyme but rather to an inhibitor he found in the particulate fraction of liver homogenates. His results also indicated that more than one gene was involved in decreasing the activity of the enzyme in DBA/1J animals. Thus, coisogenic *D/D* mice had 86% and genetically unrelated dilute black (*a/a;d/d*) mice had 65 to 75% of normal phenylalanine hydroxylase activity. These findings, along with the fact that the dilute black mice were not as susceptible as DBA/1J animals to audiogenic seizures, raised the possibility that these seizures were somehow related to the enzyme deficiency (see S. Huff and R. Huff, 1962; S. Huff and Fuller, 1964). Rauch and Yost (1963) reported that dilute lethal mice could neither transaminate nor hydroxylate phenylalanine as well as normal mice and that these defects were most pronounced during possible critical periods of postnatal development. Accordingly they suggested "that reduced enzyme activities lead to elevation in serum phenylalanine level which may in turn be responsible for those behavioral abnormalities that are related to myelin degeneration in the brain." On the other hand, Mauer and Sideman (1967) found that d^l/d^l mice displayed no difference from their normal littermates in levels of serum phenylalanine. Assays of liver phenylalanine hydroxylase activity, based on radioactive tracer methods, also failed to reveal any defect in this aspect of phenylalanine metabolism. Zannoni et al. (1966), too, did not find any evidence that phenylalanine hydroxylase activity was reduced in d^l/d^l mice. They likewise did not find any elevation in the concentration of phenylalanine in the blood or of phenylpyruvic acid in the urine of these animals (see also Woolf, 1963). Dilute-lethal has also been reported to influence both the total amount and the rate of increase of adrenal epinephrine and norepinephrine levels. Thus, according to Doolittle and Rauch (1965), by 3 weeks after birth d^l/d^l mice have about 25% more epinephrine and over twice as much norepinephrine as do normal mice.

[6]d^s was recovered from a (C57BL/6J × DBA/2J)F_1 litter, and the d^{15} mutation was radiation induced.

[7]According to R.J.S. Phillips (1962), d^{15} homozygotes are easily distinguishable from *d/d* and at about 20 days of age develop characteristic behavior abnormalities similar to those which occur in d^l/d^l mice (the larger the animal the earlier the symptoms develop). If weaned, d^{15} homozygotes usually die between 3 and 4 weeks of age. If, however, they are given wet mash, many live for a considerable time. At least one male has lived long enough to breed.

[8]A considerable number of *d*-locus mutations have also occurred in the course of

radiation experiments at Oak Ridge (L. Russell, 1971). These include (1) 10 viable mutations (known as d^x), all of which produce a somewhat darker phenotype than *d*/*d* when homozygous; (2) 3 mutations which are apparent repeats of *d*; (3) 14 mutations which when homozygous are lethal prenatally (known as d^{pl}); and (4) 95 mutations which in two doses are lethal postnatally (known as d^{op}). This last class of mutants do not differ among themselves in any easily recognizable way, nor do they differ from d^l. All are phenotypically indistinguishable from *d*/*d*, produce clonic convulsions with opisthotonus and are lethal before weaning age.

[9]The difference between $A^y/a;D/d$ and $A^y/a;D/D$ does *not* occur in *a*/*a* animals and is most pronounced on the ventrum, which appears almost white in $A^y/a;D/d$. Microscopic examination of $A^y/a;D/d$ hairs show no clumping of granules but a general reduction in their numbers along the entire shaft. This reduction is most pronounced in the ventral hairs where there may be a complete or almost complete absence of granules. This semidominant expression of *d* is also observed on the belly of $a^t/a^t;D/d$ mice (and presumably would be discernible in A^w/A^w animals as well) which have a black dorsum and a very pale yellow ventrum. It does *not* occur in recessive yellow (*e*/*e*) mice. Thus, in contrast to the situation in lethal yellow mice, *e*/*e*;*D*/*D* and *e*/*e*;*D*/*d* genotypes are indistinguishable (Poole and Silvers, unpublished).

[10]There is one mouse strain, known as the PET (for Pigmented Extraepidermal Tissues) strain, which displays an exceptionally wide disposition of melanocytes in the connective tissues of the body. This strain, which has been studied in detail by Reams and his associates (Nichols and Reams, 1960; Mayer and Reams, 1962; Reams, 1963, 1966, 1967; Rovee and Reams, 1964; Reams and Schaeffer, 1968; Reams et al., 1968, 1976) was derived from an accidental cross between C3H mice and black mice of unknown origin obtained from a local pet shop in Richmond, Virginia. A survey of these mice made shortly after the strain was established showed that, although the distribution of melanocytes was not consistent from animal to animal, within the strain as a whole they occurred in almost every tissue of the body, including cartilage, bone, serosae, and many other tissues (lungs, heart, gonads, etc.). In fact, melanocytes were consistently absent only from the connective tissues of the gut mucosa (Nichols and Reams, 1960). As time passed and a number of PET sublines were produced, it became apparent that each of them could be characterized by the particular localization of melanocytes within various tissues or body regions. Unfortunately, however, all but one of these lines, one in which the melanocytes are restricted to the skin and to certain muscles of the hind limbs, are extinct (Reams, 1963).

[11]Because melanin pigmentation of the skin involves the production of melanosomes by melanocytes and their distribution to malpighian cells, and because the latter may play an active role in controlling the rate of synthesis of melanosomes, these two cell types can be looked upon as comprising a structural as well as a functional unit. Accordingly, the term *epidermal melanin unit* has been coined (Fitzpatrick and Breathnach, 1963; Hadley and Quevedo, 1966; Quevedo, 1972). One might loosely define this unit "as a melanocyte with an associated pool of malpighian cells, the number of which may be variable" (Fitzpatrick et al., 1966). In this view, factors which influence any component in the "epidermal melanin unit" might be expected to alter the function of the entire system. Thus, the effect which UV has in stimulating melanogenesis may not be due to its direct effect on melanocytes, but rather to

its ability to induce the proliferation of malpighian cells thereby providing more vehicles for melanin transport (Fitzpatrick and Breathnach, 1963). Genetic mechanisms, such as *d/d* and *ln/ln*, also obviously influence the size of the "epidermal melanin units" by restricting the number of dendritic processes. Also "epidermal melanin units" may overlap with two or more melanocytes sharing some malpighian cells in common (Hadley and Quevedo, 1966).

[12]Following exposure to UV the epidermal melanocytes of *d/d* hairy and plantar skin are, in general, less dendritic than those of *D/D* skin but often possess more numerous and better developed dendritic processes than the pigment cells in unirradiated plantar *d/d* skin (Quevedo and J. Smith, 1963; Quevedo and McTague, 1963). To account for this, Quevedo and McTague raise the possibility that UV treatments may "increase the permeability of the cellular interstices of the epidermis for penetration by the dendrites of melanocytes." They also deem it conceivable that an increase in melanogenesis results "in the filling of pre-existing amelanotic processes with pigment granules, thus rendering them visible."

[13]*p* is linked with the albino (*c*) locus (recombination 16% in females, 12% in males) (M.C. Green, 1966). This linkage was the first to be reported in any vertebrate (Haldane et al., 1915).

[14]According to Carter (1958), the coats of $A/-;p^d/p$ mice are rather like $A/-;b/b$. p^d/p^d mice are not as dilute as p^r/p^r (Searle, 1968a) or *misty* (*m/m*) (see Chapter 5, Section VI).

[15]As pointed out by L. Russell (1964), if the mottling were the result of a somatically mutable wild type allele, then mottled (plus, perhaps, fully wild type) progeny should not exceed 25% in these matings.

[16]The heavily mottled animals usually are gonadal mosaics. Eye colors in these mottled mice range from pink to full color, and often show bilateral asymmetry (Wolfe and Coleman, 1966). Indeed, because reversions to + are so readily distinguishable in the cleared retinal pigment epithelium of p^{un}/p^{un} mice, Searle (1977) suggests that such pink-eyed mutants can provide a means of studying somatic reversions induced by chemical agents (see Chapter 12, note 32).

[17]The somatic reversion rate varied from 1.8 to 5.8×10^{-4} depending on the mating involved.

[18]Preliminary results from matings designed to test the role of meiotic recombination in the parental effect on reversion of the p^{un} allele indicate that reversion frequencies are significantly higher in p^{un}/p^{un} progeny of $+/p^{un}$ ♀ × p^{un}/p^{un}♂ matings, where recombination at the *p*-locus in the female parent is uninhibited, than in matings where recombination is inhibited or prevented by the presence of a translocation (Melvold, 1972).

[19]The effects which various *p*-alleles have on the reproductive system have received a considerable amount of attention. The p^{6H}, p^{25H}, and p^{bs} alleles when homozygous cause sterility in males and semifertility in females, whereas p^d/p^d, p^{un}/p^{un}, and *p/p* animals have normal fertility. Pituitaries from sterile males have significantly lower proportions of gonadotropic cells than pituitaries from fertile males. Pituitary gonadotropins appear likewise to be reduced in these sterile genotypes, although the lesion cannot be localized (Melvold, 1974). The ovaries of semifertile females contain large numbers of developing follicles, but no corpora

lutea or corpora hemorrhagica have been found. Semifertile females also have high numbers of polyovular follicles (Melvold, 1974). According to Wolfe (1967, 1971) *p/p* and p^{un}/p^{un} (nonmottled) females release more ova than *P/–* females when inoculated with pregnant mare serum and human chorionic gonadotropins, an enhanced response which has been traced in part to the greater endogenous gonadotropic activity of their plasma and pituitary. These females also have larger ovaries and uteri, and seem to have a shorter ovarian cycle than *P/–* females. Similar differences in reproductive function also appear to exist for males which have significantly larger seminal vesicles (but not preputial glands or testes) than *P/–* males. The evidence, based on bioassays, indicates that there is an increased synthesis and release of gonadotropin by *p/p* (and p^{un}/p^{un}) pituitaries, but neurohumoral release mechanisms of the hypothalamus, such as luteotropin releasing factor, cannot as yet be excluded. How this effect which *p* has on hormone secretion relates to its influence on pigment remains to be determined (if such a relationship exists). According to Wolfe (1971), the only possible connection would seem to be that due to the inability of pink-eyed animals to form pigment there is "an accumulation of gonadotropin precursors or a change in regulatory control in favor of gonadotropin synthesis." In this regard, it would be of interest to know whether any correlation exists between the gonadotropic activity of p^{un}/p^{un} and p^{un}/p genotypes and their degree of mottling.

[20]Hearing and his associates (1973) found that melanogenesis in *ru/ru* mice is also delayed in the choroid until after birth. They also observed a decrease in the melanization of each granule and a subsequent reorganization of fibrillar melanosomes into particulate melanin granules, especially in the choroid.

[21]Hearing et al. (1973) note that since tyrosinase levels are reduced in *ru/ru* and *p/p* mice, one must consider the possibility that one or more forms of this enzyme are inactivated as a result of these mutations.

[22]A mutation which appears to be a repeat of *ru-2* (called $ru\text{-}2^r$) occurred at Oak Ridge in the C57BL/10 strain. Although mice homozygous for this mutation may have slightly lighter eyes at birth than *ru-2/ru-2* animals, this difference is probably due to genetic background (Kelly, 1974).

Chapter 5

Grey-Lethal, Grizzled, Mocha, Pallid, Muted, Misty, and Pearl

I. Grey-Lethal (*gl*)

In considering the influence of grey-lethal on pigmentation, one is faced with the dilemma of whether to limit the discussion to those observations which bear directly on this subject or to include the considerably greater number of efforts which have been directed toward investigating the other effects of this osteopetrotic mutant. The decision to include a brief resume of all of the effects of the mutation is based on the belief that they are germane and that such treatment may help focus attention on the mutation's primary effect.

A. Origin and Influence on Pigmentation

Grey-lethal [*gl*; chromosome 10 (Lane, 1971)], which is fully recessive, was first described by Grüneberg in 1935 (see also Grüneberg, 1936a, 1938). It occurred as a spontaneous mutation in a stock segregating for c^e and was recognized by the fact that on a wild type background it appears to remove most, if not all, of the yellow pigment from the fur; the coat is a slate grey color instead of the brownish hue of the normal agouti mouse. The gene has no effect on eumelanin synthesis but can be recognized in the presence of nonagouti by its influence on the phaeomelanin-containing hairs round the ears, genital papilla, and mammae. Grey-lethal yellow mice ($A^y/-;gl/gl$) can best be described as khaki-colored.

Although originally it was believed that grey-lethal removed all of the yellow pigment from the fur (Grüneberg, 1935) this, of course, is not the case. Microscopic examination of either the yellow region of $A/A;gl/gl$

hairs, or of $A^y/-;gl/gl$ hairs, reveals that yellow pigment is indeed present but that it is distributed mostly in clumps somewhat like the distribution found in dilute and leaden animals. Evidently for some reason grey-lethal melanocytes are unable to release yellow granules in the same manner as black granules. Moreover, inasmuch as this clumping is accompanied by a considerable number of sparsely pigmented or empty hair septules, it is likely, although not proven, that grey-lethal also reduces the number of phaeomelanin granules.

The basis for the effect which grey-lethal has on the deposition and perhaps the synthesis of phaeomelanin is not known. Grüneberg (1966c) cites that the hairs themselves are structurally abnormal but, if so, it is difficult to envisage how this would selectively influence phaeomelanin synthesis.[1] The observation that grey-lethal melanocytes in all pigmented tissues except the hair bulbs produce black granules only led Markert and Silvers (1956) to conclude that this mutation probably acted "within the epidermal cells of the hair bulb in the presence of A^y, A^w, A and a^t genes to weaken and render abnormal the stimulus for yellow melanin production." Nevertheless, it is just as likely that *gl* acts within the melanocyte but produces its effect only *after* phaeomelanin synthesis is "turned on" by the appropriate *a*-allele.[2]

Also to be resolved is whether the abnormal deposition of phaeomelanin in the form of clumps into the hair of grey-lethal mice reflects a change in the morphology of the follicular melanocytes? Thus it would be of interest to know if during phaeomelanin synthesis the hair bulb melanocytes of $A/-;gl/gl$ animals switch from a nucleofugal to a nucleopetal morphology and whether $A^y/-;gl/gl$ melanocytes persist in the nucleopetal form? This information is particularly important for elucidating the relationship between melanin deposition and melanocyte morphology, including the stability of these cells in the hair follicle (see Chapter 4, Section I, D).

Finally, the fact that grey-lethal [and grizzled (Section II)] has no apparent influence on eumelanin synthesis is important because it provides further evidence that eumelanin and phaeomelanin are produced via alternative pathways.

B. Other Effects

As perplexing as grey-lethal's effect on pigmentation is, it is even more difficult to envisage how this influence ties in with the other manifestations of the mutation. Grey-lethals display profound anomalies in their skeleton and teeth (Doykos et al., 1967; H. Murphy, 1969), abnormalities which stem from their inability to resorb bone. The condition can be identified at birth, even from as early as the eighteenth day of gestation because the apex of the grey-lethal's lower incisor lies anterior to the molar region in the mandible (Hollinshead et al., 1975). It can be identified also at 2 days of age by the external appearance of amputated tibiae; the diaphyses of normal tibiae appear red under the dissecting microscope due to the presence of hemopoietic tissue occupying the central marrow cavity, whereas grey-lethal tibiae appear opaque because relatively more unresorbed bone occupies the center of the diaphysis (Hollinshead and Schneider, 1973).

Grey-lethals weigh slightly less than normals at birth but grow steadily although at a somewhat reduced rate during the first 2 weeks. Thereafter, until weaned, their weight either is stationary or increases irregularly. When weaned their weight declines rapidly, especially if they are not allowed to continue to suckle. Under normal conditions grey-lethals die between 21 and 30 days of age (Grüneberg, 1952). Although one might expect that the main cause of death is starvation as a consequence of noneruption of the teeth, this is *not* the case since hand-feeding affected animals with a liquid diet prolongs their survival only slightly (Grüneberg, 1952). Grüneberg suggests that a neuralgia of the trigeminus due to the excess formation of bone matrix makes them reluctant to move their jaws.

As might be expected from a failure of secondary bone absorption, grey-lethals are characterized by widespread anomalies of the skeleton and teeth. A general osteosclerosis prevails in the growing and developing bones, and because osteoclastic activity is deficient, a disfigurement of the osseous architecture results (H. Murphy, 1969). Hirsch (1962) reported a mean osteoclast count of only 11 (SD $\pm$ 10) cells/mm^2 in untreated sections of grey-lethal metaphyseal bone, compared with a mean of 74 (SD $\pm$ 19) cells/mm^2 in corresponding sections from untreated normal bone. The failure of the teeth to erupt is caused by a lack of resorption in the dental crypts (H. Murphy, 1969).

C. Etiology

To determine the basis of this failure of secondary bone resorption Barnicot (1941) performed a number of transplantation experiments. He found that if he transplanted grey-lethal bone to normal hosts it frequently attained a structure approximating that of normal bone whereas this was not the case if the transplant was made to another grey-lethal animal. On the other hand, normal bones transplanted into grey-lethals sometimes, but not always, became structurally similar to grey-lethal bones. Barnicot (1945) also observed that grey-lethal animals were very resistant to injections of parathyroid hormone but that massive doses of this hormone, doses which were lethal to normal mice, could induce bone absorption (see also Walker, 1966; Schneider et al., 1972). He also reported that when grey-lethal parathyroid gland and a section of grey-lethal parietal bone were transplanted intracerebrally into normal hosts, the bone adjacent to the gland displayed signs of resorption (Barnicot, 1948; see also Hirsch, 1962).

Barnicot (1945), Hirsch (1962), and Grüneberg (1963) suggested that these results were consistent with the possibility that grey-lethals produced normal amounts of parathyroid hormone but somehow inactivated or destroyed it at an increased rate. More recently, however, both thyroid and parathyroid endocrinopathies have been shown to be involved (H. Murphy, 1968, 1969, 1972, 1973; Marks and Walker, 1969), and many features of the condition can be explained by an increased secretion of the thyroid hormone, calcitonin, eliciting a compensatory increase in parathyroid hormone (H. Murphy, 1973). Nevertheless, it is not clear whether even these endocrinological changes are the *primary* cause of the osteopetrosis. Probably the best hypothesis is one proposed by Marks and his associates (Marks and Lane, 1976; Marks and Walker, 1976). They suggest that the primary lesion is in the osteoclast, i.e., that a reduction in osteoclast function is responsible for the hyperparathyroidism (and the increase in bone formation), and that the parafollicular cell hyperplasia (and increased secretion of calcitonin) is a compensatory response to elevated levels of parathormone (see also Raisz et al., 1977).

Most exciting have been the studies of Walker (1975a,b, see also 1972). He found that if irradiated grey-lethal mice are given an intravenous inoculation of cells prepared from the spleen or bone marrow of normal littermates, their capacity to resorb bone and calcified cartilage is restored. Conversely, osteopetrosis is induced when lethally irradiated, normal mice are given cell infusions prepared from the spleens of grey-lethal littermates. These results indicate either that cells from normal donors give rise to competent osteoclasts in grey-lethal recipients, or that they are the source of some humoral factor which is essential for normal osteoclast function.[3]

Although some of the problems of the grey-lethal mouse appear to stem, at least secondarily, from the malfunctioning of the thyroid and/or parathyroid, the possibility should not be overlooked that the thymus too is involved. This possibility receives support from two observations. First from Grüneberg's (1952) observation that the grey-lethal's thymus undergoes a rapid and complete degeneration of the cortex during the third week of life;[4] and second from the fact that the thymus may be involved in the pathogenesis of a genetically determined osteopetrotic condition in rats (which does not affect pigmentation) (Milhaud et al., 1977).[5] Indeed, "*op*" rats not only suffer from an early atrophy of the thymus gland, but their condition too can be cured with a single injection of normal bone marrow cells (Milhaud et al., 1977; see also Marks, 1978a,b).

While it is difficult enough to relate these disturbances in bone formation to an affect on phaeomelanin, they also must somehow be related to white spotting. This follows from the fact that microphthalmia (*mi*) homozygotes, which are "one big spot" (see Chapter 12, Section I, A, 2), likewise suffer from an osteopetrotic condition which although not as severe as grey-lethal's apparently has a similar etiology (see Chapter 12, note 5).

II. Grizzled (*gr*)

"Grizzled" (*gr*; chromosome 10) is similar to grey-lethal (*gl*) in that it too is a recessive condition which influences viability and phaeomelanin synthesis (Bloom and Falconer, 1966). In fact, the only reliable phenotypic effect of this mutation is that it dilutes yellow pigment.

A. Influence on Coat Color

$A/A;gr/gr$ mice look much like chinchilla ($A/A;c^{ch}/c^{ch}$) animals except that the dilution is usually more pronounced and, unlike chinchilla, eumelanin synthesis is unaffected. Grizzled is recognizable also in nonagouti (a/a) mice by the hairs on the ears and around the genitalia appearing white instead of yellow. When combined with yellow (A^y) grizzled produces a less pale color than does chinchilla but only because of the persistence of sootiness (eumelanin) in the coat.

The fact that *gr* and *gl* are situated about 20 cM from each other is undoubtedly a coincidence. Nevertheless, it would be of interest to know whether the dilution of yellow pigment in grizzled animals occurs via a clumping of granules, as is the case with grey-lethal, and/or whether it is due to a general reduction in the number of granules. The interaction of *gr* with recessive yellow (e/e) also is germane.

B. Other Manifestations

Grizzled mice are about three-quarters the weight of their normal littermates at birth and, although they subsequently grow relatively a little faster than normal, they are a little smaller when adult. They are also less viable, both prenatally and postnatally, than their normal littermates with the mortality being higher in males than in females. The prenatal mortality occurs at all stages of development from about 10 days to after 18 days. Some grizzled mice also have kinky tails, the degree of kinkiness being very variable (Bloom and Falconer, 1966). The cause of the "grizzled syndrome" is not known.

III. Mocha (*mh*)

Mocha (*mh*), like grizzled, with which it is very closely linked on chromosome 10[6], also dilutes yellow pigment but affects eumelanin as well (Plate 2–G). The mutation occurred in 1963 in the C57BL/6J-*pi* (pirouette) stock and was named as a result of the phenotype it produced on this nonagouti background (Lane and Deol, 1974). Mocha, which is a recessive, can be recognized at birth by the absence of visible eye pigment. The mutants are generally smaller than their normal littermates and some do not survive the first few days of life—a situation again akin to that in grizzled.

A. Influence of Pigmentation

On a nonagouti background mocha (*a/a;mh/mh*) mice are slightly lighter than the mutants ruby-eye (*a/a;ru/ru*) and white ($a/a;Mi^{wh}/+$) and a little darker than pallid (*a/a;pa/pa*). Moreover, because phaeomelanin as well as eumelanin is diluted, the hair in the pinna of the ear and around the genitalia appears white instead of yellow. The eyes of adult mocha mice are deep red (Lane and Deol, 1974).

When whole mounts of *a/a;mh/mh* and *a/a;mh/+* hairs were examined microscopically and compared, those of the mutant-type contained fewer and smaller eumelanin granules, particularly in the medullary cells (Lane and Deol, 1974).

B. Influence on Behavior

Mocha also affects behavior. According to Lane and Deol the abnormal behavior of mocha mice is similar in some respects to pallid (*pa/pa*) animals (Section IV). At weaning some mocha mice hold their head tilted to one side and some of these cannot swim on the surface. All mocha mice seem to be more active or nervous than their normal littermates and when disturbed they react quickly and violently, jumping and scampering wildly around their cage. When those mice which do not tilt their head are picked up by the tail they tend to jerk constantly, waving their front legs rapidly while quickly flexing and extending or kicking their hind legs. When animals which do tilt their head are picked up they tend to twist their body as well as constantly jerk and kick.

While there is evidence that mocha mice can hear at weaning some, and perhaps all, eventually become deaf. Mocha mice are poor breeders; most males will not breed and the few females which produce litters usually destroy them.

C. Inner Ear Abnormalities and Their Significance

Studies of the inner ear of mocha mice revealed the cochlea to be abnormal. Degenerative changes were observed in the organ of Corti, the stria vascularis, and the spiral ganglion. These changes were mild in young animals but became progressively more severe so that "in animals over 100 days old there was generally a heavy loss of hair cells in the organ of Corti, and the various types of supporting cells, although present, had largely lost their characteristic form and organization on which their identification normally depends. The tectorial membrane was frequently curled back. There was a severe loss of cells in the spiral ganglion. The stria vascularis was nearly always reduced" (Lane and Deol, 1974). Abnormalities of the otoliths in both the saccule and utricle also were observed in most of the mutants.

The casual connection between these ear abnormalities and the pigmentary changes is not evident. While all of the features of the "mocha syndrome" have been separately reported in other animals, in no other single mutant do they occur together. Thus, although the coat-color mutants pallid (Section IV) and muted (Section V) also possess abnormal otoliths and share some behavioral traits with mocha, cochlear abnormalities do not occur in these animals. On the other hand, such cochlear abnormalities do occur in the neurological mutants of the shaker-waltzer-deaf type but only one of these, varitint-waddler (*Va*) (see Chapter 11, Section IV), affects pigment (Deol, 1954, 1968). The cochlea is known to be severely affected by a number of genes associated with white spotting, e.g., piebald lethal (s^l), microphthalmia (*mi*), white (Mi^{wh}) (Deol, 1970b), some of which also cause general dilution, but the abnormalities produced by these genes follow a characteristic pattern that is quite distinct from that in mocha (Lane and Deol, 1974). Furthermore, while the spotting and inner ear abnormalities associated with these spotting genes are best explained by a defect in the differentiation of neural crest cells (Deol, 1970b), such an explanation is difficult to accept for the fully pigmented mocha phenotype.

Clearly the primary defect responsible for the various abnormalities associated with the mocha mutation remains to be elucidated.

IV. Pallid (*pa*)

While Mocha's effect on pigmentation and behavior is difficult to reconcile, there is some evidence that a subtle relationship between pigmentation and the trace element, manganese, may be responsible for the effect which pallid (*pa*) has on behavior (Erway et al., 1966, 1971).

A. Origin and Phenotypic Effects

Pallid, an autosomal recessive (chromosome 2), was described originally by E. Roberts (1931) after a mouse displaying the mutation was caught in the country and brought to his laboratory at Urbana, Illinois in 1926. According to Roberts "the eyes were pink, indistinguishable from the eyes of the common pink-eyed varieties, but the coat color, though plainly agouti, was lighter than that of a pink-eyed black agouti." Because the phenotypic effects of pallid (Plate 2–H) are similar to those of pink-eyed dilution, it has sometimes been referred to in the literature as "Roberts' pink-eye" or "pink-eye-2" (p_2) (Grüneberg, 1952).

Pallid dilutes yellow pigment as well as black and brown. Yellows homozygous for the mutation (*A^y/a;pa/pa*) are a light lemon color and animals homozgous for both pallid and pink-eyed (*pa/pa;p/p*) have very light-colored coats, significantly lighter than those produced by either of the genes separately (E. Roberts and Quisenberry, 1935; Grüneberg, 1952).

Because pallid mice have pink eyes, when combined with viable dominant spotting (*W^v/W^v*; which ordinarily produces a black-eyed white animal) a "mock-albino" phenotype results (Erway et al., 1971). However, it should be stressed that whereas albinos have amelanotic melanocytes (see Chapter 3, Section II, A), *W^v/W^v;pa/pa* mice do not (see Chapter 10, Section I, D).

B. Occurrence and Structure of Pigment Granules

Theriault and Hurley (1970) have compared the occurrence and ultrastructure of developing melanosomes in the retinal epithelium, the sensory epithelium of the inner ear, and the epidermis of pallid (*a/a;pa/pa*) mice with those of C57BL/10 (*a/a*) animals (see also Hearing et al., 1970). They found that in 14-day-old *a/a* embryos there were melanosomes in all stages of development in the retinal epithelium and pigment granules were present in the utricular epithelium, whereas in *pa/pa* embryos no melanosomes occurred in the inner ear and only premelanosomes were present in the retinal epithelium. Uniformly dense melanosomes resembling *a/a* mature melanosomes also were not observed in 1- and 3-day-old pallid retinas and in 15-day-old and adult samples there was a conspicuous absence of melanosomes in any developmental stage, although pigment granules were found in the choroid layer of the eye.[7] The inner ear epithelium of young *pa/pa* mice also lacked both mature and immature melanosomes. On the other hand, melanosomes in different stages of development were present in the cytoplasm of pallid epidermal melanocytes at 6 days postpartum. These granules, however, never became completely electron dense and their melanin was deposited in a coarsely granular way, without any coalescence. Moreover, not all of the granules were evenly electron dense, and similar differences occurred even within a single granule. Pallid granules in epidermal melanocytes were smaller than *a/a;B/B* granules, spherical, and never larger than 0.5 μm (Theriault and Hurley, 1970).[8]

C. Behavior and Its Association with Otolith Defects

In addition to its effect on pigmentation, Castle (1941) observed that when maintained under crowded conditions the viability of pallid mice was considerably reduced and that they "tended to be nervous and jumpy."[9] These behavioral defects were subsequently shown by Lyon (1951) to be associated with otolith defects within the inner ear. She noticed that some pallid animals displayed defects in their postural reflexes which were of two main types. The first type was noticeable when they were 2–3 days old and held up by the tail. Whereas a normal mouse reacts by flexing its spine and neck dorsally while stretching its forelimbs forward, pallid mice sometimes "flexed the spine and neck ventrally and stretched the limbs backward, i.e., they failed to respond normally to change of position" (Lyon, 1953). The second type of postural defect noticed by Lyon developed during the third week of life when some pallid animals, which hitherto seemed normal, displayed "an asymmetrical posture and walked about with the head constantly tilted to one side" (Lyon, 1953). Still other pallid mice displayed none of these abnormalities and remained normal.

Lyon was able to demonstrate that the anatomical basis of these postural and behavioral defects could be traced to the absence of one or more otoliths (Lyon, 1951). A normal mouse has two otoliths, one in the sacculus and one in the utriculus. "Those pallid mice which failed to respond to position change always lacked both otoliths from both ears and those with a normal response always possessed at least one otolith. Animals with asymmetrical posture showed asymmetry of the otolith defect, and tilted the affected side of the head upwards. Those with completely normal otoliths had normal behavior" (Lyon, 1953).[10]

It thus appears that whereas pallid always alters eye and coat color, its effects on the otoliths show incomplete penetrance and variable expression. To account for this variability Lyon carried out a number of experiments and demonstrated that both the genetic background upon which pallid acted, as well as environmental factors, contributed to its penetrance. She found that the proportion of pallid mice which lacked otoliths was higher in brown pallid (*b/b;pa/pa*) mice than in nonbrown pallids. Whether this is due to an effect of *b* itself or to some closely linked factor(s), and whether it is related to the fact that both *pa* and *b* are coat-color determinants is, of course, impossible to say.

Lyon (1953) observed also that the effect which pallid had on the otoliths was related both to litter size and litter order. Increasing litter size produced an increase in penetrance, especially in young and old mothers, i.e., small litters from females at their reproductive prime were least affected. Moreover, in an elegantly conceived experiment she was able to show that this effect of litter size was due to the number of fetuses surviving after implantation and not to the number of ova shed or the number implanted (Lyon, 1954). She postulated that the otolith defect "may be due to competition (*in utero*) for food substances, either general or particular, or for oxygen, space, etc." (Lyon, 1954).

D. Manganese Deficiency, Behavior, and Pigmentation: An Hypothesis

While these studies made it clear that the behavioral defects of pallid mice were due to abnormal otolith development and helped focus attention on some of the factors which contributed to the expression of this abnormality, they failed to explain how this effect might be tied in with the influence pallid had on pigmentation. The idea that perhaps such an association was related to a deficiency in manganese stemmed from a series of experiments by Hurley and her associates (Hurley et al., 1958; Hurley and Everson, 1963). These studies demonstrated that if pregnant normal animals are maintained on a manganese-deficient diet, a phenocopy of the pallid otolith defect occurs. They also showed that if pregnant pallid mice are given high concentrations of this trace metal at the appropriate time,[11] their offspring behave normally and display normal or almost normal otolith development (Erway et al., 1966, 1970, 1971). Although these treatments had no effect on pigmentation Erway and his associates (1966, 1971) propose that since the abnormality of the otoliths in pallid mice can be remedied by treating with manganese, and as melanocytes have not been found in the inner ear of these mice (see above), that the manganese requirement of this mutant is abnormal, and that the melanocytes act as a local reservoir of this trace element (see Cotzias et al., 1965; Van Woert et al., 1965). They suggest "that certain pigment mutations (such as *pa*) may produce subtle effects on trace-element metabolism, which in turn may alter the delicate balance of enzyme systems requiring these metallic co-factors. In the case of *pa*, the evidence indicates that the synthesis of mucopolysaccharides, comprising the matrix in which otoconia are formed, is affected" (Erway et al., 1970). They believe that this hypothesis not only helps explain some of the pleiotropic effects associated with coat-color determinants, but accounts for why melanocytes accumulate in certain regions of the body such as the labyrinth, harderian gland, substantia nigra, etc. (Erway et al., 1971).

To clarify the relationship between pallid's effect on pigment and otolith development, Erway and his colleagues (1971) determined the influence that albinism (*c*) and viable dominant spotting (W^v) had when combined with *pa*/*pa*. They reasoned that inasmuch as *c* presumably disrupts tyrosinase activity, it should supress "the effect of *pa* on otolith development if *pa* exerted its effect on manganese availability via an alteration of the melanin substructure (e.g., greater chelating capacity for manganese)." Similarly, they thought that inasmuch as W^v/W^v removes all melanocytes from the coat it might "be expected to suppress the effects of *pa* on otolith development if the removal of melanocytes from other regions of the body did not withdraw manganese from circulation." Because neither of these genes affected the expression of pallid on otolith development, it was concluded that pallid influences manganese metabolism not through melanin per se but via some earlier effect, presumably on the melanosomes, and that "albino animals, presumably containing amelanotic melanocytes in the inner ear, also possess the manganese reservoir." They contend also that the inability of W^v/W^v to abrogate the influence of pallid on otolith development suggests that *pa* produces a very highly localized state of manganese deficiency, a conclusion in accord with biochemical and ultrastructural comparisons of liver mitochondria from manganese-deficient and pallid mice (Hurley et al., 1970).

While this hypothesis relating the behavioral effects of pallid with its influence on pigmentation is an intriguing one, it nevertheless remains to be confirmed (see Deol, 1970b; Lane and Deol, 1974) (see also Chapter 6, note 11).

V. Muted (*mu*)

A. Origin and Influence on Pigmentation

Another coat-color mutation affecting otolith development is muted (*mu*) which arose spontaneously in a stock segregating for *t*-alleles (Lyon and Meredith, 1965a). This autosomal recessive on chromosome 13 (Lyon, 1966) slightly dilutes both eumelanin and phaeomelanin (Lyon and Meredith, 1969). Accordingly black becomes dark grey, yellow and brown become a somewhat lighter shade, and many animals have white underfur. The color of the eyes also is diluted, so that the iris ring appears very light in newborns, and the adult eye is a dark red color. When combined with other genes affecting eye color the eyes may be colorless at birth and pink in the adult. In general, the color change produced by muted (*mu/mu*) resembles that produced by ruby-eye (*ru/ru*) and beige (*bg/bg*) (Chapter 6, Section I), and is much less severe than that produced by pallid (*pa/pa*) (Lyon and Meredith, 1969).

B. Behavior and Its Association with Otolith Defects

As well as affecting pigment, muted causes changes in postural reflexes very similar to those described above for pallid mice. "Some muted homozygotes hold the head tilted to one side; some flex the spine and tuck the head under when held by the tail, and give no normal landing reaction when dropped head downwards towards a table" (Lyon and Meredith, 1969). As in the case of pallid these behavioral disorders are associated with the absence of otoliths from the sacculus and utriculus of one or both ears as well as by an absence of pigmentation in the membranous labyrinth (Erway et al., 1966). According to Lyon and Meredith (1969) of 17 *mu/mu* animals examined "12 lacked all otoliths, 3 had one otolith in each ear, one had two otoliths in the left ear and one in the right, and the remaining one had otoliths in both ears. The bony labyrinth appeared completely normal in all the animals, and none was deaf or showed circling behavior."

Although the effect of muted parallels very closely that of pallid, especially insofar as otolith development is concerned, only future studies can reveal whether these two genes act similarly. Of particular interest will be whether supplementing the mother's diet during pregnancy with manganese can alleviate the otolith defect in muted offspring, as it does in pallid mice (Erway et al., 1971).

VI. Misty(*m*)

A. Origin and Influence on Pigmentation

Misty (*m*) is a recessive mutation which arose spontaneously at the Jackson Laboratory in the DBA/J strain. It is located on the fourth chromosome, about 7 cM from brown (Woolley, 1941, 1943, 1945). *m*/*m* animals are slightly lighter than normal on both nonagouti black and nonagouti brown backgrounds. The mutation was originally described by Woolley (1941) who noted that "*m*/*m*;*b*;*b*;*a*/*a* and *m*/*m*;*B*/*B*;*a*/*a* mice are more uniform and more intense in color than *ln*/*ln*;*b*/*b*;*a*/*a*, *ln*/*ln*;*B*/*B*;*a*/*a* or *d*/*d*;*b*/*b*;*a*/*a* and *d*/*d*;*B*/*B*;*a*/*a* mice." The difference between *m*/*m* and *ln*/*ln* or *d*/*d* mice is especially noticeable at the extremeties such as the hair tip, ears, and tail where *m*/*m* seems to allow more pigment (Woolley, 1941). Misty mice also display a light underfur which is useful in classification (Grüneberg, 1952) as well as a very high incidence of white tail tips. A small white belly spot, the incidence of which depends upon the genetic backgound, may also occur. Woolley (1945) reported that whereas none of 64 *M*/– DBA/J mice displayed a white tail tip or white belly spot, 52 of 53 *m*/*m* DBA/J animals had white tail tips and two of these also had a white belly spot. On the other hand, when misty animals were outcrossed he found that all 258 F_2 *m*/*m* mice had white tail tips (versus 61% in *M*/– genotypes) and 195/275 (71%) white belly spots (versus 2% in *M*/– animals). Moreover, in all cases where a white belly spot occurred, whatever the genotype, a white tail tip also was found.

The effect misty has on phaeomelanin synthesis has not been reported.

Microscopic examination of *m*/*m* hairs reveals much more cortical pigment, and probably slightly more medullary pigment, than in *d*/*d* or *ln*/*ln* hairs on both *a*/*a*;*b*/*b* and *a*/*a*;*B*/*B* backgrounds (Woolley, 1941).

VII. Pearl (*pe*)

A. Origin and Influence on Pigmentation

Pearl (*pe*) is a simple autosomal (chromosome 13) recessive which arose spontaneously in the C3H/He strain. The mutation was described originally by Sarvella (1954), what follows is based on her description. On the agouti black (*A*/*A*;*B*/*B*) background on which pearl occurred the pigmentation of the entire hair is diluted with the base of the hair being affected to a disproportionately greater extent than the tip. The snout, ears, feet, and tail of pearl agouti mice are considerably lighter than normal and the yellowish mask about the eyes, characteristic of some C3H strains, is diluted to near-white. Pearl agouti mice can be distinguished from their typical agouti littermates at birth by the smaller amount of pigment in their eyes. This distinction, however, is not evident in adult animals.

Nonagouti black pearl (*a*/*a*;*B*/–;*pe*/*pe*) mice bear a striking resemblance to nonagouti black ruby-eye (*a*/*a*;*B*/–*ru*/*ru*) animals. In fact the two phenotypes can be distinguished readily only by blowing the fur so that the much lighter base of the pearl hair is revealed. The coat color of *a*/*a*;*b*/*b*;*pe*/*pe* mice resembles that of *a*/*a*;*b*/*b*;c^{ch}/*c* heterozygotes. The effects of *pe*/*pe* and *b*/*b* on eye color are additive so that the eyes of adult brown pearl mice have a definite ruby cast. On a yellow background (A^y/*a*;*B*/*B*) the effect of pearl depends on the amount of sootiness. When sootiness is present the coat color appears pale grey even though the bottom quarter of the hair is cream-colored. On the other hand, in the absence of sootiness, or when yellow is on a brown background (A^y/*a*;*b*/*b*), the action of pearl is similar to that of chinchilla (c^{ch}/c^{ch}), producing a cream-colored fur. As might be expected lethal yellow pearls show the same tendency toward obesity as "normal" lethal yellows (Sarvella, 1954).[12]

B. Stability

An interesting feature of pearl is its stability. While on certain backgrounds, such as its strain of origin, it is stable; on another background, strain 201(A^y/*a*;*pe*/*pe*), somatic reversions to full-color are frequent (L.B. Russell and Major, 1956). Thus, L. Russell (1964) reported in this latter strain the frequency of reversion of *pe*/*pe* to full color was about 6% in the 800 or so animals observed. Moreover, the incidence was not affected by whether one bred from mosaic or nonmosaic mice, indicating that the degree of instability was characteristic of the stock as a whole rather than being transmitted in sublines (L. Russell, 1964). While in most cases the intensely colored spots occupied from 0.1 to 1% of the coat and did not affect the germinal tissue, six animals of a total of 61 had from 50 to 100% of their coat full-colored, and one had a full-color spot covering about 5% of the surface (L. Russell, 1964). Five of the animals in which the proportions of full-colored fur were about 5, 50, 80, 100, and 100%, produced respectively 18.2, 52.7, 43.6, 38.5, and 36.2% full-colored young (L. Russell and Major, 1956). Since the two completely full-colored mice produced significantly less than 50% full-colored offspring, it was concluded that they, too, were most probably the result of somatic rather than germinal events.

Although the basis for this mottling is not known, L. Russell and Major

(1956) believe that spontaneous somatic reverse mutations at the *pe*-locus is the most likely cause and that these can occur either very early or moderately late in embryonic development. If they occur early, mutant cells participate in the fur and in the gonad and if they occur late they usually involve only the fur. Animals mosaic in other tissues would, of course, not be detected (L. Russell, 1964)[13] (see also Chapter 6, note 11).

Notes

[1]Grüneberg (1966c) maintains that the structure of the hair is abnormal in a number of "color mutants" and cites as examples dilute, varitint-waddler, and grey-lethal. Although this awaits confirmation (other than the fact that the incorporation of nucleopetal melanocytes into the hair may cause local enlargement of the hair diameter—see Chapter 4, Section I, D) it is not without precedent. It is well established that some, and probably all, of the alleles at the X-linked mottled (*Mo*) locus affect hair structure (see Chapter 8), and there is evidence that varitint-waddler (*Va*), microphthalmia (*mi*), and perhaps white (Mi^{wh}) also act both in the melanoblast and in the hair follicle (Mintz, 1971a; see Chapter 12, note 16).

[2]Inasmuch as recessive yellow (*e*), unlike lethal yellow (A^y), acts within the melanocyte (see Chapter 2, Section II, C, 3), the coat color of *e/e;gl/gl* animals might help resolve this question. If, on the one hand, the coats of such mice were indistinguishable from normal recessive yellow (*e/e*;+/+ or *e/e;gl*/+) animals, this would indicate that *gl* probably does act in concert with the *a*-locus via the follicular environment. On the other hand, if *gl/gl* has a similar effect on recessive yellow as it has on lethal yellow, this would still leave its site of action unresolved, i.e., it could act via the follicular environment to alter phaeomelanin synthesis, or, it could act intracellularly along with *e/e*.

[3]The observation of Loutit and Sansom (1976) that the osteopetrotic defect of microphthalmic (*mi/mi*) mice can be cured by inoculating them with cell suspensions containing hematopoietic stem cells from normal donors, even in the absence of X-irradiation (Loutit, 1977), has led them to conclude that osteoclastic cells are derived through circulating monocytes from hematopoietic stem cells (see Chapter 12, note 5). If correct, this would undoubtedly hold for grey-lethal as well.

[4]Because subcutaneous injections of "thymocrescin," an extract prepared from calf's thymus, had no effect on grey-lethals it was concluded that the degeneration of the thymus was most likely the consequence of the poor condition of the animals, and not its cause (Grüneberg, 1938).

[5]There are also two recessively inherited osteopetrotic conditions in mice which do not influence coat color: osteosclerotic (*oc*) and osteopetrosis (*op*). Animals homozygous for *oc* (chromosome 19) can be detected at 10 to 12 days of age by a circling behavior, odd clubbing shape of feet, and absence of teeth. Skeletal preparations show extra deposition of bone at the chondrocostal junction. There is likewise extra bone deposition in the hind limbs. Histologically there is no evidence of secondary bone resorption. The circling behavior becomes more pronounced with advancing age and most affected animals die at approximately 30 to 40 days of age. The double homozygote of this mutant and grey-lethal is viable (Dickie, 1967a). Mice

homozygous for osteopetrosis, which is linked to varitint-waddler on chromosome 12, are also recognizable at 10 days of age. They have no teeth, a short domed skull, short tail, and small body size. Compared with normal littermates, young *op* homozygotes "have excessive accumulations of bone without marrow cavities, increases in bone matrix formation and concentrations of parafollicular cells of the thyroid, and are hypophosphatemic" (Marks and Lane, 1976). Their osteoclasts are small, few in number, and have an abnormal distribution of the lysosomal enzyme acid phosphatase in their cytoplasm. Nevertheless, this osteopetrotic condition is milder than that produced by *gl/gl*, mi/mi, and *oc/oc*. The main skeletal defect seems "to be a severe restriction in bone remodeling that is capable of slowly removing the excessive skeletal mass characteristic of the disease only after bone formation has declined to one-fifth that of normal littermates" (Marks and Lane, 1976). If provided with adequate soft food *op/op* mice frequently survive weaning. Their breeding performance is, however, very poor (Marks and Lane, 1976).

[6]The two loci are less than 1.2 cM apart. In fact, of a total of 256 chromosomes tested, no recombination between *gr* and *mh* was observed (Lane and Deol, 1974).

[7]Hearing et al. (1973) also report that pallid affects melanization in the retina. However, according to these investigators although the number of granules in the adult retina is greatly reduced and their size diminished, they nevertheless display at least some melanin deposition. They also note that melanization in the choroid is affected.

[8]Another observation of Theriault and Hurley which deserves mentioning is that they found the morphology of *a/a* retinal and inner ear pigment granules to be strikingly similar; they both displayed a diversity in size and shape as well as an uneven melanization. Indeed, in their opinion, the inner ear melanocytes (melanosomes) seemed to resemble those of the retina more closely than those of the epidermis. This is somewhat surprising for whereas it is known the retinal melanocytes are formed *in situ* from the outer layer of the optic cup (see Chapter 1, note 4), those of the inner ear, like epidermal melanocytes, are of neural crest origin (Markert and Silvers, 1956).

[9]Castle (1941, 1942; see also Grüneberg, 1952) also reported that pallid reduced body size more than pink-eyed. However, as in the case of dilute (see Chapter 4, note 2) the apparent effects which these coat-color determinants have on size could be due to closely linked genes.

[10]According to Erway et al. (1971) most pallid mice behave normally under cage conditions, except that about 25% of them display various degrees of head tilting. They also report that there is a clear correlation between their ability to swim normally with their otolith defects. The seriously affected pallid mice, when initially dropped into water, frequently exhibit a momentary tonic seizure, but then become coordinated and swim "in almost any direction, often undergoing tortuous spiralling or back-circling because of the retracted position of the head." After they are removed from the water and placed on a solid surface, they often exhibit, temporarily, an exaggerated form of ataxia and head tilting, and sometimes they roll over repeatedly.

[11]Lyon (1955 a,b) reported that otolith matrix and crystals were first visible in the mouse between days 15.5 and 16.5 of gestation, and in these experiments manganese supplementation of *pa* mice was ineffective on the fourteenth day of gestation, or later, in preventing otolith defects. On the other hand, initiation of the supplement on

the tenth or eleventh day of gestation produced results comparable to supplementation throughout gestation (Erway et al., 1971).

[12]Sarvella (1954) also noted that although the viability of pearl mice until maturity is good when raised by *pe*/+ mothers, pearl females have smaller litters than *pe*/+ females, have a tendency to die during pregnancy and lactation, and are poor mothers.

[13]The behavior of *pe* in the 201 strain is very similar to the *p*-unstable (p^{un}) gene of the *p*-series (Chapter 4, Section II, C, 9) as well as to four genes at the *c*-locus (L. Russell, 1964). Two of these appear similar to *c* in phenotype (but one is homozygous lethal), and two are intermediate between *c* and c^e. All of these, when either homozygous, or heterozygous with *c*, produce "a rather high frequency of animals with small dark (possibly full-colored) patches of fur," and one has produced a uniformly full-colored animal which transmitted *C* (L. Russell, 1964). It also should be noted that at the *a*-locus somatic "reverse mutations" of the type $a \rightarrow A^w$ have been observed (Bhat, 1949; L. Russell, 1964). Both of these involved gonadal tissue. Frequent full-colored spots in mice heterozygous for Ames dominant spotting (W^a) and white (Mi^{wh}) have also been interpreted as a consequence of somatic reverse mutation (Schaible and Gowen, 1960). No breeding results have been published for these animals (see Chapter 10, note 28 and Chapter 11, note 29).

Chapter 6

Beige, Silver, Greying with Age, and Other Determinants

I. Beige (*bg*)

Beige [*bg*; chromosome 13 (Lane, 1965)] is a recessive mutation affecting both coat and eye color. The mutation is of particular interest because it produces a syndrome strikingly similar to the Chediak–Higashi syndrome in man (Blume et al., 1968; Blume and S. Wolff, 1972).[1] This syndrome is characterized by an autosomal recessive inheritance, pigment dilution, large melanin granules, large lysosomal granules, and an increased susceptibility to infections.[2] Its effects on pigmentation have been studied by Windhorst and her associates (Windhorst et al., 1966, 1968; Zelickson et al., 1967).[3]

A. Origin

Beige was reported first by Kelly in 1957 after it was probably radiation induced at Oak Ridge. However, it also occurred at Brown University in 1955 as a spontaneous mutation in Chase's YZ57/Ch strain (Chase, 1959), a mutation that was known as "slate" (*slt*) until its allelism with *bg* was demonstrated (Lyon and Meredith, 1965b; see Chase, 1965). This mutation is currently referred to as bg^{slt}. Two mutations to beige have occurred also at the Jackson Laboratory, one in a yellow stock and the other in strain C57BL/6J (Lane, 1962) (Plate 3–A). The mutation is maintained on two backgrounds; an inbred derivative of the heterogeneous stock in which it occurred at Oak Ridge, known as SB/Le (Lane and E. Murphy, 1972) and C57BL/6J (bg^J).

B. Influence on Pigmentation

In the heterogeneous agouti stock in which it was first reported, the eye color of *bg/bg* mice was light at birth and varied from ruby to almost black in adults. *A/–;B/–;bg/bg* mice display reduced ear and tail pigmentation, and the coat is lighter than the coat of wild type mice, particularly at the base of the hairs (Kelly, 1957). This basal dilution, which is first noticeable at about 15 days of age, is somewhat variable, and tends to become darker in older animals (Kelly, 1957). The overall appearance of *A/–;b/b;bg/bg* animals is "café-au-lait" and their eyes at birth are lighter than either *A/–;b/b;Bg/–* or *A/–;B/–;bg/bg* genotypes. Moreover, the relatively greater dilution effect on the base of the hair is not as pronounced in brown beige as in black beige mice (Kelly, 1957).

A more detailed account of the effect of beige on pigmentation is given by Pierro and Chase (1963) in their description of slate. This mutation was maintained in a stock segregating for yellow (A^y/a) and piebald spotting (*s*) and it was found that ruby-red eyes were restricted to $bg^{slt}/bg^{slt};s/s$ mice; $bg^{slt}/bg^{slt};S/-$ and *Bg/–;s/s* animals had dark eyes. These observations, therefore, indicate some interaction between beige and piebald in the variability in eye color.

$a/a;B/B;bg^{slt}/bg^{slt}$ mice have a slate black coat with the hairs displaying a considerable amount of medullary and cortical pigment (Figure 6-1a). The pigment in the medulla of the hair was generally very highly clumped, but individual granules could be resolved in some septae. The dimensions of individual granules along any given axis varied from 0.5 to almost 4 μm.[4] Some granules were long ovals but irregular shapes were common, especially among the larger pigment masses (compare Figure 6-lc with d). Some variation was also observed among cortical granules which were usually long ovals measuring about 1 μm long and about half as wide. However, some granules were compressed laterally and elongated to as much as 2–3 μm, while others were almost as wide as they were long (Pierro and Chase, 1963).

C. Interaction with Other Determinants

Pierro and Spiggle (1970) found that the characteristic exclusion of eumelanin granules from the proximal region of the hair shaft of B^{lt} (light) animals (see Chapter 3, Section I, G) did not occur in B^{lt} mice homozygous for bg^{slt}.[5] On the other hand, phaeomelanin was eliminated from the proximal portion of the hair shaft in yellow, beige ($A^y/a;bg^{slt}/bg^{slt}$) animals. Initially this situation was found to occur only during the later hair growth cycles in a small number of animals, but when selected for it resulted first with a lightening of the coat after the completion of the second hair growth cycle, and subsequently with the absence of yellow granules in the proximal

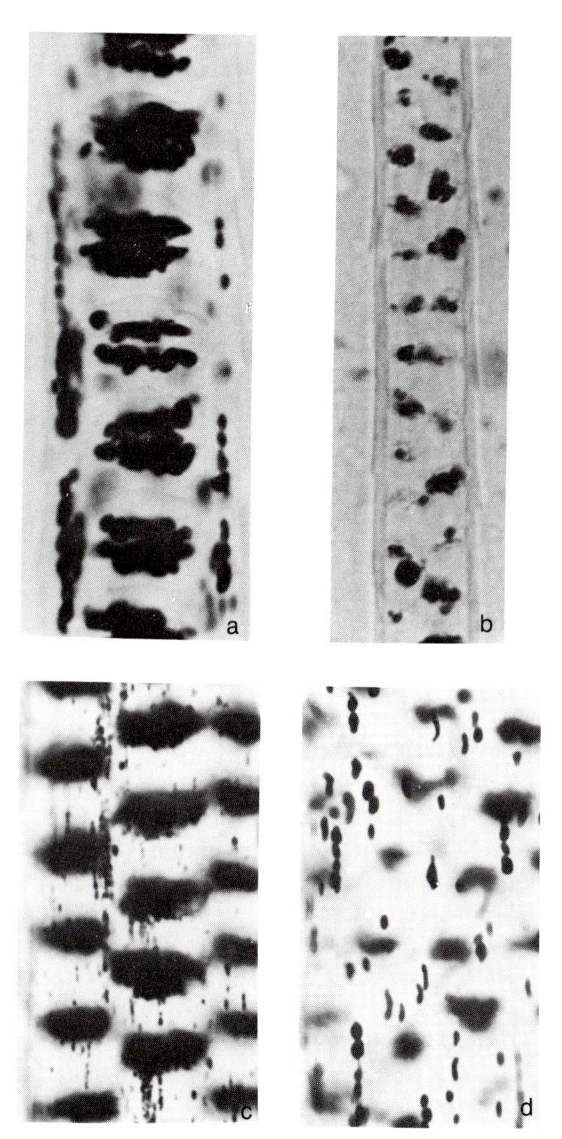

Figure 6-1. (a) Cleared whole mount of a beige ($a/a;bg^{slt}/bg^{slt}$) hair showing medullary and cortical pigment. Note clumping of granules. Approx., ×1400. (b) Cleared whole mount of a beige ruby-eye (*a/a;bg/bg;ru/ru*) hair. Note absence of cortical granules and paucity of medullary granules compared with (a). Approx., ×750. (c) Region of a hair from a C57BL/6 (*a/a;+/+*) mouse showing small, ovoid granules. (d) A similar region from a C57BL/6-bg^J/bg^J mouse showing large, irregular melanin granules. The large, out of focus densities in (c) and (d) are aggregates of melanin granules in medullary compartments. Both (c) and (d) approx., ×800. Figs. (a) and (b) from L.J. Pierro and H.B. Chase, 1963. Figs. (c) and (d) from M.A. Lutzner et al., 1967. Reproduced with permission of the authors and the American Genetic Association.

portion of hairs of the initial pelage.[6] Nevertheless, melanocytes appeared to be present in hair follicles throughout the cycle.

According to Pierro and Chase (1963) when beige is combined with *dilute* ($a/a;bg^{slt}/bg^{slt};d/d$) a "silvery-blue" coat color results and the eyes are black, or reddish if the animals are piebald. Under the light microscope cleared beige dilute hairs displayed an almost complete absence of pigment from approximately the distal third of the hair shaft, but the more proximal regions were as heavily pigmented as the hairs of nondilute beige mice.

Beige ruby-eye ($a/a;bg^{slt}/bg^{slt};ru/ru$) mice have an off-white coat and pink eye color. The cleared hairs from these mice "showed that cortical pigment was absent, and that many medullary septae had no pigment granules at all." Melanin granules were found in more medullary septae in the proximal regions of the hair shaft than in the distal portion. The granules themselves displayed the same characteristics as those in beige hairs, but were very few in numbers (Figure 6-1b).

The coats of animals homozygous for bg^{slt} and pink-eyed dilution (*p*) are only slightly lighter than those of mice homozygous for *p* alone. The eyes of these mice are of course pink. Cleared hairs revealed that cortical pigment was absent. A few medullary septae lacked pigment granules, but these were restricted to the more distal regions of the hair shaft. In general, the pigment granules themselves were very much like those found in *p/p* mice, but in a few instances larger than expected granules were observed (Pierro and Chase, 1963).

It thus appears that in animals homozygous for both beige and ruby-eye the medullary granules of the hair display the characteristic features of beige granules, whereas in pink-eyed beige mice the medullary granules generally display the features characteristic of *p* granules. These observations indicate that at least insofar as the morphology of the pigment granules of the hair is concerned, and, as noted below, those of the retina as well (Hearing et al., 1973), beige is epistatic to *ru* but hypostatic to *p* (Pierro and Chase, 1963). Nevertheless, some interaction between beige and both ruby-eye and pink-eyed is indicated by the absence of cortical pigment in both double homozygotes, as well as by the extreme reduction in medullary pigment in $bg^{slt}/bg^{slt};ru/ru$ mice (Pierro and Chase, 1963).

Although beige, like dilute (*d/d*) and leaden (*ln/ln*), appears to bring about the clumping of medullary pigment granules, it has no effect on melanocyte morphology. Presumably, therefore, this clumping effect must have a different basis.

D. Development of Pigment Granules

In a more detailed analysis of the effect of beige on melanin synthesis Pierro (1963b) studied pigment granule genesis in epithelial, uveal, harderian gland, and hair bulb melanocytes of both beige and normal (*a/a;B/B* and

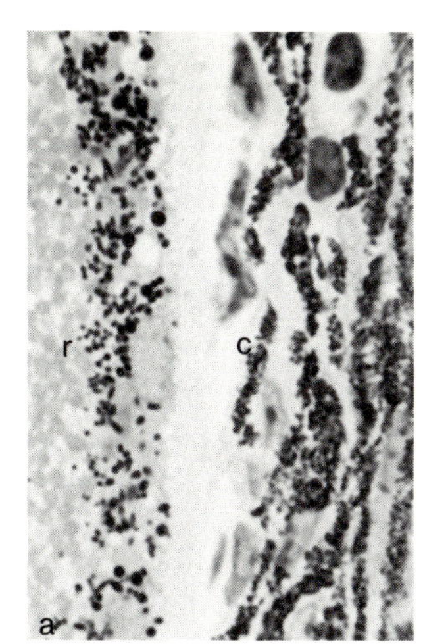

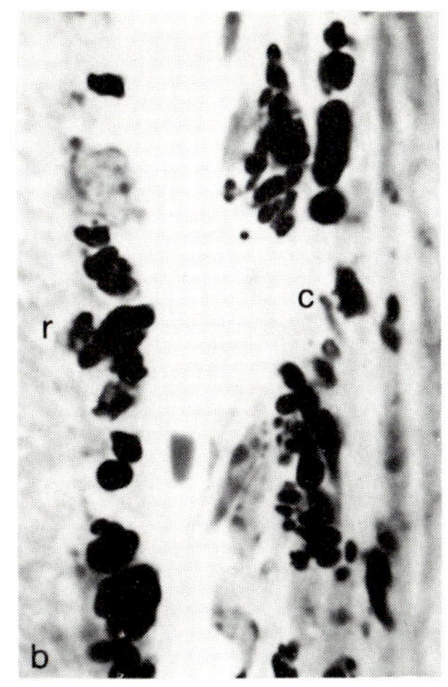

Figure 6-2. (a) From the eye of a C57BL/6J (+/+) mouse showing that normal, small ovoid granules are present in the choroid (c) and round and rod-shaped granules occur in the retina (r). (b) A similar region from the eye of a C57BL/6-bg^J/bg^J animal in which large, irregular melanin granules are present in both the choroid (c) and retina (r). Approx. ×1100. From M.A. Lutzner et al. (1967). Reproduced with permission of the authors and the American Genetic Association.

a/*a*;*B*/*B*;*S*/*s*;*Bg*/*bg*) mice. In the epithelial melanocytes of the retina (which becomes pigmented at about the eleventh day of gestation) of normal animals he found the first pigment granules to be relatively small and uniform spheroids which approximately doubled in size during the first 24 hours after their initial appearance. As the number of these granules increased their sizes and shapes became more variable: first long ovals, then rod and spindle shaped granules appeared (Figure 6-2a). Although in beige mice the first granules observed in epithelial melanocytes were likewise spheroidal, and of a size similar to the granules in normal mice, fewer granules were found within each cell. Moreover, not only did the number of granules appear to remain fairly constant for some time, but their number subsequently seemed to decline. This occurred even though granule synthesis continued and the various granule types appeared. It soon became evident that this apparent decrease in the number of granules resulted from the fusion of individual granules to form "giant granules" (Figure 6-2b). Thus in some cells "grape-like clusters and dumb-bell shaped granules" were found which appeared to represent intermediate stages leading to the formation of still larger masses. In fact, a number of cells were found in which only a single enormous pigment mass occurred (Pierro, 1963b).

As far as the neural crest-derived melanocytes of the uveal tract and harderian gland were concerned, from their first appearance the pigment granules in these cells of *normal* mice were uniform long-ovals, considerably smaller than any of those found in epithelial melanocytes. Furthermore, unlike the granules of the retina, these did not undergo any apparent change. On the other hand, while the first granules to appear in beige uveal and harderian gland pigment cells were also long-ovals, there were not as many of them and, as progressively older mice were examined, their size increased significantly. Moreover, the larger they became the more irregular was their shape. Once again it seemed apparent that in at least some cases granule size increased at the expense of granule number, indicating that the large masses of pigment were due to the fusion of granules (Pierro, 1963b).

The pigment granules of beige and nonbeige hair bulb melanocytes also were essentially similar in size and shape during the early stages (Anagen III and IV) of hair growth. In both genotypes, most of the granules were elongated ovals which did not exceed 0.5 μm in length. During later stages of hair growth (Anagen V and VI), however, there were marked differences. Not only were the pigment granules of normal hair bulbs much more numerous but they were much more uniform in size: small, long-ovals. Beige granules, on the other hand, showed considerable variation in size and shape even though a significant number of them were long-ovals approximately 1 μm in length (Pierro, 1963b).

Taken together these observations indicate that there is a reduced number of pigment granules in the melanocytes of beige mice and that this reduction is due both to the synthesis of fewer granules and to the fusion of granules into progressively larger bodies. In the hair bulb this fusion of granules occurs in both the follicular melanocytes and the recipient matrix cells and since, as the granules become larger, their shapes become more irregular, all of these attributes—granule number, size, and shape—contribute to the mutant's phenotype (Pierro, 1963b).

E. Ultrastructural Studies

The ultrastructure of the melanosomes of nonagouti beige mice, as well as those from brown (*b*/*b*), ruby-eye (*ru*/*ru*), albino (*c*/*c*), and pink-eyed dilute (*p*/*p*), nonagouti beige donors, has been examined by Hearing and his associates (1973). As might be expected they found in both the choroid and retina of bg^J/bg^J mice that giant granules appear to be formed by the random aggregation of premelanosomes and melanosomes, and their subsequent asynchronous melanization. Thus unlike normal granule formation where melanin deposition proceeds more or less uniformly, several stages of melanization often were visible within a single beige granule. This asynchronous melanization of giant granules was observed also in the eyes of brown beige animals (Figure 6-3a and b) where it was estimated that about 80% of all the granules in the adult choroid and retina were fused into giants.[7]

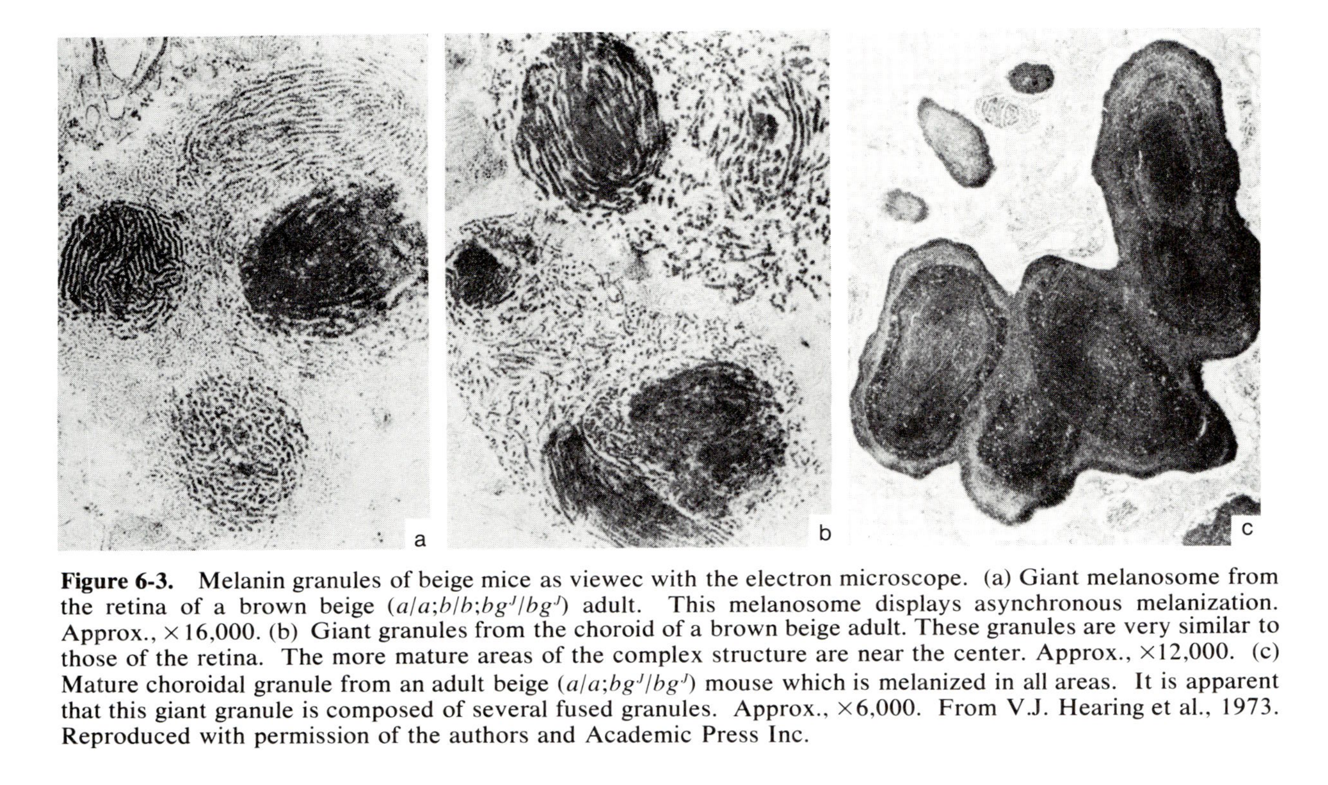

Figure 6-3. Melanin granules of beige mice as viewec with the electron microscope. (a) Giant melanosome from the retina of a brown beige ($a/a;b/b;bg^J/bg^J$) adult. This melanosome displays asynchronous melanization. Approx., ×16,000. (b) Giant granules from the choroid of a brown beige adult. These granules are very similar to those of the retina. The more mature areas of the complex structure are near the center. Approx., ×12,000. (c) Mature choroidal granule from an adult beige ($a/a;bg^J/bg^J$) mouse which is melanized in all areas. It is apparent that this giant granule is composed of several fused granules. Approx., ×6,000. From V.J. Hearing et al., 1973. Reproduced with permission of the authors and Academic Press Inc.

In ruby-eye beige animals melanogenesis in the choroid was delayed until after parturition as it is when ruby-eye is present by itself. However, unlike the premelanosomes formed in the retina of these double mutants, where approximately 80% fused to form giant melanosomes, the granules of the choroid displayed no evidence of aggregating into giants and 95% remained particulate (Hearing et al., 1973) (see Chapter 4, note 20). Thus for some reason, perhaps related to the delay in the onset of melanogenesis, in *ru/ru*;bg^J/bg^J mice, beige seems to affect the fibrillar retinal granules as well as the granules of the epidermal melanocytes (see above), but not the particulate choroidal ones.

In contrast to the effect of beige on *b/b* and *ru/ru* granules, it had no influence when combined either with *c/c* or *p/p*. In albino beige animals the premelanosomes in the retina and choroid were like those of nonbeige albino mice, and in bg^J/bg^J;*p/p* mice there was likewise no tendency for the melanosomes to aggregate into giant granules (Hearing et al., 1973).[8]

These ultrastructural observations suggest that the formation of giant granules in beige mice results from the fusion of individual, normally sized granules (Figure 6-3c), primarily those in the premelanosome stage. Whether this fusion is caused by a membrane abnormality, a modification of melanofilaments, or some other abnormality is not known (Hearing et al., 1973). However, the fact that the melanosomes of neither albino nor pink-eyed dilute animals are affected by beige raises the possibility that a threshold amount of tyrosine activity may be required before this mutant's effect in the melanocyte is expressed (Hearing et al., 1973). In this regard it would be of interest to know how the other effects of beige (see Section F) are expressed in albino and pink-eyed dilute mice.

F. Other Effects

In addition to its effect on the pigmentation of the hair, beige has been reported to be sometimes associated with temporary hair loss (Pierro and Chase, 1965). In the YZ57/Ch strain in which bg^{slt} originated, about 5% of bg^{slt}/bg^{slt} mice lost most of their dorsal hairs of the first coat during the fourth week after birth, i.e., following the cessation of the first hair growth cycle. The second coat grown by these animals was normal although somewhat delayed in some cases. Coats of affected mice usually remained normal in appearance thereafter, but in some instances the coats of older animals became somewhat sparse (Pierro and Chase, 1965). Breeding studies demonstrated that the penetrance and expression of this hair loss trait depended upon the genetic background and, since it could be readily selected for, that a small number of genes were involved. In fact the presence of *p/p*, *d/d*, and especially *b/b*, or genes closely associated with these coat-color determinants, seemed to increase greatly the number of beige animals which displayed the temporary hair loss condition.

Encouraged by Pierro and Chase's observation that beige mice possessed unusually large melanin granules, Lutzner and his colleagues (1967), who were searching for a small animal model for the Chediak–Higashi syndrome, examined the peripheral blood and bone marrow of these mice for giant leukocyte granules. Their search was not in vain as they found that lymphocytes, neutrophils, and eosinophils of all beige mice, but not of nonbeige animals, possessed giant granules. Subsequently it was demonstrated, using acid phosphatase as a marker (C. Oliver and Essner, 1973), as well as Sudan black and peroxidase stains (J. Bennett et al., 1969), that these giant granules were enlarged (anomalous) lysosomes, often in the form of aggregates, and that they occurred not only in leukocytes (Blume et al., 1969; C. Oliver and Essner, 1975) but in a variety of other tissues as well (C. Oliver and Essner, 1973). The most anomalous lysosomes occurred in liver parenchymal cells, kidney proximal tubule cells, Purkinje cells, and granulocytes (C. Oliver and Essner, 1973; Essner and C. Oliver, 1973, 1974).[9] Since there is evidence that in the Chediak–Higashi syndrome, too, there is a basic alteration in lysosomes (S. Wolff et al., 1972), these findings further strengthened the homology between the beige mouse and this disease.

This homology was made even more apparent when it was established that on some genetic backgrounds, homozygous beige mice have a significantly higher incidence of pneumonitis than nonbeige animals (Lane and E. Murphy, 1972). In the C57BL/6J strain, beige mice do not develop spontaneous pnemonitis whereas it does develop when the SB/Le background predominates (Lane and E. Murphy, 1972). While it could be argued that this increased susceptibility is caused by a closely linked susceptibility locus, the simplest hypothesis is that the situation is analogous to the Chediak–Higashi syndrome and that beige itself is the major susceptibility factor (Lane and E. Murphy, 1972).[10]

More recently it has been shown that the beige mouse also resembles the Chediak–Higashi syndrome of humans because in both situations granulocytes have decreased bactericidal activity (Gallin et al., 1974). It has likewise been shown in both bg^J/bg^J (Brandt et al., 1975) and bg/bg (Brandt and Swank, 1976) mice that there is a correlation between abnormal lysosome structure and defective lysosome function. Thus beige mice secrete much less than normal amounts of lysosomal enzymes from proximal tubule cells and, as a result, have increased lysosomal enzyme activity in their kidneys (Brandt and Swank, 1976). Indeed, this finding has led Brandt and Swank to suggest that beige mice "are defective in intracellular motility of lysosomes and/or their fusion with cellular membranes."[11]

Finally, while there is some evidence that all of the effects of the beige mutation are a consequence of some disturbance in lipid metabolism (C. Oliver et al., 1976), whether this is so, and how it specifically relates to the pigmentary disturbance, remains to be elucidated.[12]

II. Silver (*si*)

A. Origin

Silver (*si*, chromosome 10) (Plate 3–B) was reported first by Hagedoorn (1912) who claimed it occurred twice in his colony. According to Hagedoorn animals homozygous for this recessive determinant, which he called *f*, did not differ from *Ff* or *FF* animals until after they had moulted and their second coat was interspersed with white hairs. These were either distributed evenly over the entire body or were limited to patches on the back and sides.[13]

Because Hagedoorn's allele was lost, it was not possible to determine whether it was the same as the one obtained from an English fancier by Dunn and Thigpen 15 years later. This *si* allele had been represented in the English fancy for several years and was known as either "silver grey" or "silver brown." Silver grey animals appeared to be black mice displaying a mixture of white-tipped, all-white, and black and white hairs. Silver brown mice "were of a lighter shade than the standard chocolate, and also showed white-tipped and some all-white hairs, although they were never silvered to such an extent as the silver greys" (Dunn and Thigpen, 1930).[14]

B. Expression on Nonagouti Black (*B/B*) Backgrounds

The "purebred silver greys" (*a/a;B/B;si/si*) studied by Dunn and Thigpen were variable in phenotype. Indeed, dark, medium, and light silver greys were recognized by the fancy. Some young *a/a;B/B;si/si* mice were black in their first pelage, although a few silvered hairs could usually be seen on the head, behind the ears, and on the flanks, while the initial coat of others, usually those which became "light silvers," was well silvered. All the mice became progressively lighter (more silvered) with age, the males showing more silvering than the females.

Although silver greys usually exhibit the silvering effect throughout their coat, the belly is lighter than the back. The fur contains several kinds of hairs: all white, all black (these probably only in dark silvers), black hairs with white tips, and hairs with several white and grey or black bands. The all-white hairs have no pigment whereas in the all-black and black hairs with white tips the pigment granules are reduced in number and scattered. The banded hairs display white areas alternating with sparsely pigmented ones (Dunn and Thigpen, 1930). Silvering thus seems to be the consequence of a reduction in the number of melanin granules, and of their complete absence from certain areas and from some hairs. Light silvers have more completely white hairs and larger unpigmented zones than darker silvers (Dunn and Thigpen, 1930).

C. Expression on Nonagouti Brown (*b*/*b*) and Heterozygous (*B*/*b*) Backgrounds

Hairs from silver brown (*a*/*a*;*b*/*b*;*si*/*si*) mice display the same sorts of pigmented hairs as described above for silver greys, except that "the proportions of white hairs and of white in colored hairs are generally much less" (Dunn and Thigpen, 1930). Nevertheless, in black mice heterozygous for brown the effect of *si* is greatly intensified; *a*/*a*;*B*/*b*;*si*/*si* animals are *lighter* than either *B*/*B* or *b*/*b* silver mice (Dunn and Thigpen, 1930). According to Grüneberg (1952) the whole underfur of *a*/*a*;*B*/*b*;*si*/*si* mice is practically white and "the animals resemble 'reverse agoutis', the hair bases being light, the distal parts of the hairs dark." Moreover, a similar effect has been observed by Grüneberg in a^t/a^t;*B*/*b*;*si*/*si* animals. Aside from the fact that this heterozygous effect of *b* on the expression of silver represents an exception to the rule that *B* is always dominant to *b*, it is interesting because its effect on the phenotype is unexpected, i.e., since *a*/*a*;*b*/*b*;*si*/*si* mice usually display *less* silvering than *a*/*a*;*B*/*B*;*si*/*si* animals, one would anticipate any heterozygous effect of *b* likewise to be in the direction of less silvering.

D. Expression in Agouti and Lethal Yellow Animals

Silver agoutis (*A*/–;*B*/*B*;*si*/*si*), like silver greys, are reported by Dunn and Thigpen to be "variable in appearance, some being well silvered, others being very similar to black agouti, except for white or light bases in the hairs which serves as a distinguishing mark of silver agouti." The yellow band on the agouti hair is not visibly influenced. Silver yellow (A^y/–;*si*/*si*) mice also display the same variation in intensity of silvering as silver blacks. Moreover, as in the case of agouti, the predominant effect of silver on yellow seems to be at the base of the fur which becomes very light, although white tips and silvered shafts likewise occur. Grüneberg (1952) notes that both agouti and yellow silver mice "have a more or less white underfur." Whether the *B*/*b* heterozygous condition influences the expression of silver on agouti and yellow backgrounds is not stated.

E. Influence of Age on Expression

One of the most interesting observations of Dunn and Thigpen (1930) is that in contrast to the situation in nonagouti silver mice, where the animals became progressively more silvered, in agouti and yellow silver mice the silvering *decreases* markedly as the animals get older. In fact, Dunn and Thigpen report that some silver agoutis became almost indistinguishable from black agoutis. This finding is of particular importance insofar as the etiology of silvering is concerned. If silvering merely was a consequence of a loss of pigment cells, as is thought to be the case in man, one would expect that its severity would increase with age on all backgrounds. The fact that it does not and, indeed, the fact that its severity *decreases* with age on some backgrounds, indicates that the condition cannot be explained simply by the demise of pigment cells. Furthermore, inasmuch as the genetic background affecting this expression of silver is determined primarily by the agouti locus, a locus known to influence the follicular environment (see Chapter 2, Section I, E), it appears that whatever affect this locus has on this environment must also somehow influence the expression of *si/si*.

F. Comparison with "Light"

Superficially the expression of silver resembles that of light (B^{lt}; see Chapter 3, Section I, G) not only because both mutations produce an overall dilution in pigmentation[15] but because at the end of the hair cycle all light hair bulbs and many *a/a;B/B;si/si* follicles (e.g., those which are not pigmented) do not possess clear cells and are indistinguishable from those in white spotted areas (Chase, personal communication). Nevertheless, these coat-color mutants behave differently. In a very simple but significant experiment Chase (see Quevedo and Chase, 1958) was able to deplete the pigment cell populations of *a/a;B/B;si/si* hair follicles through repeatedly inducing hair cycles by replucking immediately at the end of each cycle. Eventually completely white hairs were regrown. In contrast, attempts to deplete the melanocyte population of light mice by this means proved unsuccessful (Quevedo and Chase, 1958). Of course one wonders what the effect would be of subjecting agouti and yellow silver mice to this plucking treatment? On the basis of Dunn and Thigpen's observation that these mice get darker with age, one might anticipate a *decrease* rather than an *increase* in silvering.

G. Influence of X-Irradiation and Possible Etiology

Chase and Rauch (1950) reported that "a moderate grade of silver resembles in appearance an animal treated with 300 r on inactive follicles." Thus, a silver animal treated with 400 r responded as if it already had a 300-r treatment. The net effect was the production of more grey than results from the treatment of a black mouse with 400 r. Since the silver mice employed in this study were also on a nonagouti black background (Chase, personal communication), it again raises the question whether $A^y/-;si/si$ and $A/-;si/si$ genotypes would respond similarly.

Clearly the *si* mutation has not received the attention it deserves. On the basis of the available evidence its effect appears to stem from some breakdown in the intimate relationship which normally prevails between pigment cells and the epithelial cells of the follicle; a breakdown which seems to result in the inability of melanocytes to differentiate fully.[16] A similar hypothesis was initially suggested as being responsible for the greying effects of low dosages of X-irradiation (Chase, personal communication) and could be the basis of greying in man as well.

Finally it should be noted that very little is known about the factors, genetic or otherwise, which are responsible for the variations in the intensity of silvering. Thus, it remains uncertain whether modifying genes in the strict sense, and/or "minor" silvering genes influence the expression of the *si/si* genotype (Grüneberg, 1952).

III. Greying with Age

Greying with age (*Ga*), an autosomal dominant, produces a phenotype similar to silver. The mutation was found in wild mice from several localities in South Australia by Kirby and the following is based on his account (Kirby, 1974). *Ga* has not been mapped nor have crosses been made to rule out the possibility that it is allelic with *si*.

Two wild grey mice, one from Yorke Peninsula and one from Adelaide, were bred in Kirby's laboratory. Although the stocks derived from these animals have been maintained separately, animals of both stocks look alike and it is assumed that the greying is produced by the same factor.

A. Maternal Influence

Unlike silver, greying with age does not manifest itself until relatively late in life. Moreover, its expression seems to be influenced by maternal effects. When *Ga/–* mice were crossed reciprocally with C57BL (*ga/ga*) animals, different results were obtained. When the female parent was C57BL, the proportion of grey mice in the progeny was considerably lower than when the male parent was C57BL. Only 7 of 65 of the progeny of 8 C57BL females were detected as being grey after they were more than 10 months old, indicating that more of these offspring undoubtedly would have become grey if they had been maintained longer. On the other hand, all 12 of the progeny derived from 3 *Ga/–* females became grey in less than 10 months. Indeed, the fact that all of these progeny became grey indicates that one or more of the mothers was probably homozygous for *Ga*, and that such animals are viable and fertile.

B. Influence of Genetic Background on Expression

The age at which *Ga/–* mice begin to lose their hair pigment is variable and depends upon their genetic background. In animals which had been selected for three generations for early greying it was noticeable as early as 3 months of age, whereas it did not appear until after 1 year of age in animals outcrossed to C57BL.

Ga/– mice on an *a*/*a*;*B*/*B* background become grey first on their belly, then usually on their face and hind quarters. The last area affected is the mid-dorsum. Furthermore, the greying is variable; some mice are only faintly grey while others are nearly white. Microscopic examination of *Ga*/– hairs revealed most of them to be either fully pigmented or completely lacking in melanin; only rarely were partially pigmented hairs found.

C. Comparison with Silver and Other Considerations

Although there is a marked resemblance between greying with age mice and silver mice, *Ga* differs from *si* in that it is inherited as a dominant, is recognized later in life, produces fewer mosaic hairs, and is influenced by maternal factors. The etiology of this condition remains to be elucidated. Nevertheless, the observation that, unlike silver, the greying intensified with age in the presumably agouti (or white bellied agouti) wild population in which it occurred suggests that it may have a different etiology. Information on its expression in A^y/– animals as well as whether repeated plucking can hasten its expression is awaited.

Ga is also of interest because it evidently has been present in widespread local populations of wild mice in South Australia for some time. Its exact frequency in the wild is difficult to establish because most *Ga* mice in the laboratory do not express the gene until about 6 months of age and few wild mice reach this age (Newsome, 1969a,b). Hence, one wonders what the selective forces, if any, are for maintaining this polymorphism?

IV. Other Determinants

In addition to the above coat-color determinants there are a number of other color mutants in mice (given below) which have been described only briefly.

A. Ashen (*ash*)

This recessive mutation arose in strain C3H/DiSn (Lane and Womack, 1977). The coat color of these mice mimics that of dilute (*d*) and leaden (*ln*). It has been assigned to chromosome 9.

B. Buff (*bf*)

This mutation occurred spontaneously in the C57BL/6J strain and was originally described by Dickie (1964a). The mutation is inherited as an autosomal recessive and has been assigned to chromosome 5 (M.C. Green, 1966a). Nonagouti buff (*a*/*a*;*bf*/*bf*) animals are khaki colored and look somewhat like misty brown (*a*/*a*;*b*/*b*;*m*/*m*) mice. The eye color does not seem to be affected (see Chapter 4, Section IV, B).

Recently this mutation has been shown to be associated with the activities of three lysosomal glycosidases: β-galactosidase, β-glucuronidase, and *N*-acetyl-β-hexosaminidase. Thus *bf*/*bf* mice usually have much higher activities of these enzymes in their kidney cells than *bf*/+ mice. The significance of this, however, remains to be determined (Håkansson and Lundin, 1977).

C. Dark Footpads (*Dfp*)

This dominant mutation, which has not as yet been mapped, was first observed in a descendent of an irradiated female (Kelly, 1968). The mutation is lethal when homozygous; *Dfp*/*Dfp* embryos die before implantation (Kelly, 1970). *Dfp* heterozygotes are distinguished by the fact that the pads at the base of the digits, as well as the more proximal pads, are either black or significantly darker than the rest of the foot. The trait is somewhat variable since occasionally one foot may be unaffected, or some pads on all feet may be unaffected. In extreme cases the pads on distal phalanges as well as those on the feet are black (Kelly, 1968).

D. Dilution-Peru (*dp*)

This autosomal (linkage unknown) recessive arose in the descendants of a cross between a CBA/FaCam male and a wild Peru–Harland female (Wallace, 1971) derived from specimens trapped in a Peruvian yard used to dry maize (Wallace, 1970). When homozygous *dp* is fully penetrant and in combination with extreme nonagouti (a^e/a^e) produces a phenotype which is very similar to misty (although it has no belly spot or white tail tip), but nearer black. Animals expressing this mutation are somewhat similar in color to pale ear (*ep*), light ear (*le*), pearl (*pe*), and taupe (*tp*) mice. Because their wild Peru ancestry causes segregation of agouti modifiers, dilution-Peru mice are difficult to classify in *A*/– animals.

E. Freckled (*Fkl*)

This mutation, an autosomal dominant about 6 cM from piebald (*s*) on chromosome 14 (Lyon and Glenister, 1978), is probably an allele of Roan (Rn) (see Section J). It arose spontaneously in a random bred stock carrying *t*-alleles. Heterozygotes are variegated with small white or light colored patches in the coat and whitish ear hairs (J. Butler and Lyon, 1969). In young animals these patches are clearly discrete, but as they get older more and more unpigmented hairs appear in the coat generally so that an uneven roan effect is produced. Animals homozygous for freckled are viable and fertile and a whitish or cream color, though again uneven (J. Butler and Lyon, 1969). A dominant modifier, which has provisionally been called fawn (*Fw*), makes freckled heterozygotes considerably lighter but has no effect on normal mice (J. Butler and Lyon, 1969).

F. Gunmetal (*gm*)

Gunmetal, an autosomal recessive [chromosome 14 (M.C. Green, 1973)], occurred spontaneously in the C57BL/6J strain (Dickie, 1964a). Homozygotes have a diluted coat color somewhat similar to that of dilute (*d*/*d*). Eye color does not appear to be affected. Gunmetal (*gm*/*gm*) mice have a high mortality of unknown etiology and do not breed well (Dickie, 1964a).

G. Light (*Li*)

This mutation, not to be confused with B^{lt}, has provisionally been given the symbol *Li*. The mutation occurred at Oak Ridge in the offspring of an irradiated female. The coat of heterozygotes is slightly but definitely diluted and homozygotes are lighter than chinchilla (c^{ch}/c^{ch}). Both eumelanin and phaeomelanin are affected by this mutation which has been shown to be inherited independently, and probably assorts independently, of the *c*-locus. This mutation has an additive effect with c^{ch}/c^{ch} (Bangham, 1968).

H. Ochre (*Och*)

Ochre is a semidominant which was found in the progeny of an irradiated male. According to R.J.S. Phillips and her colleagues (1973) who first reported this mutation "*Och*/+ mice look very light at weaning age (similar to Mi^{wh}/+ except that their ears are not lightened) but darken with age, becoming more *d*/*d* like in color." Animals homozygous for *Och* are yellower and smaller than heterozygotes and many die before or soon after weaning. From about 8 days of age homozygotes also display an abnormal landing reaction; when held by the tail and lowered fairly quickly toward a table they tend to tip their heads ventrally. Moreover, when handled similarly as adults their whole body gives a convulsive jerk (R.J.S. Phillips, 1975). Although this behavior is somewhat reminiscent of the behavior of pallid (*pa*) mice, the gross structure of the ear and the otoliths of ochre homozygotes are normal (Lyon, 1975).

Ochre interacts with white (Mi^{wh}) producing a cream-colored animal. It also interacts with viable dominant spotting (W^{v}) but not obviously with steel (Sl^{d}) or light (B^{lt}) ((R.J.S. Phillips *et al.*, 1973). Ochre is on chromosome 4 (R.J.S. Phillips and Hawker, 1973), about 3 cM from brown (*b*) (R.J.S. Phillips, 1974).

I. Pale Ear (*ep*) and Light Ear (*le*)

There are two autosomal recessive color mutations in the mouse, pale ear (*ep*) and light ear (*le*), which mimic each other but which have been shown to be inherited independently. Light ear is on chromosome 5, about 10 cM from viable dominant spotting (W^v) and pale ear is about 1.5 cM from ruby eye (*ru*) on chromosome 19. Both mutations occurred within a year of each other at the Jackson Laboratory, *le* in C3H/HeJ and *ep* in C3HeB/FeJ (Lane and E. Green, 1967).

Neonatal *le/le* and *ep/ep* mice can be distinguished from wild type animals by a lack of eye pigment. The pigment is not completely absent from the eye but it is reduced considerably for the first day or two after birth. At 3 to 4 days of age, light ear and pale ear mice display less skin pigment than normal. This becomes most striking in the pinna of the ear and on the tail and feet. The coat of juvenile *ep/ep* and *le/le* mice also is lighter than normal and this combined with the light ears and tail makes each mutant clearly recognizable (Lane and E. Green, 1967).

When combined with nonagouti (*a*), brown (*b*), and dilute (*d*), *ep/ep* (and presumably *le/le*) mice retain their lighter colored ears, tail, and coat. *a/a;b/b;D/D;ep/ep* mice have dark red or ruby eyes in contrast to the light red or pink eyes of *a/a;b/b;d/d;ep/ep* animals. According to Lane and E. Green it may be possible to differentiate *B/B* from *B/b* and *D/D* from *D/d* in *ep/ep* animals.

When cleared hairs from the juvenile coats of nonagouti light ear and pale ear mice were examined microscopically, and compared with the corresponding normal genotypes, no gross differences in the amount of pigment were apparent. However, the eumelanin granules of the *le/le* and *ep/ep* hairs were smaller, particularly near the tip of the hair. Slices of skin from 5-day-old pale ear and normal mice did not differ in their ability to incorporate ^{14}C-labeled tyrosine into melanin (Lane and E. Green, 1967).[17]

J. Roan (*Rn*)

Roan is another autosomal (chromosome 14) semidominant characterized in the heterozygote by white hairs distributed through the coat (M.C. Green, 1966b, 1971). Other hairs may have both pigmented and nonpigmented areas with no regular pattern in their distribution. In *Rn*/+ animals the expression of the mutation is variable, some heterozygotes possessing only a very few white hairs. This variability is in part genetically determined. Roan homozygotes appear as very light grey, with a much higher proportion of white hairs than heterozygotes. In spite of the fact that the roan phenotype is characterized by the occurrence of white hairs, in neither heterozygotes nor homozygotes have discrete areas of white spotting been observed (M.C. Green, 1966b).

Roan strongly resembles freckled (*Fkl*) (Section E) with which it is very closely linked (M.C. Green, 1971). Indeed, Green suggests they may be alleles (see Lyon and Glenister, 1978). Nevertheless, it is clear that these mutations are not identical. When crossed on to strain 101/H, *Fkl* produces a very light-colored coat with almost no wild type patches, whereas *Rn* gives a near wild type coat with a few small light-colored spots. "*Rn*/+;*Fkl*/+ double heterozygotes are very variable, and not certainly distinguishable from single heterozygotes, but among segregating litters from *Rn*/+;*Fkl*/+, *Rn* and *Fkl* young can be distinguished" (Lyon and Glenister, 1973). No evidence of crossing-over between *Rn* and *Fkl* has yet been observed, but because of the normal overlapping of *Rn* and variability of *Rn*/+;*Fkl*/+ this evidence would be difficult to obtain. It thus remains uncertain whether *Rn* and *Fkl* are alleles or part of a gene cluster (Lyon and Glenister, 1973,1978).

K. Sepia (*sea*)

Sepia (chromosome 1) is another recessive mutation which has only very recently been described (H. Sweet and Lane, 1977). It occurred in the C57BL/6J strain and closely resembles beige (*bg*) but the coat, ears, and tail are not quite as light as in *bg*/*bg* mice, and the eyes are not light enough to recognize this mutant at birth.

L. Slaty (*slt*)

This recessive autosomal [chromosome 14; about 5 cM from piebald (*s*)] mutation occurred in a heterogeneous stock carrying limb-deformity (ld^J) and mahogany (*mg*). On a nonagouti background slaty homozygotes have a slightly diluted coat and slightly yellowish ears (M.C. Green, 1972).

M. Taupe (*tp*)

Taupe, an autosomal recessive about 2 cM from albinism (*c*) on chromosome 7, arose spontaneously in the C57BL/10 strain (M.C. Green, 1966a). Nonagouti taupe (*a*/*a*;*tp*/*tp*) animals have a diluted (slate grey) coat color which resembles the coat of nonagouti ruby-eye (*a*/*a*;*ru*/*ru*) mice except that the underfur is lighter, with distinct yellow at its margins. Besides, unlike ruby eye, the color of the eye is not affected (Fielder, 1952).

The viability of taupe homozygotes is normal and the fertility of taupe males also is normal, but taupe females may have difficulty gestating their young (M.C. Green, 1966a). Thus while many taupe females become pregnant, litter members often either die *in utero* and are resorbed, are born dead, or die soon after birth. Furthermore, a number of young, of both those born alive and those born dead, are unusually small, though other animals in the same litters are of normal size (Fielder, 1952). Taupe females sometimes have such great difficulty in delivering their young that they may die within several days afterwards. Moreover, even when they successfully deliver a viable litter and survive they are unable to nurse it. Although the ducts and alveoli of their mamary glands are normal, most or all of their nipples are so underdeveloped that they are difficult to detect (Fielder, 1952). The primary lesion(s) responsible for the diverse effects of this mutation is not known.

N. Underwhite (*uw*)

Underwhite is an autosomal (chromosome 15) recessive mutation which arose spontaneously in the C57BL/6J strain (Dickie, 1964a). The dorsum of *uw* homozygotes is a light buff color whereas the ventrum is white. The eyes of *uw*/*uw* mice are unpigmented at birth but darken to a dark reddish color at maturity (M.C. Green, 1966a). According to Dickie mice carrying underwhite and agouti (*A*), black and tan (a^t), or such spotting genes as viable dominant spotting (W^v) and white (Mi^{wh}) are almost impossible to classify because their coat color is so dilute.

O. White Underfur (Provisional Symbol; *wuf*)

This recessive mutation (linkage unknown) occurred in a stock carrying rosette (*rst*). It is impenetrant on brown (*b*) and makes extreme nonagouti (a^e) difficult to classify. *wuf*/*wuf* is not completely penetrant and, unlike underwhite (*uw*), does not affect the eyes (Ferguson and Wallace, 1977).[18]

Notes

[1]Similar syndromes have been reported in mink (Padgett et al., 1964), cattle (Padgett et al., 1964), cats (Kramer et al., 1977), and in a killer whale (Taylor and Farrell, 1973). In mink and cattle the condition has been conclusively shown to be inherited as an autosomal recessive and this appears to be the case in cats, too. (For reviews see Padgett 1968; Padgett et al., 1970; Davis and Douglas, 1972; Windhorst and Padgett, 1973; Prieur and Collier, 1978.)

[2]Patients with Chediak–Higashi syndrome also often display a widespread mononuclear infiltration of organs resembling lymphoma that is referred to as the accelerated or malignant phase. They also may display peripheral granulocytopenia, defective granulocyte regulation, and intramedullary granulocyte destruction (Blume et al., 1968).

[3]Children suffering from the Chediak–Higashi syndrome display either a decrease or absence of uveal pigment and there is a marked photophobia. "The hair is pale

gray, blond, or brunette with a distinctive overcast and steaks of gray. The skin is generally pale, but on the exposed parts of children born of dark-skinned parents it may have "slate-gray" coloration or hyperpigmentation" (Windhorst et al., 1966). Electron microscopy of hair bulb melanocytes disclosed that "the granules were large and numerous. Each granule was limited by a single membrane and contained a number of strands which made up the granule matrix. The melanosome granules were 2 to 3 times as thick and approximately twice as long as normal melanosomes. The large granules contained approximately twice as many strands as are normally found within a matrix of pigment granules, and in some instances the strands appeared to be more irregular than those found in the typical matrix of melanosomes. Melanin was present, although its relative concentration on the larger melanosomes may have been less than it is in fully developed, normal melanosomes" (Windhorst et al., 1966). In a study aimed at examining the formation of the massive melanosomes of the disease, and the basis of the hypopigmentation, Zelickson et al. (1967) concluded that the giant particles appear to originate from defective premelanosomes, and that the continued growth and/or fusion of these premelanosomes eventually results in enormous particles which degenerate. It is believed that although the destruction of these giant melanosomes may dilute the color of the skin, that a more likely basis for the hypopigmentation is the abnormal packaging of normal sized melanosomes into large lysosome-like structures in the epidermal cells. An examination of the eye of affected individuals "revealed relatively normal (though fewer) granules in the choroid, with grossly deficient melanin granules of the retinal pigmented epithelium." Moreover, in these granule-deficient areas of the retina the melanin granules were abnormally large (Windhorst et al., 1968). As a consequence of these observations, Windhorst and her colleagues (1968) conclude "that the basis for the pigmentary anomaly of the Chediak–Higashi syndrome can best be regarded in terms of a structural abnormality of the melanosome, perhaps in the lipo-protein building blocks that are thought to be common to all membranes."

[4]Lutzner et al. (1967) report that the giant melanin granules of the beige mouse range from 2 to 10 μm.

[5]The coats of $B^{lt}/B^{lt};bg/bg$ mice resemble the "blue" coat of animals homozygous for both *bg* and *d*, but have a somewhat less silvery appearance, presumably due to greater amounts of pigment in the hair tips (Pierro, 1965).

[6]Poole and Silvers (unpublished) also have observed that $A^y/a;bg^J/bg^J$ mice can be distinguished from normal yellow animals because their coats are slightly more dilute, especially on the belly and at the base of the hairs.

[7]Lutzner (1970) and Lutzner and Lowrie (1972) also have studied the fine structural development of beige granules and their observations are similar to those of Hearing et al. They found that the earliest observable melanin granules in the retina and choroid of beige embryos were larger than those in C57BL/6 (+/+) "controls." "Granule membranes were disrupted and melanosome fibers were surrounded by vacuolar spaces. In older beige embryos giant granules were seen with a central nidus of normally formed cylinders surrounded by layers of melanin fibers, exhibiting a spectrum of maturity with the least mature fibers towards the periphery [see Figure 6-3d]. Premelanosomes appeared to be fusing to the outside of these giant granules" (Lutzner, 1970). Lutzner and Lowrie also observed that the giant granules in adult beige melanocytes are composed predominantly of fully melanized structures. Con-

centric rings were often seen in abnormal granules which could represent waves of melanization (see Figure 6-3a). While in some adult melanocytes both normal premelanosomes and melanosomes occur along with fully mature giant granules, in others almost the entire cytoplasm appears filled with maturing giant granules. From these observations Lutzner and Lowrie suggest "that the giant beige melanin granule is formed through the continual deposition of melanofilaments by granule fusion and/or failure of size-control mechanisms."

[8]In $bg^J/bg^J;p/p$ mice, no giant granules were observed in the retina and 2% of the choroidal granules were giants (Hearing et al., 1973).

[9]The mast cells of beige mice also display giant granules (Chi and Lagunoff, 1975; Chi et al., 1978). Indeed, the fact that beige mice possess giant granules in a variety of cells has been employed as a cell marker. Thus, it has been used as a neutrophil marker to study the kinetics of bone marrow replacement in anemic W/W^v [see Chapter 10, section I(G)] mice (E. Murphy et al., 1973; Maloney et al., 1978; Patt and Maloney, 1978). It also has been used to investigate the origin of mast cells (Kitamura et al., 1977; see Chapter 10, note 27).

[10]J. Vassalli et al. (1978) have reported that the polymorphonuclear leukocytes of patients with Chediak–Higashi syndrome are profoundly deficient in elastase and that the corresponding murine protease is similarly decreased in the leukocytes of beige mice. They suggest in the case of the human syndrome that this deficiency may account, at least in part, for the high incidence of infections.

[11]Excretion into the urine is a major mechanism by which lysosomal enzymes are removed from the normal mouse kidney and Brandt and his associates propose that this process is impaired in *bg/bg* mice because of an abnormal fusion of lysosomes with the plasma membrane of the kidney proximal tubule cells. Recently Novak and Swank (1979) have reported that other coat-color determinants also affect kidney lysosomal enzyme concentrations. Thus, pale ear (*ep/ep*), pearl (*pe/pe*), and pallid (*pa/pa*), like beige (bg^J/bg^J), produce a 2.5-fold increase in the concentration of kidney β-glucuronidase; maroon ($ru\text{-}2^{mr}/ru\text{-}2^{mr}$) and ruby eye (*ru/ru*) produce a 1.6- to 1.8-fold increase; and light ($B^{lt}/+$), dominant spotting (*W/+*), extreme nonagouti (a^e/a^e), slight dilution (d^s/d^s), and himalayan (c^h/c^h) produce a smaller (about 1.3-fold) but still significant increase in the concentration of this enzyme. Further characterization of the three mutants, *ep/ep*, *pe/pe*, and *pa/pa*, indicated that, like bg^J/bg^J, they have a generalized effect on lysosomal enzymes since they display coordinate increases in kidney β-galactosidase and α-mannosidase (these enzymes also are increased in the kidneys of maroon and ruby-eye mice). The effects of these three deviants appear to be lysosome specific since rates of kidney protein synthesis and activities of three nonlysosomal kidney enzymes proved normal. The influence of these genes also appears to be relatively tissue specific since all were associated with normal liver lysosomal enzyme concentrations. A common dysfunction in all three mutants was a decreased rate of lysosomal enzyme secretion from kidney into urine. Whereas normal C57BL/6J animals daily secreted 27–30% of total β-glucuronidase and β-galactosidase, secretion of both these enzymes was coordinately depressed to 1–2, 8–9, and 4–5% of total kidney enzyme in *ep/ep*, *pe/pe* and *pa/pa* mice, respectively. Although depressed lysosomal enzyme secretion appears to be a major influence of these coat-color determinants, Novak and Swank also found that the synthetic rate of kidney glucuronidase was *increased* 1.4- to 1.5-fold in pearl and pallid

animals. As pointed out by Novak and Swank, taken together these observations "reinforce the histochemical and biochemical similarities reported between lysosomes and melanosomes and suggest that there may be many steps in common in the control of subcellular organelles."

[12]The basic biochemical defect in the Chediak–Higashi syndrome also remains to be elucidated. However, recent studies indicate that in this condition, too, there is a microtubular defect which appears correctable by treatment with cyclic guanosine monophosphate, cholinergic agonists (J. Oliver, 1976), or ascorbate (Boxer et al., 1976). For other investigations on the beige syndrome see Prieur et al. (1972), Chi et al. (1975, 1976), Holland (1976), J. Oliver et al. (1976), Frankel et al. (1978), Guy-Grand et al. (1978), Kaplan et al. (1978), Lyons and Pitot (1978), and Swank and Brandt (1978).

[13]In addition to silver, Hagedoorn studied the interactions of a number of other coat-color factors. His 1912 paper is particularly entertaining because it includes a color plate of 24 charmingly drawn mice, each displaying a different phenotype.

[14]In outcrosses, silver is nearly, though not quite, recessive as some heterozygotes may display a few white hairs, mainly on the belly (Dunn and Thigpen, 1930; Grüneberg, 1952).

[15]However, whereas all light hairs are pigmented at their tips many hairs in silver animals are unpigmented or banded with pigment.

[16]On the basis of the pigment patterns displayed by allophenic mice (see Chapter 7), Mintz (1971a) suggests that *si/si* genotypes express hair follicle clonal patterns. If this is the case it would place the site of action of *si* in the skin and not in the melanocyte.

[17]LaVail and Sidman (1974) report that the slightly diluted pigmentation of the juvenile coat of *le/le* mice becomes darker in adults. They also found the eye pigmentation of these mutants to be altered. The melanosomes of the retinal epithelial cells were fewer and usually smaller than those in +/*le* mice. There also were fewer melanosomes in the choroidal melanocytes of homozygotes, and these were, on the average, larger, much more heterogeneous in size, and less densely packed than those in +/*le* melanocytes. The diameters of some of these *le/le* granules were three to four times those in +/*le* eyes. *le* homozygotes also appear to have fewer choroidal melanocytes and in newborns these are restricted to the anterior regions of the eye. During the first 2 weeks after birth, however, melanocytes come to occupy progressively more posterior parts of the eye. Finally, it should be noted that light ear, like pale ear (see note 11), has a striking effect on lysosomes in the kidney. Thus, males homozygous for *le* display a 4-fold elevated concentration of kidney β-galactosidase. They also have a decreased secretion of kidney β-hexosaminidase but in this case there is no significant accumulation of this enzyme in the kidney (Meisler, 1978).

[18]In 1957 Pizarro briefly described a recessive gene which he called platino (*pl*) and which appeared to be a mimic of extreme dilution (c^e). This determinant, which was not allelic to *c*, produced a very light-colored phenotype with dark eyes. The number of pigment granules in the hair was greatly reduced. As far as I am aware this mutation is no longer maintained.

Chapter 7

The Pigment Patterns of Allophenic Mice and Their Significance

There is no endeavor that has contributed more to our understanding of how specific coat-color determinants produce their phenotypic effects than Mintz's work on allophenic mice (see Mintz, 1967, 1970, 1971a). Indeed, because her observations on these mice are so intimately related to the mode of action of the X-linked coat-color determinants, as well as those which produce white spotting, it is necessary to review this remarkable work in some detail before considering these subjects.

Allophenic mice are produced (Figure 7-1; for specific details see Mintz, 1971b) by explanting *in vitro* two cleavage-stage embryos of dissimilar genotypes and placing them in contact after their membranes are removed. Following this procedure the cells migrate in the composite mass, and a double size, genetically mosaic blastocyst develops which is transferred surgically to the uterus of a pseudopregnant mother. Normal embryo size subsequently is restored and many of the composites are viable at birth and live a "normal" life.

I. The Standard Pattern

Because the two genotypes from which these mice are derived jointly participate in their devlopment (and in this *very important* respect allophenic mice are unlike experimental radiation or other graft chimeras) they have been invaluable in investigating a wide range of questions (see Mintz, 1974), not the least of which are those involving pigmentation. Thus, when cells from

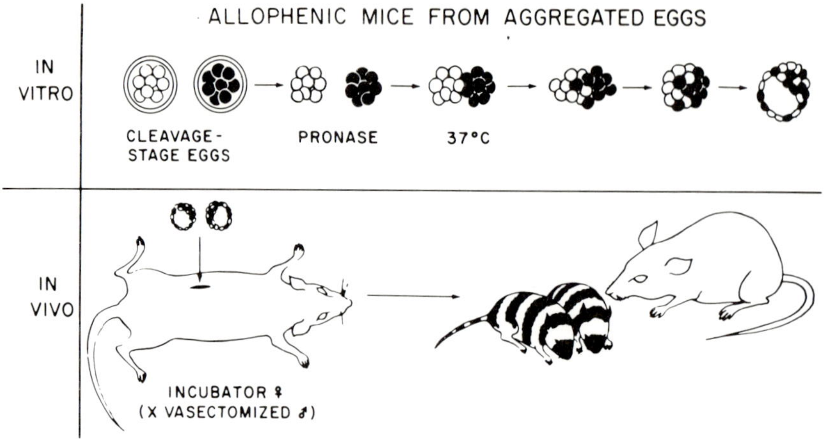

Figure 7-1. Diagram of the experimental procedures involved in producing allophenic mice. From B. Mintz (1967). Reproduced with permission of the author.

embryos of two different coat-color genotypes, e.g., black (*B*/*B*) and brown (*b*/*b*), are combined the resultant bicolored animals display a wide range of phenotypes, spanning from those with only a small proportion of black (*B*/*B*) to those which are almost completely of this color (Figure 7-2). Nevertheless, these different phenotypes all represent variations on the same theme or pattern and it is precisely this pattern, the so-called *archetypal* or "standard pattern" (Mintz, 1967), and its variations, which reveal the mechanisms by which most, if not all, coat-color phenotypes are produced (not only in mice but in all mammals). The "standard" or archetypal pattern of the allophenic mouse occurs when approximately equal proportions of melanoblasts of the two colored strains contribute to its phenotype and, by chance, alternate in their original positions. When this occurs in the *B*/*B* ↔ *b*/*b* situation an animal is produced with transverse black and brown bands of similar width, extending from the mid-dorsum to the mid-ventrum. This striping pattern is continued down the length of the head, body, and tail with the banding on one side of the mid-dorsum being completely independent from that on the other side. Mintz (1967) proposes that each band represents a single clone on that side, descended from *one primordial melanoblast* in the neural crest. Thus, inasmuch as mice displaying this standard pattern possess 17 stripes—the head has three, the body six and the tail eight—and those on one side of the mid-dorsum are derived independently from those on the other, she contends that the entire population of melanocytes is derived from a total of 34 clonal initiator cells (Mintz, 1967).[1] This interpretation not only is completely in accord with the fact that the primordial melanoblasts originate in the neural crest longitudinally flanking the dorsal midline on each side, and progressively proliferate and migrate laterally towards the mid-ventrum (Rawles, 1947), but it is also in accord with transplantation studies which have shown that the areas of different color in the coats of these mice are genotypically specific and express their mutually exclusive allelic phenotypes (Mintz and Silvers, 1970).

II. Modified Patterns

The variability which is observed in the standard allophenic pattern can be readily explained simply by assuming different total proportions of clones of the two genotypes and the permutations and combinations of how the progenitor cells are arranged on the mid-dorsal line. Thus, if by chance in the example we have chosen, three *b*/*b* stem cells occurred in succession, the resulting brown band would be three times as wide (and it should be noted its origin would be inferable only from a knowledge of the standard pattern). Also contributing to pattern variability is clonal selection upon which factors such as differences in cell viability and differences in rates of clonal expansion may be superimposed. Closure of skin areas and regional disparities in growth rates, etc., also contribute to modifying the pattern. For example, in some allophenic mice the entire complement of clones of

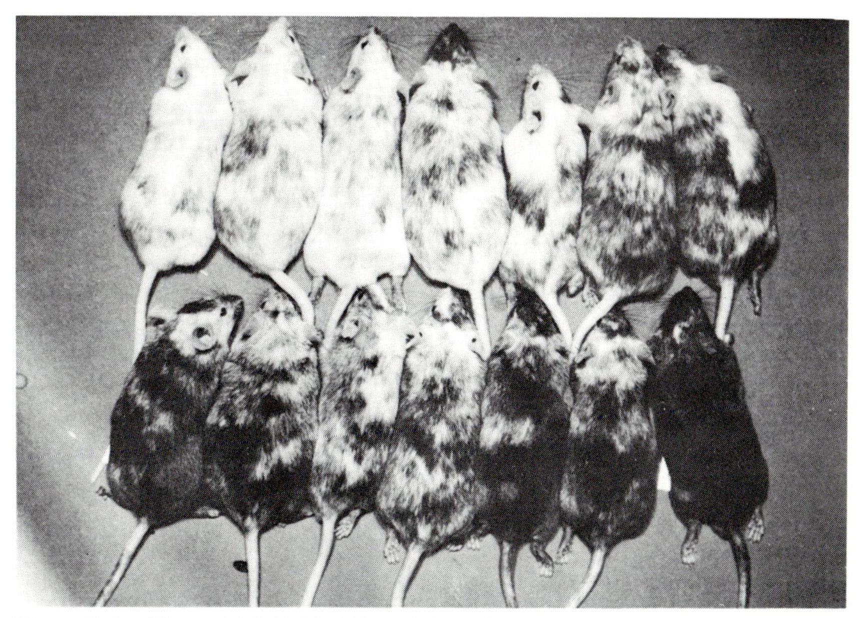

Figure 7-2. These 14 B10CBAF$_1$ (*C*/*C*) ↔ SJL (*c*/*c*) allophenics demonstrate the variation that can occur between animals of the same genotype. From R.J. Mullen and W.K. Whitten (1971). Reproduced with permission of the authors and The Wistar Institute Press.

one of the colors produces narrower bands relative to the other color, and this almost certainly indicates that the melanoblasts of one type have a greater proliferative capacity and therefore can encroach upon the territory of neighboring clones derived from cells of the other genotype (Mintz, 1970). Indeed, there are so many factors which can contribute to varying the standard pattern that it is not surprising that it is observed relatively rarely.

III. Inception of Gene Activity

In addition to demonstrating that all the melanocytes of mice are derived from 34 clonal initiator cells, Mintz's observations also demonstrate at approximately what time in development the specific gene loci responsible for melanoblast determination first become active. This follows from the fact that as the left and right sides of the animal can be of different colors, and hence must be derived from different clones, the clonal-initiator cells must arise *before* the neural crest cells can pass from one side to the other, i.e., before the neural folds on the two sides start to come into physical contact and fuse. This occurs on day 8 of embryonic life in the mouse, so that day 7 would presumably be the latest time at which clonal initiation of melanoblasts can occur (Mintz, 1970). On the other hand, inasmuch as there must be a sufficient number of cells present in the embryo to produce not only the pigmentary system but other systems as well, it is doubtful if the loci which are responsible for initiating melanoblast differentiation can become active earlier than the fifth day of development. Thus, it appears that it is sometime between day 5 and day 7 that melanoblast differentiation is initiated (Mintz, 1970, 1971a).

IV. Expression of Albinism and Occurrence of Bicolored Hairs

In considering the pigment patterns of allophenic mice it should be noted that Mintz's findings also are compatible with the fact that albino animals possess a population of amelanotic melanocytes since when allophenics are produced from merging an albino embryo with a pigmented one, the standard pattern is exactly like the one produced from two differently *pigmented* genotypes except that half of the bands are white (Figure 7-3). This, as we will see, is quite different from the case when white-spotting is involved.

It is also important to mention that at the interfaces between phenotypically different clones and, to graded extents, toward the clonal interiors

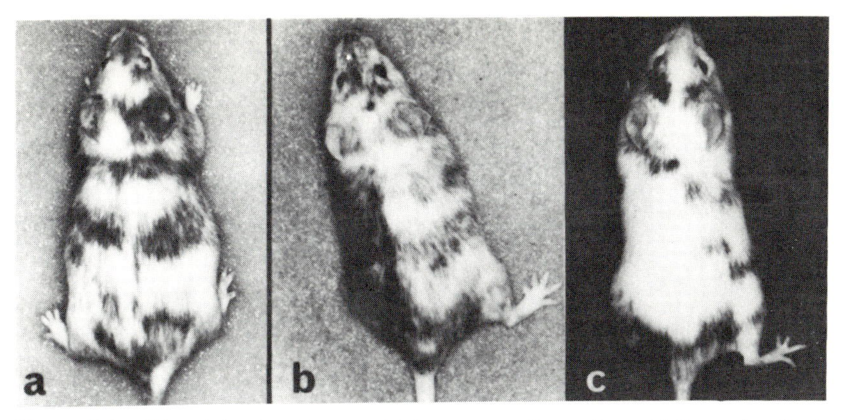

Figure 7-3. The striped patterns which occur in these three $C/C \leftrightarrow c/c$ allophenics are the same as those which occur when two pigmented genotypes are employed except that, on the average, half the bands are white. The transverse bands are believed to represent melanoblast clones and the fact that there is bilateral asymmetry indicates separate clones on each side. From B. Mintz (1971a). Reproduced with permission of the author and Cambridge University Press.

of allophenic mice not only does an intermingling of colors occur, but single hairs sometimes are populated with two colors. Thus in $B/B \leftrightarrow b/b$ allophenic animals both black and brown pigment granules occur within single hairs, and in $Ln/Ln \leftrightarrow ln/ln$ mice the septules of some hairs display clumped granules characteristic of leaden mice, whereas other regions of the same hairs possess a normal distribution of granules. These observations are important because they demonstrate that more than one melanoblast contributes toward the pigmentation of single hairs (Mintz, 1967; Mintz and Silvers, 1970; Cattanach et al., 1972).

V. Hair Follicle Clones

Inasmuch as in mice melanoblasts must usually enter a hair follicle to produce any melanin at all, it is not surprising that the clonal history of the coat also contributes to the pigment patterns of allophenic animals. Indeed, the interaction between the *independently derived* melanoblast and hair follicle clones is yet another factor which increases the variablility of allophenic phenotypes.

As in the case of the allophenic pigment pattern the archetypal and modified developmental hair patterns can be discerned only when the two contributing strains differ with respect to a suitable genetic marker, in this case one which affects either the morphology of the hair, such as fuzzy (fz/fz) and normal (Fz/Fz), or by a melanocyte marker which acts via the hair follicle, such as agouti (*A*/*A*) and nonagouti (*a*/*a*). When such a marker is present the pattern produced, in its maximal or standard form, is similar to the melanoblast pattern in that it too is characterized by a number of transverse bands with frequent mid-dorsal and mid-ventral asymmetries on the left and right side. However, these bands are noticeably narrower and much more numerous than the melanoblast bands. The left–right asymmetry again indicates that the bands on each side of the mid-dorsal line are independently derived and therefore must have originated before the fusion of the neural fold. Although it is difficult to accurately determine the number of bands (clonal-initiator cells) there are approximately 170, 85 on each side. The numbers of these bands on each side of the trunk correspond to the somite number (approximately 30 per side) and, assuming a similar situation for the tail, it is judged to possess about 35 clones on each side. Eighteen pairs of bands occur in the head region (Mintz, 1974).[2] As in the case of the melanoblast clones, these hair follicle clones also can vary in their proportions, positions, proliferative capacities, etc. and by so doing provide a continuous and almost unlimited number of phenotypes.[3]

A. Subclones

The interaction between melanoblast clones and hair follicle clones can be observed in allophenic mice produced from embryos known to bear *both* melanocyte and follicular cell markers, such as brown nonagouti (*b*/*b*;*a*/*a*) and black agouti (*B*/*B*;*A*/*A*), and in such animals it is found that each melanoblast clone physically overlaps about six hair follicle clones (Mintz, 1969b). Consequently, in the standard pattern of *a*/*a*;*b*/*b* ↔ *A*/*A*;*B*/*B* allophenics a *b*/*b* melanoblast clone includes about three brown pigmented hair clones (*a*/*a*) and about the same number which display the brown agouti (*A*/*A*) pattern. These subdivisions of the melanoblast clone are known as "subclones" because they can be shown by graft tests to be reversible modifications (Mintz and Silvers, 1970). Figure 7-4 illustrates the tremendous range of phenotypic variablility which is realized in single allophenic animals as a result of the interaction between the independently originating melanoblast clonal and hair follicle clonal patterns.

VI. Expression of Allophenic Patterns in Single Genotype Mice

Although Mintz's interpretations of allophenic patterns have been criticized (e.g., Mystkowska and Tarkowski, 1968; McLaren and Bowman, 1969; Lyon, 1968, 1970, 1972a, Schaible 1972),[4] they are consistent with almost everything that is known about the behavior of pigment cells. Moreover,

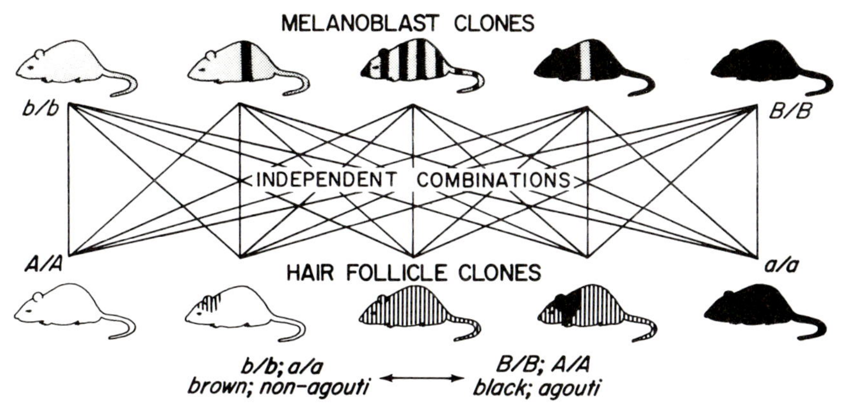

Figure 7-4. Patterns realized in allophenic animals as a consequence of independently originating melanoblast and hair follicle clones. From B. Mintz (1971a). Reproduced with permission of the author and Cambridge University Press.

they are especially attractive because they readily explain the patterns displayed by single genotype mice. Thus, if one assumes that mechanisms are available in normal mice which are capable of producing melanoblast or hair follicle clones that differ phenotypically from each other, *all* of the pigment patterns of these animals appear to conform either to the archetypal or to the modified pigment patterns observed in allophenic subjects (Mintz, 1971a). That such an assumption is not difficult to accept derives from the fact that some pigment patterns of ordinary mice have in fact been shown to be due to the occurrence of different clonal populations. Such clones, which are known as "phenoclones" (see Mintz, 1971c), are exemplified best in the pigment patterns of animals bearing an X-linked coat-color determinant. However, before dealing with these patterns, let us refocus on some of the autosomally associated bicolored patterns already considered to see if they can be interpreted in terms of the phenoclonal expression of allophenic patterns.

When the effects of the various *p*-locus alleles were discussed it was noted that there was one allele at this locus, known as *p*-unstable (p^{un}), which, when homozygous, produces a phenotype in which both dilute (p/p) and intensely pigmented ($P/-$) areas occur (see Chapter 4, Section II, C, 9). It was noted also that this bicolored pattern was very variable and that it seemed to be due to a spontaneous somatic reversion to wild type (Melvold, 1971). An examination of many p^{un}/p^{un} mice clearly indicates that their pigment patterns conform to the patterns of two-color allophenics and that their spots are essentially abbreviated parts of transverse bands. In fact, the most highly patterned p^{un}/p^{un} mice are indistinguishable from allophenic mice displaying maximal differentiation among melanoblast bands (Mintz, 1970). Clearly, as a result of somatic mutation(s) two populations of primordial cells have been produced in these mice and the behavior of these popula-

Figure 7-5. A p^{un}/p^{un} mouse which, as a result of spontaneous somatic reversion to wild type, possesses two populations of melanoblasts. The fact that the pigment patterns formed by these populations mimic those of allophenic animals indicates a similar clonal history. From B. Mintz (1971a). Reproduced with permission of the author and Cambridge University Press.

tions is exactly what one would predict from their behavior in the allophenic animal (Figure 7-5).

Two other alleles at the *p*-locus, p^{m1} (pink-eyed mottled-1) and p^{m2} (pink-eyed mottled-2), when heterozygous with *p* also produce phenotypes which from their descriptions (see Chapter 4, Section II, C, 8) are like those in p^{un}/p^{un} mice and consequently mimic the melanoblast clonal patterns of allophenic animals. Although there is no evidence that somatic mutation is responsible for the reversion to intense pigmentation in these genotypes (L. Russell, 1964), whatever the mechanism is, it appears to produce two phenoclones which again behave as Mintz's hypothesis predicts.

Chinchilla-mottled (c^m/c^m) mice (see Chapter 3, Section II, B, 7) apparently are also characterized by patterns which resemble those of allophenic mice and, therefore, almost surely are produced via a similar mechanism.

The bicolored patterns produced by other autosomally inherited coat-color factors mimic the *hair follicle* clonal patterns of allophenics and this is consistent with their known activity in the hair bulb. Some of these ex-

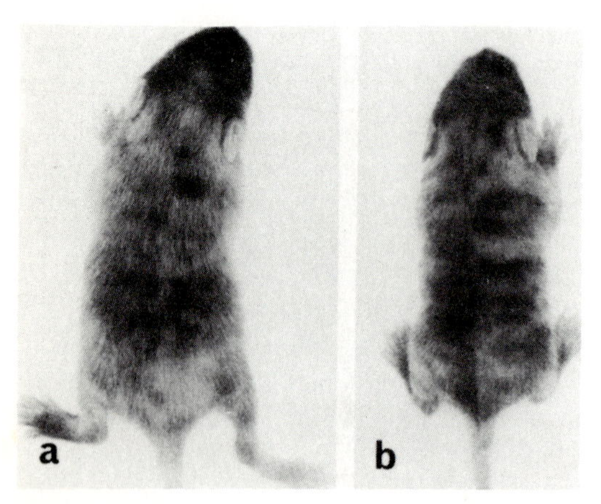

Figure 7-6. (a) $A/A \leftrightarrow a/a$ allophenic showing a typical hair-follicle clonal pattern and (b) viable yellow (A^{vy}/a) single-genotype mouse displaying the same pattern. From B. Mintz (1971a). Reproduced with permission of the author and Cambridge University Press.

amples involve *a*-locus alleles. Thus, viable yellow (A^{vy}/a)(see Chapter 2, Section I, A, 2), mottled agouti (a^m/a^m) (Chapter 2, Section I, A, 12), and agouti suppressor (A^sA^w/A^{sAw}) (Chapter 2, Section I, A, 17) all produce phenotypes which are included in the array of patterns seen in allophenics derived from embryos of different *a*-locus constitutions (Figure 7-6) (Mintz, 1971a). Another coat-color determinant which may behave similarly is silver (*si*/*si*) (Mintz, 1971a) (see Chapter 6, Section II).

While some heterozygous bicolored phenotypes (e.g., p^{m1}/p, A^{vy}/a, c^m/c^{ch}) could be due to the activity of one or the other allele in each cell, this does not seem to be a likely mechanism for the phenoclonal origin of most of the alleles with which we are dealing since many of them (e.g., A^{vy}, a^m, c^m) produce mottled phenotypes when *homozygous*. Nevertheless, this in no way invalidates the Mintz hypothesis since there is no reason to assume that the phenoclones of single genotype mice are all produced via the same mechanism. Indeed, we already have noted that whereas in p^{un}/p^{un} animals somatic mutation is responsible for the two populations of cells, this is *not* the case in p^{m1}/p mice (L. Russell, 1964).

VII. Allophenic Patterns and the Etiology of White Spotting

In addition to the significance of the similarities between allophenic pigment patterns and those produced by a few autosomally inherited and all X-linked coat-color determinants (see Chapter 8), gene expression in allophenics also relates to the subject of white spotting. Inasmuch as it has been shown that white spotting, in contrast to albinism, results from an absence of melanocytes, it is hardly surprising that the allophenic pigment patterns which are produced when potentially white spotted and fully pigmented embryos are merged are usually quite different from the patterns which occur when blastomeres from albino and pigmented genotypes are aggregated.

In the mouse there are a number of genes [e.g., white (Mi^{wh}), dominant spotting (*W*)] which when heterozygous produce white spotting but when homozygous result in animals with no pigment in their coats at all. Such homozygotes may be considered as "one big spot" since, like the white spotted areas of pigmented mice, and unlike the situation in albinos, their hair follicles do not possess any melanocytes (clear cells) (Silvers, 1956). When allophenics are produced from merging the embryos of such all-white animals with those of a fully pigmented strain, they *never* resemble the archetypal pattern of $c/c \leftrightarrow +/+$ allophenics. Such allophenics either resemble the phenotypes of ordinary white-spotted mice or they are completely pigmented (Mintz, 1971a). For example, when $a/a;Mi^{wh}/Mi^{wh}$ ("one big spot") blastomeres are aggregated with $A/A;+/+$ (pigmented) blastomeres about 50% of the allophenics which are skin mosaics, i.e., display areas of both agouti and nonagouti pigmentation, are white-spotted while the remainder are fully pigmented. In contrast, in $A/A;C/C \leftrightarrow a/a;c/c$ allophenics both pigmented and nonpigmented areas usually are present whenever hair follicle mosaicism occurs. It thus appears that for some reason Mi^{wh}/Mi^{wh} melanoblast clones are much less likely than albino melanoblast clones to lead to any phenotypic manifestation in allophenic animals. To account for this Mintz (1970, 1971a) proposes that the Mi^{wh}/Mi^{wh} melanoblast clones are inviable and that, depending upon when they die, they may or may not be replaced by an invasion of melanoblasts from a viable (+/+) clone. If they die early in development, the areas they occupy can usually be invaded and replaced by +/+ clones. On the other hand, if their death occurs late in development, the +/+ cells are unable to proliferate and migrate sufficiently to "fill the gaps."[5]

Another factor which can influence the ability of viable clones to migrate into and pigment regions previously occupied by nonviable clones is the growth rate of the area. For example, it is less likely that a +/+ clone will be able to *completely* replace an inviable clone in a region which grows relatively rapidly, such as the rump, than to replace such a clone in a more slowly growing region such as the neck (Mintz, 1970).

Still another factor which influences the amount and location of the white spotted areas is the late closure of some regions and the distal locations of others. Thus, some distal areas, such as the feet and the tip of the tail, are often white and this Mintz (1970) attributes to the inability of +/+ clones to reach these areas. The comparatively late closure of the umbilical region, or of the anterior neural folds, may also lead to a white belly spot or a spot on the top of the head as a consequence of viable melanoblasts not being able to migrate into these areas.

Once again all of the white spotting patterns which one finds in these allophenic animals mimic those produced in single genotype mice by known spotting genes (Figure 7-7). This includes the fact that many white spotted genotypes are exemplified by a belly spot, by a head spot, by white feet and a white tail tip, or by a white band (belt) around the trunk (which presumably represents an inviable clone(s) which failed to be replaced). Mintz (1969a,b) therefore suggests that white spotting results from *preprogrammed clonal death*, i.e., that white spotted genotypes such as $Mi^{wh}/+$ and $W/+$, like $Mi^{wh}/Mi^{wh} \leftrightarrow +/+$ allophenics, possess some melanoblast clones which yield normal populations of functioning pigment cells and others which give rise to inviable populations of these cells, and that the spotting patterns produced by these clones are a consequence of exactly the *same* factors which operate in the allophenic model.

To support her contention that white spotting results from genotype-specific melanoblast cell death and that this lethality is entirely a consequence of gene expression within the melanoblasts, and not due to an influence of

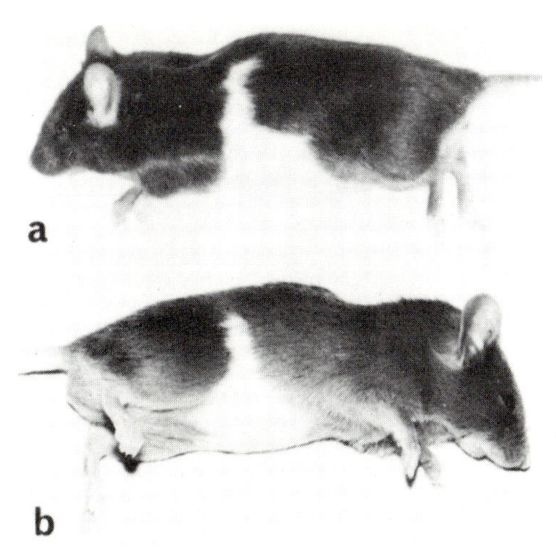

Figure 7-7. (a) $Mi^{wh}/Mi^{wh} \leftrightarrow +/+$ allophenic and (b) $Mi^{wh}/+$ single-genotype animal showing the same spotting pattern. From B. Mintz (1971a). Reproduced with permission of the author and Cambridge University Press.

the skin environment, Mintz carried out the following experiment: she produced $W/W \leftrightarrow +/+$ allophenics from animals known to bear different alleles at the major histocompatibility locus, known as the H-2 complex (J. Klein, 1975), and then demonstrated that both the white and pigmented areas of these animals possessed cells of *both* H-2 types. Thus, when pieces of white skin or of pigmented skin were grafted separately from these allophenics to members of the two isogenic strains from which they were derived, in every case a semirejection occurred. This indicates clearly that despite the fact that the melanoblasts in these grafts must have been derived from only one of the strains, the skin in which they reside (and do not reside) nevertheless is comprised of cells from both (Mintz, 1970).[6]

Notes

[1]In an analysis of the position effect variegation in female mice heterozygous for the *flecked* X-autosome translocation (see Chapter 3, Section II, H), Cattanach (1974) provides data which are in broad agreement with the "standard" pattern as derived by Mintz from allophenics. He too concludes that three bands can be allotted to the head (and neck) but that perhaps seven, rather than six, occur on the body, four in the thoracic and lumbar regions, and probably three more in the rump area. The uncertainty about the exact number stems from the great amount of cell mingling in this latter area. In agreement with Schaible (1969; see Chapter 9, Section I), Cattanach also draws attention to a twin head-spot. Taken together, then, according to his reckoning, including the head spot, there are at least 10 and more probably 11 pairs of sites on the head and body which are colonized by single clones of cells.

[2]It is interesting to note that the occurrence of bands in the head region provides

concrete evidence for the long held theory that mammalian embryos might have anatomically "invisible" head somites (Mintz, 1970).

[3]Since in *fz/fz* ↔ *Fz/Fz* allophenic mice some hairs, especially at the clonal boundaries, are obviously morphologically intermediate between abnormal *fz/fz* hairs and normal *Fz/Fz* hairs, and since in *A/A* ↔ *a/a* allophenics some hairs with yellow pigment display less than the usual amount of yellow banding characteristic of agouti hairs (Mintz and Silvers, 1970; Mintz, 1974), the dermal component of a single hair follicle apparently must be derived from at least two cells (Mintz, 1974; see also McLaren and Bowman, 1969).

[4]Some of this criticism (Lyon, 1968) was aimed at Mintz's original assertion that the two color types of primordial melanoblasts were not arranged randomly (Mintz, 1967), an assertion which Mintz now agrees is *not* the case (Mintz, 1970). Indeed, in a more recent paper in which Dr. Lyon is a coauthor (Cattanach *et al.*, 1972) she apparently accepts Mintz's interpretation of the archetypal pattern.

[5]It seems most likely that if, e.g., Mi^{wh} affected only melanoblast differentiation and *not* melanoblast survival, a situation which would result in the continued occurrence of melanoblasts in the skin (and elsewhere), that the patterns displayed by +/+ ↔ Mi^{wh}/Mi^{wh} allophenics would more closely resemble the banded patterns which exemplify +/+ ↔ *c/c* allophenics. This is especially the case since Mayer (1970) observed that the prior presence of albino melanoblasts in the skin prevented the entrance of normal (+/+) melanoblasts (see Chapter 11, note 11).

[6]Although the coat color patterns of allophenics do not necessarily reflect the relative proportions of each strain in the other organs (see Mintz and Palm, 1969; Mintz, 1974; Gordon, 1976), there are situations where such a correlation has been found. Employing the combination $B1OCBAF_1$ (*A/a;C/C*) ↔ SJL (*A/A;c/c*), Mullen and Whitten (1971) observed that the allophenics which produced only SJL sperm had the greatest proportion of white in the coat (42% average); those which produced only $B1OCBAF_1$ sperm had the lowest (12%); and those which produced both sperm were intermediate (23% white coats). They also observed that whereas normal sex ratios occurred in multicolored allophenics in which one genotype predominated, in those in which neither component predominated, the sex ratio approached 3♂ :1 ♀. R. Roberts and his colleagues (1976) have made allophenics from mice selected for large (L) and small (S) body size, marked with different coat colors. Their observation that the growth of these allophenics was related to the proportion of L pigmentation in the coat indicates a correlation between the proportion of melanocytes of L and S origin and the proportions of L and S cells in whatever tissue (or tissues) regulate growth. Finally, Markert and Petters have produced some hexaparental mice, i.e., animals derived from aggregates of three genetically marked 8-cell embryos, and one of these, as well as one involving the aggregation of four genetically marked embryos, displayed three of the markers (albinism, black, and recessive yellow). These tricolored mice, which exhibited patterns in accord with the supposition that they had been derived from clonal initiator cells arranged linearly from head to tail, are important because they demonstrate that, in the 64-celled blastocyst, at least three cells, and probably only three, are the source of all adult tissues (Mintz, 1970; Markert and Petters, 1978).

Chapter 8

X-Linked Determinants

I. X-Chromosome Inactivation and the Allophenic Model

One of the most fundamental discoveries in mammalian genetics stems from the pigment patterns of female mice heterozygous for an X-linked coat-color determinant. Thus, the single-active-X hypothesis was originally suggested to explain the variegated or banded coat-color patterns of such females (Lyon, 1961; L. Russell, 1961). It was proposed that in the early embryo one X-chromosome, chosen at random, was inactivated in each mammalian somatic cell and that the affected chromosome remained inactive. Thus, in the adult all the cells descended from any particular embryonic cell *after* the time of decision constituted a clone characterized by the presence of the same active X-chromosome (with the exception of the germ cells in which both X-chromosomes remained active). The wild type colored patches of the coat of heterozygous females were therefore attributed to the activity of cells in which the X-chromosome bearing the mutant allele had been inactivated, whereas the mutant colored patches were attributed to cells in which the X-chromosome carrying the gene for wild type color was inert (Figure 8-1). This inactivation hypothesis, which is now accepted as dogma (see Lyon, 1970, 1971, 1972a, 1974),[1] is of particular interest because, like the allophenic patterns described in the previous chapter, it too can result in a situation in which melanocytes and/or hair follicle cells are derived from clonal-initiator cells bearing different markers. Consequently, a careful comparison of the patterns produced by X-linked coat-color determinants with those of the appropriate allophenics provides a means of assessing

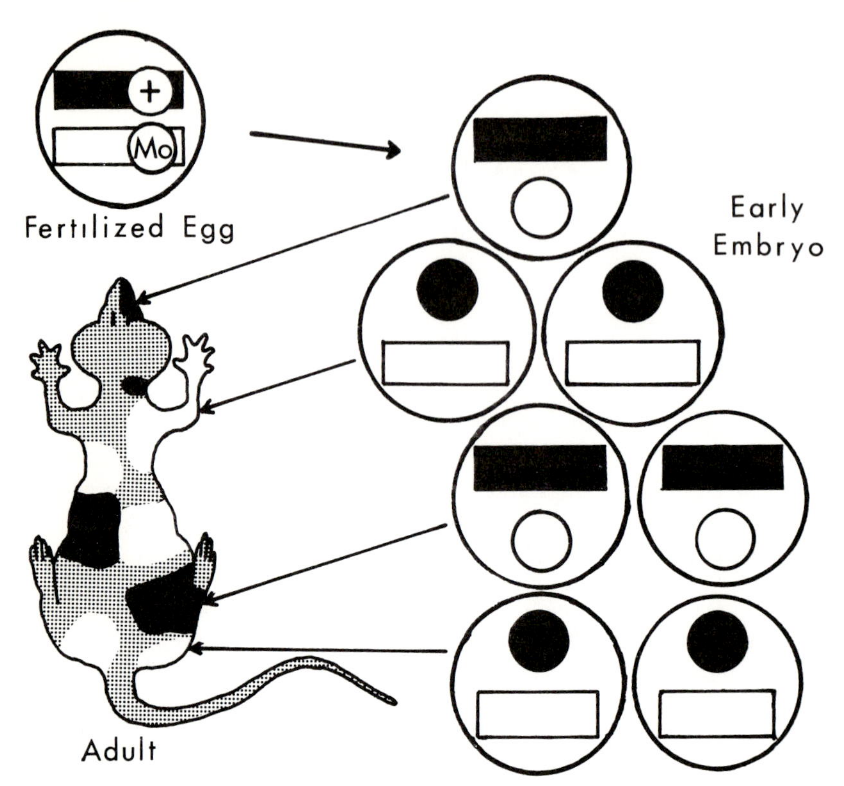

Figure 8-1. A diagrammatic representation of X-inactivation as applied to coat-color mosaicism in the mouse. The inactive X-chromosome in cells of the early embryo is represented by a circle and the active one by a rod. Light areas of the coat stem from regions in which the X-chromosome bearing the mutant gene *Mo* is active within melanocytes [and the X-chromosome bearing the normal (+) allele is inactive]. In the dark areas, the reverse situation prevails and the X-chromosome bearing the + allele is active within melanocytes. The grey area designate regions where a significant intermingling of the two types of melanoblasts has occurred during development. Adapted from M.F. Lyon (1963).

whether Mintz's interpretation of the latter also applies to these *naturally* occurring phenotypes. Such a comparison has been made (Cattanach et al., 1972) and it clearly indicates that all of the phenotypic features of X-linked heterozygotes can be found in their allophenic counterparts.[2] Indeed, the pigment patterns displayed by females heterozygous for an X-linked coat-color factor are so similar to those of certain allophenics that *all* the postulates which Mintz has formulated to explain the latter almost certainly apply to the activity of these genes.

What is most surprising however is that some of the pigment patterns of females heterozygous for X-linked color genes stem from *both* melanoblast and hair follicle phenoclones (Mintz, 1970; Cattanach et al., 1972). This observation is unexpected because it implies that some X-linked coat-color determinants are complex genetic units, of which different parts are expressed in different cells that give rise to different clones (Mintz, 1970). Indeed, Mintz (1970) raises the possibility that "perhaps many (or even all) other genes once thought of as acting only in melanoblasts, or only in hair follicles, may be complex or highly polycistronic loci."

Finally, before focusing our attention on the effects of specific X-linked coat-color factors, it should be noted that because the patterns produced by these factors are replicas of those found in allophenic mice, they must be determined at a similar development stage. Thus the same line of reasoning that led to the conclusion that the melanoblast and hair follicle patterns of allophenics were established sometime between the fifth and seventh day of gestation, undoubtedly applies to these X-linked heterozygotes as well (Mintz, 1974). In accord with this supposition are the results of injecting blastocysts with single cells from embryo donors heterozygous for an X-linked coat-color marker. The phenotypes which result from this procedure show that at 4.5 days of gestation X-chromosome inactivation has not yet occurred (Gardner and Lyon, 971; Lyon, 1972a).

II. The Mottled Locus

Although a number of X-linked mutations have been reported which affect coat color, all of these, with the exception of yellow mottling (*Ym*) and Pewter (*Pew*) (Section III), probably represent mutations at one locus, the so-called mottled or *Mo* locus. Because some of these mutations are lethal in hemizygous males, it has been difficult sometimes to demonstrate allelism directly, but all the information which is available is consistent with the notion that they all represent changes at the *same* locus (Falconer, 1953, 1956b, Lyon, 1960, 1972b, Grahn et al., 1971, 1972; Cattanach and Williams, 1972). Thus there is sufficient evidence to consider five of the six X-linked mutations, mottled (*Mo*), blotchy (Mo^{blo}), dappled (Mo^{dp}), brindled (Mo^{br}), and viable-brindled (Mo^{vbr}) as representing a multiple series of X-linked alleles and they have accordingly been so designated. Only in the case of tortoiseshell has a separate locus designation been maintained. However, there is no evidence that this mutation represents another X-linked coat-color locus (Lane, 1960b), and much evidence to support the notion that it too is another *Mo* allele (Grahn et al., 1969a,b). Indeed, its effects are so similar to the alleles at the *Mo*-locus that there is much to be said for changing its designation from *To* to Mo^{to}. Nevertheless, tortoiseshell will be referred to as it is currently designated.

A. Mottled (*Mo*)

1. Origin and Description

The initial mottled (*Mo*) female occurred among the progeny of a cross-segregating for albinism (*c*), pied (*s*), brown (*b*), and hairlessness (*hr*) (Fraser et al., 1953). The female was *B*/−;*S*/− and had many regions of light-colored (off white) hair scattered in an apparently patternless fashion over the entire body. The depth of color of the hairs in these regions varied between regions. All of the subsequent mottled females derived from this initial female displayed essentially the same pattern although the diffuse areas of very lightly pigmented hairs varied in extent and sometimes seemed to be arranged in an irregular pattern of transverse bars (Falconer, 1953) which, when well defined, very rarely crossed the mid-dorsal or mid-ventral line (M. C. Green, 1966a).

Another feature of the mottled mutation is that it produces curling of the vibrissae, a characteristic which allows the condition to be recognized as early as 1 or 2 days after birth (Falconer, 1953).

2. Survival of Males

While it was immediately apparent that mottled expressed itself when heterozygous, it was also evident that it never occurred in males and that some females bearing the mutation died before 2 weeks of age (Fraser et al., 1953). To determine the fate of the mottled males, Falconer (1953) killed and dissected mottled females at various stages of pregnancy. He found a significant group of dead embryos at the 11-day stage of development and by the twelfth day of gestation their frequency (23%) agreed well with the expected frequency of 25% of mottled males. It therefore appeared that although these dead embryos did not display any obvious external abnormality, they must in fact have been mottled males.

B. Dappled-2

In 1960 Lyon described another sex-linked mutation which she provisionally called dappled-2. However, because she subsequently judged that this mutation was not clearly distinguishable from mottled, she withdrew this name and no formal designation has been given this deviant.[3] Nevertheless, this mutation deserves special consideration because some of its features, as described by Lyon (1960), are not mentioned either by Fraser and his colleagues (1953) or by Falconer (1953). Indeed, this raises the question as to whether Lyon's mutation is in fact a remutation to *Mo* or whether it represents a different allele? If it is *Mo*, it illustrates that different descriptions of the same mutation can often be misleading. It also indicates that there are more features associated with the *Mo* mutation than had previously been described.

1. Origin and Pigmentation

The original mutant animal of Lyon's was a *male* which upon reflection must have been a mosaic with some cells, both of his soma and his gonads, carrying a normal X-chromosome and some carrying an X-chromosome bearing the mutation. Phenotypically this male resembled his female decendants which possessed, to varying degrees, a mottled coat with patches of white, light-colored and full-colored hairs, as well as intermingled hairs of different colors.[4] Females which were heterozygous for this mutation also sometimes exhibited patches with well-defined edges and when these occurred they rarely crossed the mid-dorsal or mid-ventral line of the animal (Figure 8-2). This was especially obvious on the head, which was often divided by a line from the ears to the tip of the nose into dark and light halves. Thus, insofar as their coat colors were concerned these females were identical in every respect to the descriptions noted above for mottled. They also resembled *Mo*/+ in that they had curly vibrissae, although the degree of curliness was variable. In situations where the color of the hair was very different on the two sides of the snout, the whiskers were curlier on the lighter side.

2. Other Abnormalities

In addition to these features, however, Lyon also noted that "in some of these females the fur was slightly wavy and of a finer texture than normal." Moreover, she noted that some heterozygous females were distinguished by incoordination of the limbs which was first noticeable as a tremor when they were a few days old. This condition progressed rapidly until the affected animals lost all coordination of the limbs and died at about 2 weeks of age.

Lyon also reported that some of her heterozygous mutants when a few months old displayed calcified lumps either attached to their bones or free in their muscles or tendons. These lumps were frequently seen attached to the vertebrae or the bones of the feet, but were also common in the muscles of the limbs. The frequency of these calcifications increased as the animals got older.

3. Expression in Hemizygous Males

As in the case of mottled, males hemizygous for Lyon's mutation also fail to reach term and apparently die *in utero* at about the same stage of development. Thus, at 12.5 days of gestation Lyon found some recently dead, normal looking embryos. However, these embryos were white with yolk sac circulations which appeared empty. Indeed, this observation, together with the fact that at 11.5 days some living embryos were found with hemorrhages inside the yolk sac, is undoubtedly significant. Since other *Mo*-locus alleles

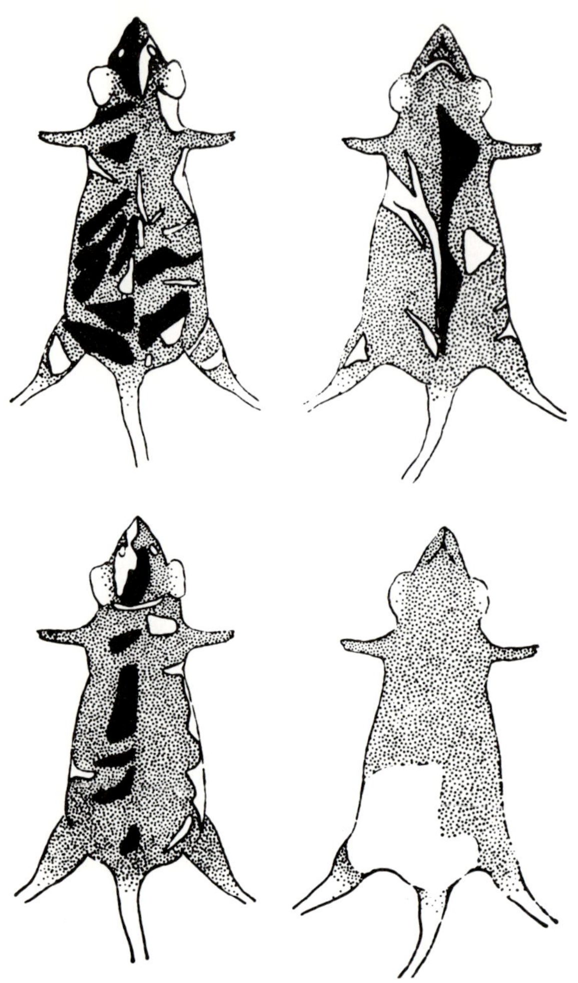

Figure 8-2. Diagrams showing dorsal and ventral sides of two typical *dappled*-2 females. The black areas denote full-colored and the stippled areas light-colored patches. Note that patches rarely cross the mid-line. From M.F. Lyon (1960). Reproduced with permission of the author and the American Genetic Association.

are known to be associated with vascular disorders (see Table 8-1) and there is every reason to believe that these defective embryos were hemizygous for Lyon's mutation, it is very likely that they too were afflicted with some disorder of the circulatory system.

C. Brindled (Mo^{br})

Brindled arose spontaneously in the C57BL strain (Fraser et al., 1953). The phenotypes associated with this mutation have been described in detail by Grüneberg (1969) and most of what follows is based on his description.

Table 8-1. Frequency of Aortic Lesions in Various *Mo*-Locus Genotypes[a]

Genotype	Total number	Normal (%)	Aortic lesions: Aneurysm (%)	S-curve[b] (%)
$Mo^{blo}/+$	19	63	32	5
Mo^{blo}/Mo^{blo}	13	15	85	0
Mo^{blo}/Y	63	2	98	0
Mo^{blo}/Mo^{br}	12	100	0	0
$Mo^{br}/+$	32	100	0	0
Mo^{br}/Y	13	100	0	0
$Mo^{dp}/+$	47	90	4	6
$To/+$	328	58	25	17
To/Y	26	19	81	0

[a]Modified from Rowe et al. (1974).
[b]Sometimes aneurysm is accompanied by a defined S-curve in the thoracic aorta or an S-curve may be seen alone.

1. Pigmentation and Pelage of Males

The coat of Mo^{br} males is usually almost white with sooty tips, but sometimes it is light grey in color (Figure 8-3). The eyes are dark and the ears, tail, and scrotum all display normal levels of pigmentation. Although the color of the brindled male has been compared with Himalayan albinism (Falconer, 1953) this resemblance is only superficial for unlike himalayan

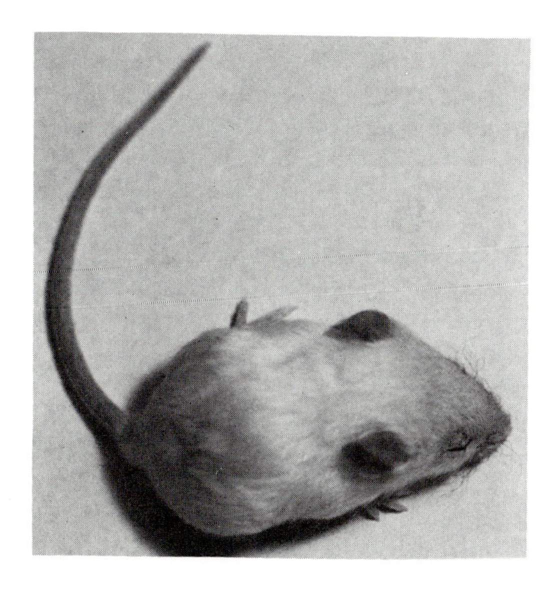

Figure 8-3. a young brindled (Mo^{br}) male. Note light color of coat as compared with the ears and tail. Note also the curliness of the vibrissae and the texture of the coat. Courtesy of D.M. Hunt.

animals (see Chapter 3, Section II, B, 6), where the pigmentation of the extremities involves the hair, in brindled males only the *skin* of the ears, tail, and scrotum is pigmented normally. The vibrissae of Mo^{br} males are very curly and irregular and the coat, which as a whole is short, possesses many finely curved or undulated hairs (see Figure 8-3). Whereas in the normal coat a halo of guard hairs usually projects above the rest of the fur, these hairs are almost completely absent in Mo^{br}/Y mice. Moreover the tail hairs of these animals are reduced in number and less regularly arranged than in normal mice.

A careful examination of the coat of brindled males reveals that it includes both overhairs (awls, auchenes, and probably some guard hairs) and underfur (zigzags), but these hairs are reduced in caliber and structurally abnormal (Figure 8-4). Most of the hairs are unpigmented, but some have a little, mostly cortical, pigment in the tip. The first few septules may also contain melanin granules.

2. Pigmentation and Pelage of Females

The coat of $Mo^{br}/+$ females displays a wide variation in its pigmentation (Figure 8-5). Some females are very light, while others are phenotypically almost normal, but all display at least some irregular mottled or brindled regions in which ill-defined lighter and darker areas tend to grade imperceptibly into each other. Hair bases are much lighter than the tips and the hairs on the tail are reduced and less regular in arrangement than in normal animals.

In addition to mottling some $Mo^{br}/+$ females display more regular patterns. Thus, as pointed out by Falconer (1953) and Grüneberg (1969), the light hairs sometimes form an irregular pattern of transverse stripes which is most conspicuous in the juvenile coat and tends to decrease with age. This banding pattern resembles those found in allophenics whose components differ with respect to a hair follicle marker, and like some allophenic patterns is occasionally characterized by a "midline effect" in which dark and light areas meet sharply along the midline. There is no color mottling on the ears or tail of brindled females and, as in mottled ($Mo/+$) and dappled ($Mo^{dp}/+$) females, there seems to be a correlation between the color and curliness of the whiskers: they are straight in dark, but a little curled in light individuals.

Females heterozygous for Mo^{br} sometimes change in color with age and when this occurs they tend to get darker.

According to Grüneberg the juvenile coat of $Mo^{br}/+$ females includes some morphologically normal hairs which are pigmented completely, and some which are abnormal and contain as little pigment as the hairs of brindled males (Figures 8-4 and 8-6). However, these latter hairs are usually not nearly as structurally abnormal as those of brindled hemizygotes. Most of the hairs are intermediate between these extremes. They usually start normal both with regard to structure and pigmentation, but the pigmentation

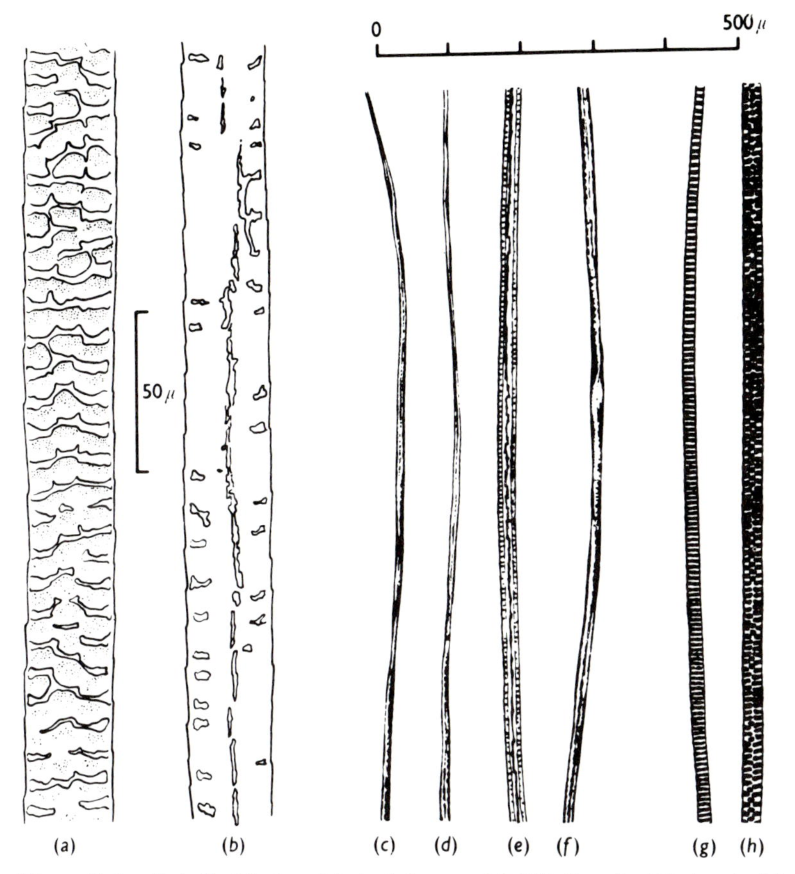

Figure 8-4. Brindled hairs. (a) Awl from a $Mo^{br}/+$ female (19 days), (b) awl from Mo^{br} male (littermate, 12 days), (c–f) other hairs from the same animal at lower magnification (all completely devoid of pigment in the segments shown). (g and h) Normal male (19 days). (a,b,d,e, and h) awls (d a distal tip), (c and g) zigzags, (f) presumed guard hair with a long nonmedulated tip. From H. Grüneberg (1969). Reproduced with permission of the author and Cambridge University Press.

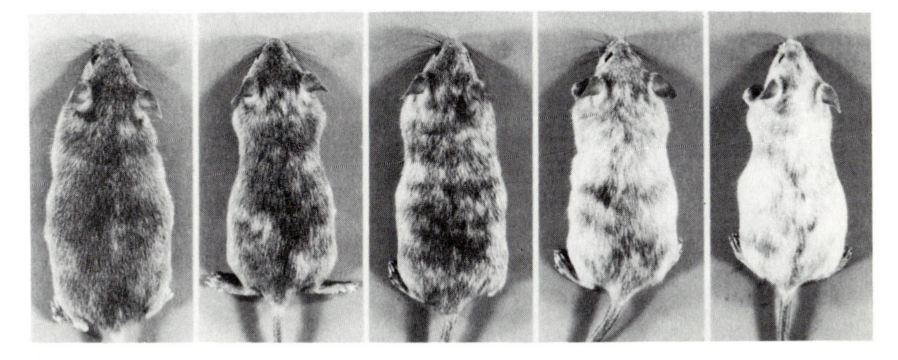

Figure 8-5. Variation in the degree of expression of brindled (Mo^{br}) in heterozygotes. From D.S. Falconer and J.H. Isaacson (1972). Reproduced with permission of the authors and Cambridge University Press.

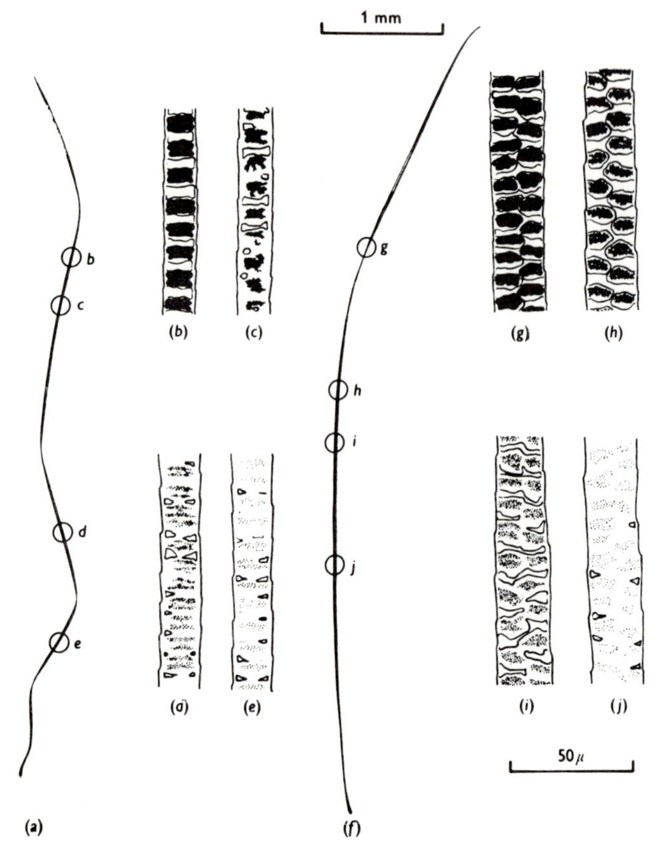

Figure 8-6. Brindled ($Mo^{br}/+$) female, 19 days (same as in Figure 8-4). (a) A zigzag, (f) an auchene, both with polarized pigmentation as shown in the enlarged regions (b–e) and (g–j), respectively. From H. Grüneberg (1969). Reproduced with permission of the author and Cambridge University Press.

decreases and eventually ceases altogether, and their structure becomes increasingly abnormal. Although the same polarization of pigment occurs in $Mo^{br}/+$ females as in brindled males, a greater portion of the hair shaft contains pigment and the level of pigmentation is significantly heavier. In the morphologically abnormal hairs the deterioration in hair morphology may begin early or late in the hair cycle. While in some instances only a middle portion of the hair may appear abnormal, there are no hairs which start abnormal and gradually become normal.

In the *adult* coat of heterozygous brindled females hairs as extreme as those found in the juvenile coat are all but absent, and changes in the degree of mottling are evidently largely the result of differences in the amount of pigment deposited in different hair generations.

The exact relationship between the structure and pigmentation of the hair in brindled mice is not clear. While it is conceivable that the effect of hair structure is primary, and that melanocytes cannot function properly in follicles which give rise to abnormal hairs, it is also possible that Mo^{br} acts via the melanocyte and that these cells can somehow alter the morphology of the hair. This latter possibility, however, seems highly unlikely because the pigmentation of the *skin* is unaffected, and because it is inconsistent with the fact that some of the pigment patterns of $Mo^{br}/+$ females simulate the hair follicle banding patterns of allophenics. Indeed, the most likely possibility, and one to which we will return, is that while the major effect of Mo^{br} is on the hair follicle (Grüneberg, 1969) it also is able to influence directly, albeit in a relatively more minor way, the melanocyte.

3. Survival and Behavior of Hemizygous Males

Brindled is usually lethal in males. As originally described all Mo^{br} males died between 10 and 14 days of age (Fraser et al., 1953). However, Grüneberg (1969) reported that they sometimes live up to 3 weeks and occasionally longer and a few have lived and been fertile (Falconer, 1956b; M.C. Green, 1966a). In fact the survival of such males enabled Falconer (1956b) to produce homozygous brindled females which were in every way indistinguishable from hemizygous males. He also was able to produce females which carried both mottled and brindled. These were indistinguishable from Mo^{br}/Mo^{br} females or Mo^{br}/Y males. One of these brindled males was also crossed with one of Lyon's "mottled" females, producing one litter in which there was one black-eyed white female which died at 11 days of age (Lyon, 1960). These findings, along with the phenotypic resemblances of *Mo* and Mo^{br}, and their similar linkage relationships to tabby (*Ta*),[5] make it reasonable to assume that they are alleles.

Brindled males suffer from a neurological impairment (Sidman et al., 1965; D. Hunt and D. Johnson, 1972a,b) characterized by "slight tremor, uncoordinated gait, and characteristic reflex of the hind legs when the animal is held up by the tail, the feet clasping each other instead of being spread out" (Falconer, 1956b).[6]

4. Selection for the Expression of Brindled

Because there is a great deal of variation in the variegated patterns of $Mo^{br}/+$ females, Falconer and Isaacson (1969, 1972) attempted to determine whether these patterns could be selected for. They reasoned that since this variegation is due to X-chromosome inactivation, if it could be modified by selection this would show that the inactivation process, or some property of the derived cell populations, is under genetic control. $Mo^{br}/+$ females were accordingly selected for on the basis of the amount of mutant color they displayed. They found that although selection based on *individual* phenotypes was ineffective in influencing the amount of variegation, four cycles of reciprocal recurrent selection based on progeny-means did produce a

"High" line with 64% mutant area and a "Low" line with 30% mutant area, from a base population with 53% mutant area (Falconer and Isaacson, 1972). They went on to demonstrate that the difference between the selected lines was not due to autosomal modifiers but was due entirely to properties of the X-chromosomes bearing the Mo^{br} allele, i.e., apparently there are X-linked modifiers of X-linked loci which operate only upon the allele located on the same chromosome. They also showed that the altered properties of the X-chromosomes were not restricted to the *Mo*-locus but extended at least as far as tabby (*Ta*). Although there was no evidence that the chromosomes carrying the + allele of brindled were altered by the selection procedure, normal X-chromosomes from other strains did influence the degree of variegation. Falconer and Isaacson (1972) concluded from their experiments "that the difference between the selected lines was due either to non-random inactivation or to somatic cell selection." Their protocol did not allow them to distinguish between these possible mechanisms.

Employing an X-linked gene, mosaic (*Ms*) (see Section J), which produces a variegated phenotype very much like that of brindled, Krzanowska and Wabik (1971, 1973) also attempted to select for high and low proportions of mutant areas. In contrast to Falconer and Isaacson, these investigators were successful in establishing such lines on the basis of individual selection. Thus they obtained lines with 73 and 35% mutant areas after four generations of within family selection. To account for this success Falconer and Isaacson (1972) suggest that it was due to variations in the properties of the X-chromosome arising by recombination. This variation was presumably either not present in their strain, or at least no suitable crossovers occurred.[7]

D. Viable-Brindled (Mo^{vbr})

1. Pigmentation and Pelage

This mutation has been described only briefly (Cattanach, et al., 1969) but from all indications the range of coat-color phenotypes it produces when heterozygous, as well as the coat color of hemizygous males, is indistinguishable from those produced by brindled. As in the case of $Mo^{br}/+$ females, the banding patterns of $Mo^{vbr}/+$ mice more closely resemble those associated with hair follicle phenoclones than with melanoblast phenoclones, the bands being composed of whitish hairs like those which comprise the coat of the hemizygous Mo^{vbr} male. Like Mo^{br}, Mo^{vbr} also causes a rippling of the coat and a curling of the whiskers which is more pronounced in hemizygous males. This influence on the texture of the coat suggests that the primary effect of the allele is on hair structure. Nevertheless, it seems likely that this allele has a direct effect on melanoblasts as well since the banding in $Mo^{vbr}/+$ heterozygotes is broader than the banding in tabby heterozygotes (*Ta*/+) (Cattanach et al., 1969) which, as already noted (see note 5), is believed to be produced solely by hair follicle phenoclones.

2. Other Characteristics

$Mo^{vbr}/+$ females are not good mothers and about 50% of their offspring, regardless of genotype, do not survive to weaning. This, together with the additional loss of 50% of Mo^{vbr}/Y males before they are a month old, results in an overall early mortality of about 75% for these males. Although the Mo^{vbr}/Y males which succumb early in life have a mean survival of 22.7 days, and most die between days 20 and 30, an appreciable number die at 12–14 days of age (Grahn, personal communication).

Viable brindled, like brindled, males suffer from a neurological disturbance (D. Hunt and D. Johnson, 1972a) but unlike surviving brindled males they are sterile although spermiogenesis appears normal at the ultrastructural level (D. Hunt and D. Johnson, 1972a). Many Mo^{vbr} males which survive into adulthood also display aortic aneurysms, which are *not* observed in Mo^{br} males (Rowe et al., 1974). Indeed, many viable-brindled males succumb to blood vessel rupture between 50 and 100 days of age (Rowe et al., 1974).

E. Dappled (Mo^{dp})

1. Origin and Evidence of Allelism

Dappled (Mo^{dp}) occurred in an F_1 male from a low-dosage γ-irradiation experiment (Carter et al., 1958). The only affected male sired 10 dappled females which phenotypically resembled himself, as well as 278 normal females and 290 normal males (R.J.S. Phillips, 1961b). Inasmuch as dappled has proved to be lethal in the hemizygous condition, it is apparent that the original male who displayed the mutation, like the male who originally manifested the "mottled" mutation described by Lyon (1960), carried a sectorial mutation with at least the coat and part of the testes being affected.

The supposition that dappled is a member of the *Mo*-series is supported both by linkage tests and by the fact that dappled females mated to a brindled (Mo^{br}) male produced six females which resembled brindled males. These females, which died before weaning, presumably carried both dappled and brindled, i.e., were Mo^{dp}/Mo^{br} (R.J.S. Phillips, 1961b).

2. Pigmentation and Pelage

Most dappled females can be identified at birth by a curling of the vibrissae, but the degree of curling is very variable and some animals' whiskers are nearly straight (R.J.S. Phillips, 1961b). The adult coat has a variegated pattern similar to that of mottled.[8] Although the hairs have not been studied in the same detail as brindled, and dappled males die prenatally and hence cannot be studied, some dappled hairs possess abnormalities which resemble those found in $Mo^{br}/+$ females (Cattanach et al., 1972). Medullary cells appear to be fused, perhaps as a result of the presence of liquid (absence of air) as suggested by Grüneberg (1969), and a flattening of the hair is also apparent (Cattanach et al., 1972). In addition to the structural abnormalities, the hairs also tend to be deficient in pigment. However, there is not a complete correlation between the structural abnormalities of the hair and lack of pigmentation (Cattanach et al., 1972).

Although the variegated phenotypes of heterozygous dappled females appear to resemble more closely those produced by melanoblast phenoclones than those which result from hair follicle phenoclones, it seems apparent from the structural abnormalities of the hair, as well as from a close examination of the pelage of an $Mo^{dp}/+ \leftrightarrow Ta/Ta$ allophenic female (Cattanach et al., 1972), that this allele too most likely acts both in the melanocyte and in the hair follicle.

3. Interaction with Piebald or Belted

A most fascinating effect on the pigmentation of dappled females occurs in the presence of either piebald (*s*) (Chapter 9, Section II, A) or belted (*bt*) (Chapter 9, Section IV) spotting. In $Mo^{dp}/+;s/s$ and $Mo^{dp}/+;bt/bt$ mice the two hair colors which constitute the "normal" variegated pattern are separated out to give an overall tricolored animal possessing normally (+) pigmented, lightly (Mo^{dp}) pigmented, and nonpigmented (*s*/*s* or *bt*/*bt*) areas (Cattanach et al., 1972). This effect, which undoubtedly is associated with the fact that the major influence of Mo^{dp} is on the melanoblast,[9] is of interest not only in its own right but because a similar situation occurs in guinea pigs, rabbits, tortoiseshell cats, and other animals as well (Searle, 1968a). For example, the extension series in the guinea pig (see Chase, 1939), which is autosomal, includes *E* (self black) which is dominant over e^p (tortoiseshell) and *e* (self yellow). In the absence of white spotting, e^p/e^p quinea pigs display a brindling pattern of yellow hairs on a black background. However, in the presence of white spotting (*s*/*s*) the black and yellow areas become segregated from each other and a tricolored yellow, black, and white phenotype is produced. Indeed, the two inbred strains of guinea pigs which are currently most utilized, strains 2 and 13, are $e^p/e^p;s/s$ and represent such tricolored phenotypes. How white spotting produces this effect and whether the mechanism is the same in mice, guinea pigs, cats, and rabbits is not known, although it probably is.[10]

4. Other Abnormalities

In addition to its influence on pigmentation, Mo^{dp} sometimes influences the morphology of the feet. Some $Mo^{dp}/+$ females show clubbing of one or both fore feet at birth, or, at weaning, have a tendency to walk on the dorsal surface of the hind feet. Moreover, the occurrence of abnormal feet seems to be associated with a greater degree of curling of the whiskers (R.J.S. Phillips, 1961b). The amount of curling of the vibrissae is likewise associated with the degree of lightness of the coat at weaning: the lighter the coat, the more pronounced the curling (R.J.S. Phillips, 1961b).

Another condition associated with some $Mo^{dp}/+$ females is the development, with age, of calcified lumps in the region of the periosteum especially on the vertebral column, the lumbar and thoracic regions of which are principally affected. Whether these bodies are outgrowth from the bone is not clear. None has been observed in mice under 8 weeks of age (R.J.S. Phillips, 1961b). This anomaly also occurred in Lyon's (1960) "mottled" mutation.

Finally, about 10% of $Mo^{dp}/+$ females have been reported to have aortic lesions (Rowe et al., 1974) (see Table 8-1).

5. Expression in Hemizygous Males

When the offspring of the dappled daughters of the original dappled male were sexed at birth, there were about half as many males as females. This of course suggested that, as in the case of mottled, males hemizygous for dappled die before birth. R.J.S. Phillips (1961b) found that among embryos 15 days old or more some had ribs which were very white, thickened, and bent and some also displayed distortion of the pelvic and pectoral girdles and the limb bones. She also observed that most of these abnormal embryos either were dead already or were moribund at 17–18 days of gestation.[11] These defective embryos were assumed to be the hemizygous males and this was supported by data which showed that half the males and none of the females were affected. Males with abnormal ribs have also occasionally been found dead at birth. The specific cause of death is not known.

F. Blotchy (Mo^{blo})

1. Pigmentation and Pelage of Heterozygotes

Blotchy (Mo^{blo}) arose spontaneously at Oak Ridge (L. Russell, 1960).[12] As in the case of the other *Mo*-alleles, females heterozygous for this mutation also display irregular patches of light colored fur. Its expression is occasionally poor at weaning (heterozygous females are sometimes misclassified as wild type) but is complete by adulthood (L. Russell, 1960). The pigment patterns which occur in $Mo^{blo}/+$ females resemble those produced by melanoblast phenoclones (Mintz, 1971a). However, inasmuch as Mo^{blo} causes no lightening of the pigment in the skin of the ears and tail, its hair-lightening effect could be due to some effect on the hair follicle (Lyon, 1970). Thus there is some evidence that this allele, too, acts *both* in the melanocyte and in the hair.

2. Manifestations in Hemizygotes and Homozygotes

Blotchy when hemizygous in either males or females (X/O), and when homozygous in females (Mo^{blo}/Mo^{blo}), produces a light phenotype (with no blotching) somewhat resembling c^{ch}/c in intensity. These genotypes usually also are of small size, and occasionally have deformed hind legs. Their whiskers are kinked at birth but straighten out by the time they are weaned. Whereas the viability and fertility of heterozygotes is normal, and Grahn and his colleagues (1969b) note that Mo^{blo}/Mo^{blo} females also are of normal viability, L. Russell (1960) reports that hemizygous and homozygous blotchy mice have reduced viability and many are infertile.

3. Evidence of Allelism

Evidence that blotchy is a member of the *Mo*-series of alleles (and hence that its designation should be changed from *Blo* to Mo^{blo}) was provided by the fact that it displays the same crossover frequency with tabby (~ 4%) as the other members of the series (L. Russell, 1960; L. Russell and Saylors, 1962). Moreover, no crossovers between Mo^{blo} and Mo^{vbr} have been observed (Cattanach and Williams, 1972) and both Mo^{blo}/Mo^{br} and Mo^{blo}/Mo^{vbr} females are lightly pigmented. Animals of the former genotype are of a uniform light coat color (Lyon, 1972b) and those of the latter are described as tending to show a white (Mo^{vbr})–brownish white (Mo^{blo}) variegated coat although the two types of pigmentation are not readily distinguishable (Cattanach and Williams, 1972).[13]

4. Other Abnormalities

In addition to these originally described features of blotchy the mutation is also associated with spontaneous aortic and abdominal aneurysms (Grahn et al., 1969b; Rowe et al., 1974; Andrews et al., 1975). According to Rowe and his colleagues about 95% of hemizygous blotchy males, 85% of Mo^{blo}/Mo^{blo} females, and 35% of $Mo^{blo}/+$ females have aortic aneurysms (see Table 8-1). Indeed, many blotchy males succumb to blood vessel rupture when more than 150 days of age (Rowe et al., 1974). Their median age at death is about 200 days and over 90% are dead by 1 year of age (Grahn et al., 1971). Blotchy males also have recently been reported to develop osteoarthrosis at a much earlier age than "normal" animals. Silberberg (1977) found that from the age of 3.5 months on such males develop this condition in their knee joints and by the time they die 88% have it. Emphysema-like changes have also been noted in the lungs of these males (Silberberg, 1977).

G. Tortoiseshell (*To*)

1. Pigmentation and Pelage of Heterozygotes

This mutation arose spontaneously in an obese stock at the Jackson Laboratory (Dickie, 1954). Heterozygous tortoiseshell (*To*/+) females (Plate 3–C) are variegated and resemble *Mo*/+ females in color. They also, as expected, display a wide variety of phenotypes so that a large number of individuals must be examined to determine the general pattern trend. Such an examination indicates that the basic *To*/+ pattern is composed of both melanoblast and hair follicle phenoclones, one superimposed upon the other (Mintz, 1971a). Indeed, it is because of this that Mintz (1970) suggests that the *To* locus may be a complex one, with one portion controlling and being expressed in melanoblasts, and another in hair follicle clones. Further support for this contention is provided by the fact that *To* affects the texture of the hair which is silkier than normal. The vibrissae are also slightly waved (Dickie, 1954).

Unlike the situation in dappled females, piebald (*s*/*s*) spotting has no effect on the tortoiseshell pattern (Dickie, 1954).

2. Other Abnormalities

Some tortoiseshell females also suffer from skeletal abnormalities in the fore and hind limbs (Dickie, 1954) and an appreciable number of them have aortic lesions, including aneurysms (see Table 8-1) (Fry et al., 1967; Grahn et al., 1969a,b; Rowe et al., 1974).

3. Hemizygous Males and Evidence of Allelism

Although a few *To*/*Y* males are stillborn, many of them die *in utero* with blood vessel aneurysms (Grahn et al., 1969a,b; Rowe et al., 1974) (see Table 8-1).[14] A presumed mosaic male carrying *To* (see Grahn and Craggs, 1967) was utilized to produce Mo^{blo}/To, Mo^{br}/To, Mo^{dp}/To, and *To*/*To* females and it was found that whereas animals of the first two genotypes reach term but usually die by 15 days,[15] the latter two genotypes are presumably prenatal lethals (Grahn et al., 1969a,b). Such information provides further evidence that *To* is a member of the *Mo*-series of alleles.

H. Summary of Phenotypic Characteristics of the Mo-Series of Alleles

The phenotypic effects of *Mo*-locus alleles are summarized in Table 8-2 and the relative viability of various *Mo*-locus genotypes is given in Table 8-3. While it is evident that all of these determinants produce banded patterns when heterozygous with the wild type allele, these patterns seem to vary from those which are similar to the melanoblast bands of allophenics to those which bear a much closer resemblance to the hair follicle clonal stripes of these mice. Whereas the coat-color patterns of *Mo*/+, $Mo^{blo}/+$, and $Mo^{dp}/+$ females resemble those of allophenics whose components differ with regard to melanocyte pigment, the patterns found in $Mo^{br}/+$ and $Mo^{vbr}/+$ heterozygotes more closely resemble the striped patterns produced by hair follicle clones. Nevertheless, evidence has been presented which indicates that *all* of these alleles may act *both* in the melanocyte and the hair follicle but to quite different degrees. This evidence includes the following: (1) all these genes appear to be associated with curly whiskers; (2) although the $Mo^{blo}/+$ pattern appears to resemble that produced by melanoblast pheno-

Table 8-2. Phenotypic Characteristics of the *Mo*-Locus Series of Alleles

Allele	♂ Homozygote	♀ Heterozygote
Mottled (*Mo*)	Lethal *in utero* at ~ 11 days gestation	Irregularly mottled or brindled coat, curly whiskers; reduced-viability; survivors usually fertile
Dappled-2[a]	Lethal *in utero* at 11–12 days gestation (blood vessel aneurysms?)	Phenotypes similar to *Mo*; curly vibrissae; coat maybe slightly wavy and of fine texture; incoordination of hind limbs in some; calcified lumps in muscles, tendons, or attached to vertebrae or bones of feet; reduced viability and fertile
Brindled (Mo^{br})	Dirty white to light grey in color; curly whiskers; abnormal hairs and coat wavy; neurological disturbance; usually die at ~ 14 days postpartum but a few have lived and been fertile	Phenotypes similar to *Mo*; curly whiskers; some hairs abnormal but not as structurally altered as in♂♂; slight rippling of coat; viable and fertile
Viable-brindled (Mo^{vbr})	Light-colored fur; curly whiskers; neurological disturbance; reduced viability (many succumb to blood vessel rupture), and sterile	Like Mo^{br}
Dappled (Mo^{dp})[a]	Lethal at ~ 17 days gestation but some survive to about time of birth; skeletal defects	Phenotypes similar to *Mo*; curly whiskers; some hairs abnormal and similar to those in Mo^{br}/+ ♀♀; some have clubbing of fore feet at birth; calcareous deposits (especially on vertebral column); low incidence of aortic lesions; viable and fertile

Table 8-2. *continued*

Allele	♂ Homozygote	♀ Heterozygote
Blotchy (Mo^{blo})	Light-colored fur (somewhat resembling c^{ch}/c); usually of small size; occasionally have deformed hind legs; vibris sae kinky at birth but straighten; high incidence aortic aneurysm and osteoarthrosis; reduced viability (many succumb to blood vessel rupture) and many infertile	Irregularly mottled or brindled coat; curly whiskers; significant incidence of aortic aneurysm; viable and fertile
Tortoiseshell (*To*)	Lethal *in utero*; often with macroscopic aortic lesions	Phenotypes similar to *Mo*; whiskers slightly curled; silky textured coat; some display skeletal abnormalities in fore and hind limbs; viable and fertile but many die of aortic aneurysm rupture

[a]This mutation was described by Lyon (1960) and was provisionally named dappled-2. However, since it was not judged to be clearly distinguishable from mottled, *Mo*, this name was withdrawn.

Table 8-3. Relative Viability of Some *Mo*-Locus Genotypes[a]

Genotypes	Prenatal	Postnatal
Mo/Y, Mo^{dp}/Y, To/Y, Mo^{dp}/To, To/To	Lethal	—
Mo^{blo}/To, Mo^{br}/Y, Mo^{br}/To, Mo^{dp}/Mo^{blo}, To/Mo^{blo}, Mo^{dp}/Mo^{br}	Sublethal	Lethal
$To/+$	Sublethal	Sublethal
$Mo^{dp}/+$	Sublethal	Viable
Mo^{br}/Mo^{blo}, $Mo/+$, Mo^{vbr}/Y	Viable	Sublethal
$Mo^{blo}/+$, Mo^{blo}/Mo^{blo},[b] Mo^{blo}/Y,[b] $Mo^{br}/+$, $Mo^{vbr}/+$	Viable	Viable

[a]Much of this table is based on information from Grahn et al. (1969b).
[b]According to L. Russell (1960) these genotypes have reduced viability.

clones (Mintz, 1971a), Mo^{blo} causes no lightening of the skin pigment in the ears and tail, and therefore its hair-lightening effect could be attributed to some effect on hair structure (Lyon, 1970); (3) Mo^{vbr} is known to affect the structure of the hair and yet the banding in $Mo^{vbr}/+$ heterozyotes is broader than in tabby (*Ta*/+) heterozygotes (Cattanach et al., 1969): (4) the variegated phenotypes of $Mo^{dp}/+$ females appear to resemble those produced by melanoblast phenoclones and yet the hairs of these mice are structurally abnormal; and (5) in tortoiseshell females (*To*/+) both melanoblast and hair follicle clonal patterns are clearly visible (Mintz, 1971a).

I. Etiology of the Mottled Syndrome

It is apparent from the above that the alleles at the *Mo*-locus not only affect pigmentation and hair texture but are associated with a number of other more serious conditions including neurological disturbances, aortic aneurysm, and skeletal abnormalities. To account for these seemingly unrelated effects D. Hunt (1974a) has suggested that the mottled syndrome is caused by a primary defect in copper transport. His arguments include the following:

1. The mottled syndrome bears a striking similarity to an inherited progressive human brain disease known as Menkes' kinky hair syndrome, a disease which also is X-linked (Menkes et al., 1962; Danks et al., 1972a,b). Besides retarded mental development and kinky hair, this disease is characterized by abnormal white hair, changes in the elastic fibers in the arterial walls, scorbutic bone changes, and hypothermia. A defect in the intestinal absorption of copper has been observed in babies with this syndrome and it is believed that this is responsible for the disease (see Goka et al., 1976). Indeed, the neurological effects and blood vessel aneurysms which occur in patients with this disorder resemble those found in animals experimentally deprived of copper (Carnes, 1968; Dankes et al., 1972a,b).
2. The observation that many mottled locus mutants exhibit aneurysms of the aorta and its branches indicates that their connective tissue is not normal and this has been confirmed by Rowe and his associates (1974). Thus, mice bearing the *To*, Mo^{blo}, and Mo^{vbr} alleles all display a defect in the cross-linking of both collagen and elastin. The defect appears to be localized to the step at which lysine residues are converted to aldehydes and copper is the cofactor for lysyl oxidase the enzyme required for this aldehyde-generating step.[16]
3. The skeletal abnormalities displayed by some of these X-linked coat-color mutants can also be explained from the defective synthesis of bone collagen.
4. The hair abnormalities, including the curliness of the whiskers, also can

Plate 1. (A) Lethal yellow (A^y/a); (B) viable yellow (A^{vy}/a); (C) agouti (A/A); (D) black-and-tan (a^t/a); (E) nonagouti (a/a); (F) nonagouti brown ($a/a;b/b$); (G) nonagouti sombre ($a/a;E^{so}/E^{so}$); (H) nonagouti light ($a/a;B^{lt}/B^{lt}$). (A and B) courtesy of George C. McKay, Jr., (C–F) courtesy of Earl L. Green, and (G and H) courtesy of Timothy W. Poole.

Plate 2. (A) Albino (c/c); (B) agouti chinchilla (c^{ch}/c^{ch}); (C) lethal yellow platinum ($A^y/a;c^p/c^p$) (D) nonagouti himalayan ($a/a;c^h/c^h$); (E) nonagouti brown dilute piebald spotting ($a/a;b/b;d/d;s/s$); (F) nonagouti pink-eyed dilution ($a/a;p/p$); (G) agouti mocha (mh/mh); (H) nonagouti pallid ($a/a;pa/pa$). (A, B, D, E, and F) courtesy of Earl L. Green and (G and H) courtesy of George C. McKay, Jr.

Plate 3. (A) Nonagouti beige ($a/a;bg^J/bg^J$); (B) nonagouti silver ($a/a;si/si$); (C) nonagouti tortoiseshell ($a/a;To/+$); (D) nonagouti viable dominant spotting ($a/a;W^v/+$); (E) viable dominant spotting homozygote (W^v/W^v); (F) nonagouti steel ($a/a;Sl/+$); (G) varitint-waddler ($Va/+$); (H) nonagouti white ($a/a;Mi^{wh}/+$). (A, B, E, and F) courtesy of George C. McKay, Jr. and (C, D, G, and H) courtesy of Earl L. Green.

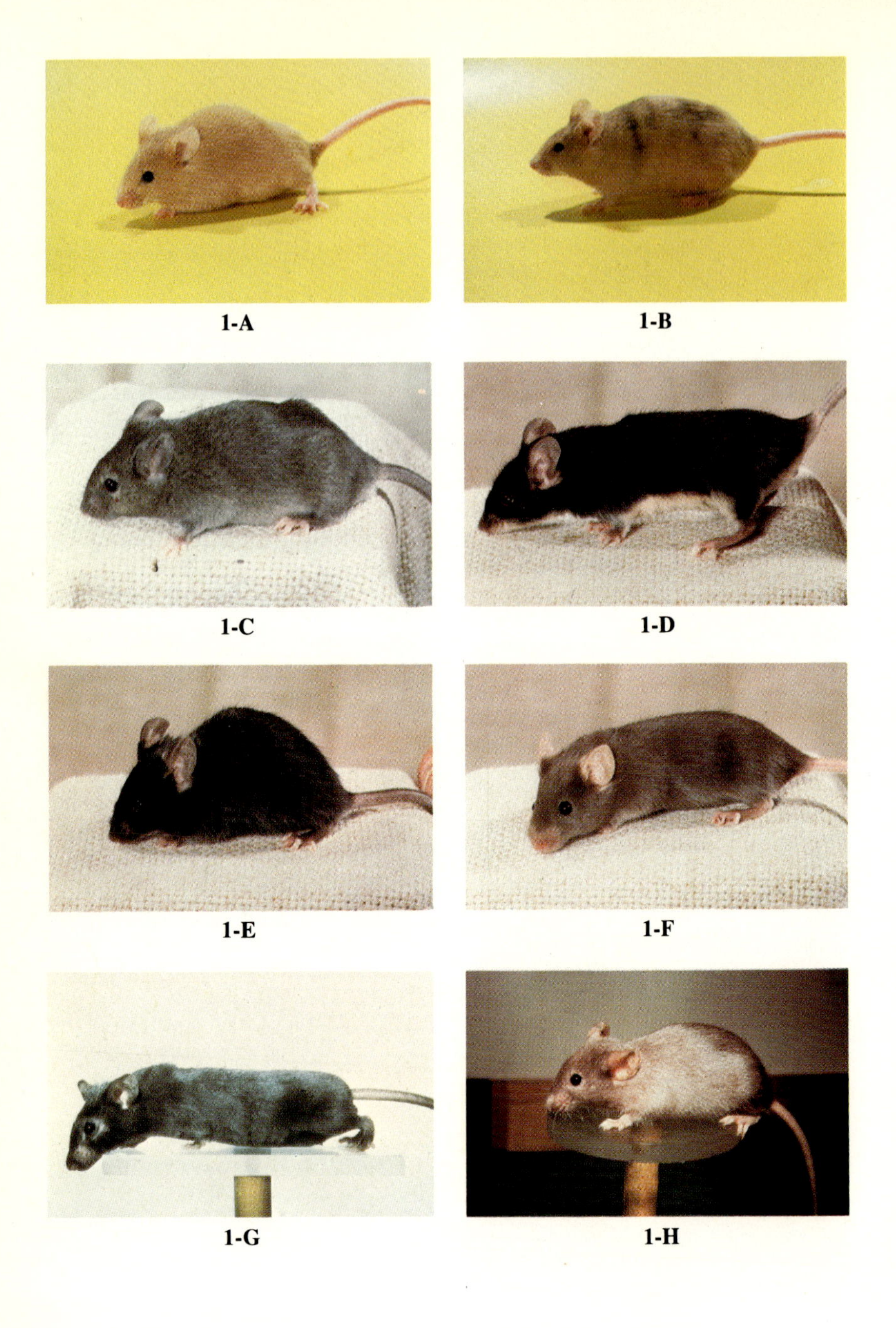
1-A 1-B 1-C 1-D 1-E 1-F 1-G 1-H

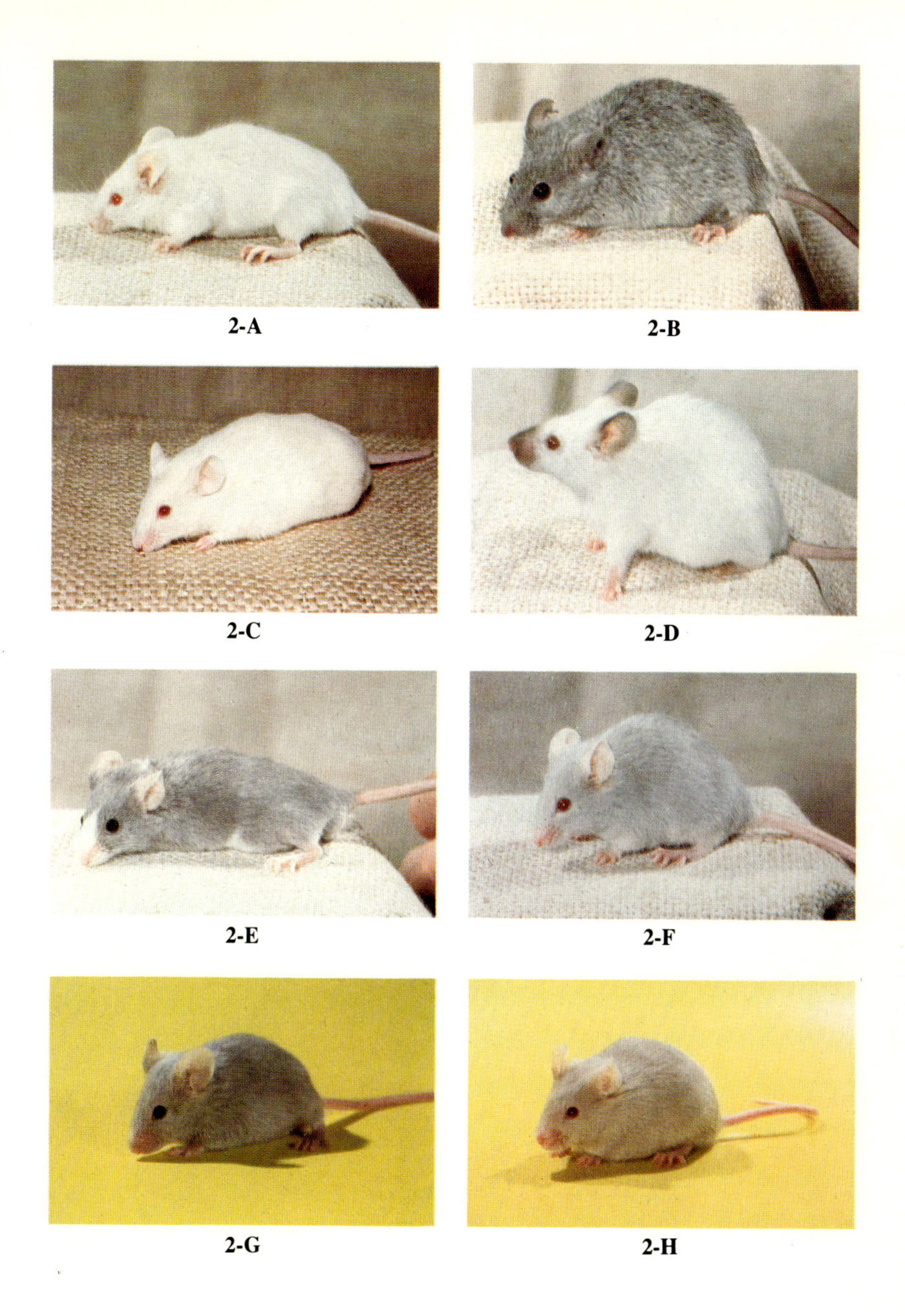
2-A 2-B
2-C 2-D
2-E 2-F
2-G 2-H

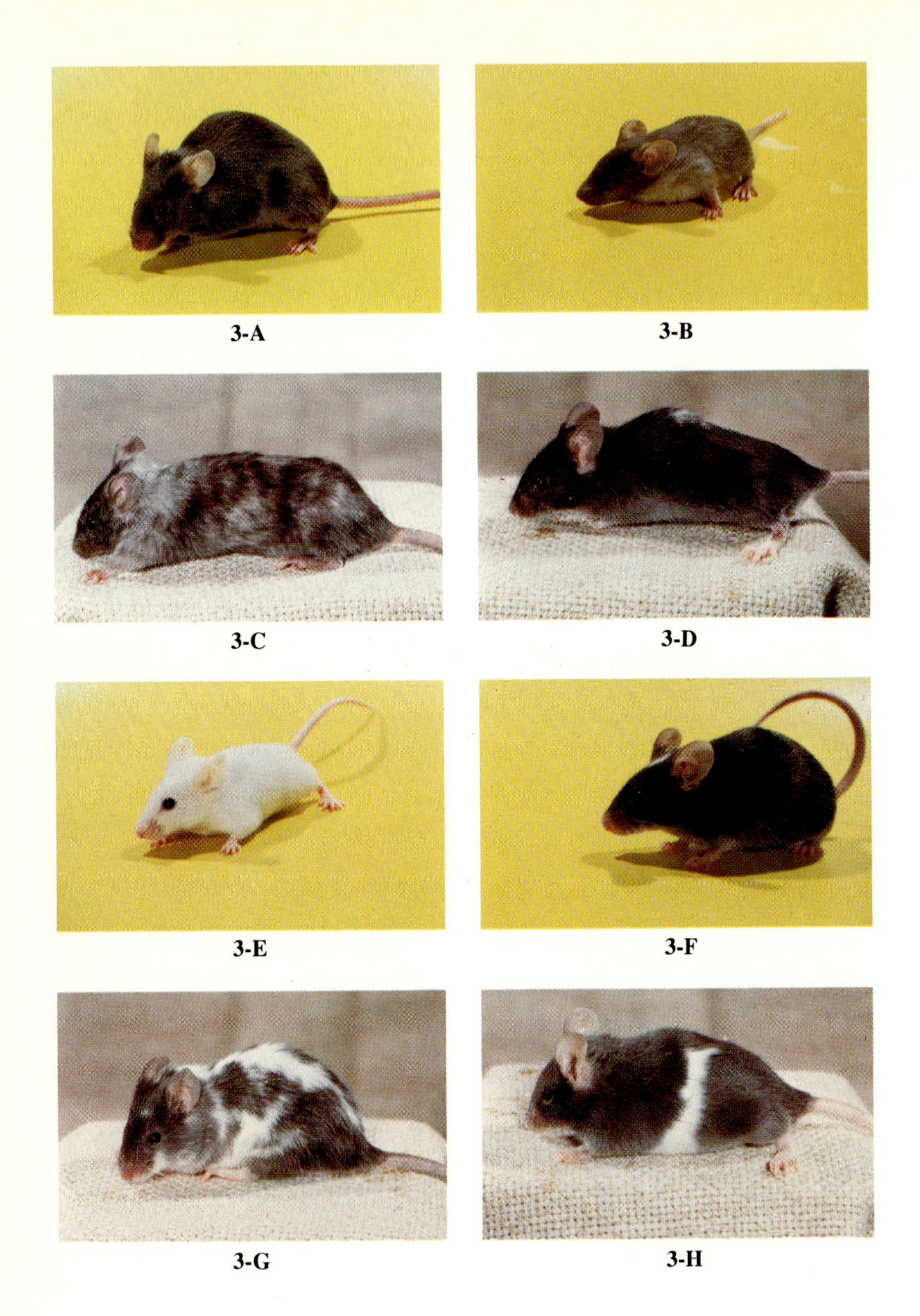

3-A

3-B

3-C

3-D

3-E

3-F

3-G

3-H

be related to a copper deficiency since it is known that the formation of disulfide bonds in keratin is copper dependent.

5. The neurological symptoms of brindled, viable-brindled, and blotchy males are associated with a severe deficiency in central norepinephrine levels and this has been traced to the reduced *in vivo* conversion of dopamine to norepinephrine by the copper-containing enzyme with a cofactor requirement for cupric ion, dopamine-β-hydroxylase (D. Hunt, 1974a,b).[17]
6. There is a copper deficiency in the liver and brain of Mo^{br} mice, a deficiency which appears not to be due to the uptake of the metal (its accumulation in the intestinal wall indicates normal uptake from the gut lumen) but to its transport (D. Hunt, 1974a).[18]
7. The virtual absence of hair pigment in males hemizygous for *Mo*-locus mutations is readily explained by a direct effect of a copper deficiency on the activity of tyrosinase. Indeed, the fact that the tips of the hairs of these males are sometimes pigmented could be due to an initial but rapidly depleted copper store in the hair bulb (D. Hunt, 1974a).

To further support the thesis of copper involvement D. Hunt and Skinner (1976) have recently reported that the activity of the copper-dependent enzymes cytochrome *c* oxidase and superoxidase dismutase are reduced in $Mo^{br}/-$ brain tissue. They have also found that the lethality and to some extent the pigment deficiency of brindled males can be overcome by administering copper chloride (see note 18).

On the other hand Rowe and his associates (1977) note that there are a number of inconsistencies with the assumption that the primary effect of all *Mo*-alleles is on the transport of copper. These include the fact that:

1. The activity of lysyl oxidase in blotchy males is 25% and in $Mo^{blo}/+$ females 50% of normal and such a gene dosage relationship indicates that the enzyme rather than its cofactor is affected.
2. Although brindled males have been shown to have reduced copper levels and reduced dopamine-β-hydroxylase activity, lysyl oxidase is not reduced sufficiently to cause a cross-linking abnormality in these mice. Thus the Mo^{br} allele does not appear to affect all copper-dependent enzymes to the same extent. Moreover, if the *Mo*-locus was concerned only with determining copper levels, then one would expect Mo^{blo}/Mo^{br} females to be affected similarly as blotchy and brindled males and they are not. While such heterozygotes have a reduced viability they do not display vascular lesions and are often fertile (Grahn et al., 1969a,b, 1971).
3. Inasmuch as the mottled coat of $Mo/+$ heterozygotes results from areas in which affected cells lack and normal cells have pigment, any defect in copper transport would have to occur at the cellular level.

Because of these inconsistencies, Rowe et al. (1977) suggest that "a possible explanation for the multiple defects of the mottled alleles are overlapping deletions with a common defect in pigmentation." Thus, according to this scheme there are genes adjacent to the "mottled locus" which code for an enzyme necessary for copper transport, for lysyl oxidase and for dopamine-β-hydroxylase, and these are affected to varying degrees.

J. Mosaic (*Ms*)

This X-linked mutation, which we have already referred to (Section C, 4, notes 7 and 18), has been described and studied by Krzanowska (1966, 1968), Radochońska (1970), Wabik (1971), and Styrna (1976, 1977a,b). It arose spontaneously in an outbred strain maintained in the Department of Animal Genetics, Jagellonian University, in Krakow.

1. Pigmentation and Pelage

Males hemizygous for this mutation display a notably lighter coat. According to Radochońska, agouti males carrying the mutation are a light shade of silvery-grey, whereas black mosaic males are brownish black with hair which is cream-colored at the base and darker at the tip. The head and the midline of the trunk of hemizygous males are darker than other parts of the body. When combined with pink-eyed dilution (*p*/*p*), agouti mosaic males (*A*/*A*;*p*/*p*;*Ms*/*Y*) are ivory and nonagouti mosaic males (*a*/*a*;*p*/*p*;*Ms*/*Y*) are completely white (Krzanowska, 1966). Heterozygous females frequently display pigmented patches arranged in an irregular pattern of transverse bars. This effect is most conspicuous in *p*/*p* females when white and wild type stripes occur (Krzanowska, 1966).

The coat of males and females bearing the mutation also is silkier than normal and grows thinner. The specific changes in the hairs have been described by Radochońska. One interesting feature is that the zigzag hairs of hemizygous males usually have two instead of three constrictions. The structure of the awls also is abnormal. The vibrissae of *Ms* mice are curly. The greatest degree of curliness occurs in hemizygous males and homozygous females. In heterozygous females there appears to be a significant correlation between the lightness of the coat and the curliness of the whiskers. In the more darkly pigmented females the vibrissae are curly at birth but gradually straighten and often appear nearly normal by the third week of life. Very light females, however, display twisted vibrissae even when mature. The lightness of the coat of heterozygous females also is associated with reduced body weight and lowered viability (Radochońska, 1970).

2. Other Characteristics

Hemizygous *Ms* males grow normally up to about 10 days of age but then suddenly stop growing, develop a progressive paresis of the hind limbs

(Radochońska, 1970), and die when about 12–20 days old.[19] However, a few males, about 4%, survive this crisis, resume growing, and eventually are of near normal size. Some have been fertile and have sired litters containing *Ms*/*Ms* females. Such females are identical phenotypically with hemizygous males and also die when about 15 days old (Krzanowska, 1968).

Although this mutation undoubtedly belongs to the *Mo* locus series of alleles (and accordingly might be noted as Mo^{ms}), to my knowledge no tests have been undertaken to determine its relationship to this locus. From its description it appears most similar to brindled and, indeed, could represent a remutation to this allele.

III. Yellow Mottling (*Ym*) and Pewter (*Pew*)

The only other X-linked coat-color determinants are yellow mottling (*Ym*) and pewter (*Pew*). *Ym* arose in an experiment at Oak Ridge in which (101 × C3H)F_1 males received a partial body exposure of 500 R acute X-ray and, 24 hours later, an additional 100 R. These males were mated to females homozygous for seven recessives and scored for mutations. An F_1 male from this cross produced a single *Ym*/+ daughter out of a large number of progeny. Hence it appears that the mutation was spontaneous (Hunsicker, personal communication).

Ym/+ females differ from females of the *Mo*-series in that their mottling is yellowish. Moreover, the vibrissae are not curly (Hunsicker, 1968). *Ym*/*Y* is lethal. The locus is closely linked to blotchy (Mo^{blo}) but how close remains to be determined (Hunsicker, personal communication).

Pewter (provisional symbol *Pew*), which appeared in the CBA/J strain, has only recently been described (Fox and Eicher, 1978). Affected males are light grey and affected females are slightly mottled. Both sexes are fertile. Its relationship to *Mo* and *Ym* is currently under investigation.

Notes

[1]Grüneberg (1966b,c, 1967a,b) has raised a number of objections to this hypothesis and his arguments, as well as Lyon's (1968) retorts, are concisely summarized and discussed by Eicher (1970b).

[2]It should be noted, however, that Cattanach et al. (1972) observed that there was a tendency for allophenics to display "phenotypes that were at the extremes or possibly beyond the range" of those found in the equivalent heterozygotes. Indeed, a detailed analysis of the variability displayed by allophenics and X-inactivated mosaics by Falconer and Avery (1978) has confirmed this observation. These investigators found that although allophenics and mosaics are alike in appearance, there is a significant difference in the percentage of marker cells likely to be found in their coats. Whereas mosaics display a binomial (bell-shaped) distribution

when the percentage of animals is plotted against the percentage of marker cells in the coat, allophenics show a flat distribution, with all percentages of marker cells being equally likely, including pure classes. Falconer and Avery believe the greater variability in allophenic patterns is caused by two sampling events which occur when cellular heterogeneity is present in allophenics but not in mosaics. These are (1) the separation of the inner-cell mass from the trophectoderm and (2) the division of the inner cell mass into primary ectoderm and primary endoderm. This paper also includes an ingenious mathematical analysis of the variation in allophenic pigment patterns, a treatment which although based on certain assumptions, some of which remain to be confirmed, nevertheless leads to results in accord with Mintz's hypothesis of 34 progenitor cells. (see *J. Embryol. Exp. Morphol.* **43:**195–219, 1978).

[3]At least one other X-linked deviant has been reported which also could represent a mutation to *Mo*. It occurred at Oak Ridge and is known as 26*K* (Welshons and L. Russell, 1959; L. Russell, 1960).

[4]When sections of skin from 7-day-old heterozygous females were examined microscopically they revealed that all the hair follicles possessed either pigment cells or amelanotic melanocytes. The follicles which contained pigment displayed a range of variation from completely normal amounts of melanin, to almost none. This histological picture therefore paralleled the gross appearance of the coat.

[5]Tabby (*Ta*) is an X-linked mutation which when homozygous or hemizygous eliminates both guard hairs and zigzags from the coat. The fibers which are present have been regarded as awls by Falconer (1953) though many of them are atypical (Grüneberg, 1969). *Ta*/*Ta* females and *Ta*/*Y* males also possess a bald patch behind each ear, bald tail with a few kinks near the tip, reduced aperture of the eyelids, a respiratory disorder, and, in combination with agouti, a modified agouti pattern. *Ta*/+;*A*/– females display transverse dark stripes. These dark stripes, which according to Mintz (1970, 1971a) are the result of hair follicle phenoclones, are deficient in zigzags and are composed mainly of normal and abnormal awls which lack the yellow band. The mutation was described originally by Falconer (1953) who demonstrated that it was situated about 4 cM from both *Mo* and Mo^{br}. Tabby has been described in detail by Grüneberg (1966b,c, 1969, 1971a,b) and his colleagues (1972), by Kindred (1967) and by McLaren et al. (1973).

[6]D. Hunt and D. Johnson (1972a) have shown that the neurological involvement of Mo^{br} is associated with significant differences in brain tyrosine and tryptophan levels which cannot be accounted for by the observed changes in the systems of synthesis and degradation of these amino acids. They suggest that the accumulation of tryptophan may be a contributory factor in the lethality of Mo^{br} since elevated levels of brain tyrosine but not of tryptophan occur in viable-brindled (Mo^{vbr}) males (D. Hunt and D. Johnson, 1972). These investigators also have provided evidence that the neurological impairment of both Mo^{br} and Mo^{vbr} males is associated with decreased central norepinephrine levels (D. Hunt and D. Johnson, 1972a,b).

[7]Falconer and Isaacson (1972) also observed that brindled males which originated from mothers of the light line were not as viable as those from mothers of the dark line and suggest that autosomal modifiers are the cause of this difference. They state that "these autosomal differences might be in the viability of the Mo^{br} males themselves, or in a maternal effect." Styrna (1975) too found that male mutants from the *Ms* dark line were clearly more viable than those from the light line. This difference

was, however, obscured when $Ms/+$ mothers were replaced by $+/+$ foster mothers from either the light or dark line. It thus appears, at least in this latter situation, that a maternal effect(s) does play an important role in the survival of mutant males (see note 18).

[8]There is no significant correlation between the color of the mother's coat and that of the offspring, and attempts to select for color intensity have failed (R.J.S. Phillips, 1961b).

[9]This effect does *not* occur in $Mo^{br}/+$ (Cattanach et al., 1972) or in $Mo^{vbr}/+$ heterozygotes (Cattanach, personal communication) and, as already noted, the major influence of Mo^{br} and Mo^{vbr} seems to be in the hair follicle.

[10]Although the basis for the variegated pattern produced by e^p remains to be determined there is some evidence which suggests that, although autosomal, it may be due to a mechanism similar to that responsible for X-linked variegation. Thus not only does white spotting have a similar influence on the e^p/e^p genotype as it has on *dappled* and on the X-linked tortoiseshell pattern of the domestic cat, but in both situations there is considerable variation in the amount of normal and mutant hair (for further information, including possible mechanisms, see Searle, 1968a).

[11]Phillips notes that death at this stage of gestation is unusual in the mouse as most prenatal deaths occur prior to the fifteenth day of development.

[12]In 1972 Eicher reported a new sex-linked mutation affecting coat color and hair which she called "silver-grey." This mutation resembles blotchy and could be a remutation to this allele.

[13]According to Grahn et al. (1969b), Mo^{blo}/Mo^{br} is sublethal postnatally and Mo^{blo}/Mo^{dp} is essentially a postnatal lethal.

[14]About half of To/Y males die before implantation (Grahn et al., 1969b).

[15]One Mo^{blo}/To female survived for 104 days. A faint mosaic pattern was visible in the light grey coat (Grahn et al., 1969b, 1971).

[16]Rowe and his colleagues (1977) recently reported that lysyl oxidase activity is markedly reduced in Mo^{vbr}/Y and Mo^{blo}/Y skin and that its activity in $Mo^{blo}/+$ skin is intermediate between Mo^{blo}/Y and normal (C3H) skin. On the other hand, the lysyl oxidase activity in Mo^{br}/Y skin is only moderately reduced from that of C3H skin and the activity in $Mo^{br}/+$ skin is normal. These findings correlate with the fact that whereas viable-brindled and blotchy males have a connective tissue defect which results in weakness of skin and of blood vessels, brindled males succumb to a neurological disorder rather than to connective tissue weakness.

[17]In contrast, *in vitro* assays of dopamine-β-hydroxylase in the presence of exogenous copper resulted in increased activity, suggesting that, *in vivo*, there is an increased synthesis of enzyme protein but a reduced availability of copper cofactor (D. Hunt, 1974b).

[18]The main defect in mosaic (*Ms*) (Section J) males may likewise be related to a deficient absorption of copper from the intestines. Styrna (1976, 1977b) found a higher than normal content of copper in the intestine and in the feces, and a lowered content in the brain and liver of Ms/Y males. He also found the mutants to possess a lower level of ceruloplasmin, which is one of the main cuproproteins in the body, and to have reduced levels of *o*-diphenol oxidase, an enzyme involved in the synthesis of

melanin which requires the presence of copper ions for its activity. Styrna likewise found higher amounts of cysteine in the brains and liver of mosaic males. Furthermore, he has reported that the life of these males can be prolonged if beginning at 4 days of age they are inoculated with an aqueous solution of $CuCl_2$. Following this treatment some darkening of the coat occurs too (Styrna, 1977a). Finally it should be noted that Styrna (1977b) found that the mammary glands of *Ms*/+ females of the light line (see Section C, 4) contained less copper than the mammary glands of *Ms*/+ females of the dark line and of +/+ females. This is especially interesting since it could explain the maternal effect noted previously (see note 7). Thus the milk of *Ms*/+ mothers of the light line may supply less copper than *Ms*/+ females of the dark line.

[19]Because these symptoms, namely, light pigmentation of hair and neurologic disorders (characteristic bending of extremities on lifting *Ms*/*Y* males by the tail), suggested that they might result from changes in the metabolism of aromatic amino acids which are precursors of melanins and of cerebral amines necessary for the normal function of the central nervous system, Klein and his colleagues (A. Klein and Sitarz, 1971; A. Klein and Styrna, 1971) analyzed the occurrence of free amino acids in the blood and urine of these males. They found that the blood of mutants contained no tryptophan, and smaller amounts of tyrosine and phenylalanine than normals. In the urine no differences in the composition of aromatic amino acids were noted but (particularly in adult *Ms* males) the levels of glycine were nearly twice as high as in normal males. Nevertheless, since there are a number of nonlethal genes which exert an influence on the content and levels of amino acids excreted in the urine (Hrubant, 1963), Styrna (1977b) believes these observed changes are probably secondary effects of the mutation.

Chapter 9

White Spotting: Piebald, Lethal Spotting, and Belted

I. Introduction

The preceding chapters have been concerned almost entirely with the genetic control of melanin synthesis in melanocytes which are distributed uniformly, at least insofar as the coat is concerned. Attention now will be focused on genetic factors which can eliminate melanocytes either partially or completely.

Evidence already has been presented (see Chapter 3, Section II, A) which indicates that white spotting results from the complete absence of melanocytes from certain areas of the coat. The cause of these unpigmented areas in terms of Mintz's allophenic model has also been considered (see Chapter 7, Section VII). Thus, according to this model, all-white spotting results from the "preprogrammed" death of one or more melanoblast clones. There are, however, a number of other hypotheses which attempt to explain piebaldism. Although these will be discussed in the sections to which they most appropriately relate, the hypothesis proposed by Schaible (1969, 1972) deserves to be considered at this time since, like Mintz's, it proposes that all spotting patterns have a similar etiology.

Schaible contends that the complete pigmentation of mammals and birds develops from the expansion and merger of clones of pigment cells from the same centers that appear as pigmented spots against a white background in white spotted mutants. Employing different mouse mutations which cause white spotting but which are associated with nearly normal pigment coverage in certain parts of the body, such as belted (*bt*/*bt*) and piebald (*s*/*s*), either alone or in combination with other spotting factors, e.g., $Mi^{wh}/+$ (white) or $W^a/+$ (Ames dominant spotting), Schaible selected for progressive restriction of these pigmented areas until the center of each was located. Employing this method he was able to locate 14 such centers (see, however, Lyon and Glenister, 1971). He contends therefore that the integument of the mouse is pigmented by the proliferation of 14 primordial melanoblasts which migrate from the neural crest and which locate themselves singly in bilateral centers in the nasal, temporal, aural, costal, lumbar, and sacral areas and in medial centers in the coronal and caudal areas (see Figure 9-1).[1] In fully pigmented mice these 14 clones proliferate and expand until they merge. However, in white spotted animals one or more of these clones fails to proliferate either entirely or completely. According to this model white spotted areas not only lack melanocytes but at no time are they populated with melanoblasts. Nevertheless, pigment cells can migrate into these areas secondarily from neighboring pigmented areas during postnatal growth.

While the method employed by Schaible to form this model was indeed an ingenious one, it does not seem to make as much biological sense as Mintz's. Unlike her's, which is completely in accord with Rawles' (1947) embryological observations, many of his clonal centers have absolutely no physical connection with the neural crest (see Mintz, 1974). Moreover, contrary to his hypothesis there is evidence (see note 11) that melanoblasts reach all areas of the integument in at least some white spotted genotypes (Mayer, 1967a).

II. Piebald Alleles

A. Piebald (*s*)

Piebald spotting (*s*) and its associated "*k*" complex has been covered in some detail by Grüneberg (1952) and much of what follows is based on his excellent account.

1. Origin

Although few mutations have been described at this locus on chromosome 14, it is very likely that some piebalds in existing stocks are of independent origin. This possibility stems from the fact that piebald spotting is carried in many fanciers' stocks, and has evidently occurred both in European laboratory mice and in stocks of Japanese waltzing mice. It has been isolated from wild populations in Turkey (Keeler, 1933), Germany (K. Zimmerman, 1941), and the United States (Littleford, 1946).

2. Spotting Patterns: External and Internal

The most common allele at this locus, *s*, when homozygous[2] usually produces distinct white spots, i.e., unpigmented and pigmented areas are

Figure 9-1. A diagram of the 14 primary areas proposed by Schaible (see text). Each area has been restricted to mark its boundaries. Based on R. H. Schaible (1969).

clearly defined by sharp borders, without any intermingling of white and pigmented hairs (Plate 2–E). In general, the ventrum displays more white than the dorsum and according to Grüneberg "'centers of depigmentation' are the feet, the tail, particularly its distal part, an area round the umbilicus on the belly, a blaze between the eyes, and the tip of the nose." White shoulder spots may also occur and unite to form a "collar." Lumbar spots likewise often join to form a "belt." As a consequence of modifying factors, known as the "*k*" complex (see Section D), the extent of all these centers is highly variable, and as they increase in size, they merge in numerous ways. The last areas to lose their pigmentation are the region of the eyes, cheeks, ears, and haunches. Thus *s*/*s* may produce entirely or almost entirely unpigmented phenotypes, or it may produce only a few white spots on the belly, feet, and forehead (Dunn, 1920b).

The amount of "internal" spotting may likewise vary and reflects to some extent the amount of external spotting. Thus whereas a predominantly white *a/a;s/s* stock described by Mayer (1965) and two almost completely white *W/+;s/s* genotypes examined by Markert and Silvers (1956) did not possess any melanocytes in the choroid[3] and harderian gland, almost normal numbers of melanocytes, with only an occasional white spot, occurred in the choroid, and normal numbers of melanocytes were observed in the harderian gland, of predominantly pigmented *s/s*, *a/a;s/s*, A^y*/a;s/s*, *a/a;b/b;s/s*, and A^y*/a;b/b;s/s* animals (Markert and Silvers, 1956) (see Table 10-1). Deol (1971) reported that in his heavily spotted *s/s* stock there was a variable degree of spotting of the choroid even though, on average, the unpigmented and pigmented areas were about equal in extent.[4] Moreover, he also observed a great deal of variability in the amount of pigment in the harderian glands of these animals. In the majority of cases this gland seemed to be totally unpigmented but in some the number of melanocytes was severely reduced with the remaining cells unevenly distributed and considerably smaller and less dendritic than normal. Deol also examined 20 *s/s* ears and observed that one was "lacking in pigment in the region of the lateral crista, another two in the region of the posterior crista, and one in the regions of all three cristae." (Deol's observations on the pigmentation of *s/s* mice are summarized in Tables 12-1 and 12-2.)

3. Other Effects

In addition to its effects on pigmentation piebald spotting is also associated with megacolon. Thus Bielschowsky and Schofield (1960, 1962) reported that about 10% of their *s/s* animals had this condition. Since this same affliction is invariably associated with piebald-lethal (s^l) it is considered below.

B. Piebald-Lethal (s^l)

1. Origin and Description

This allele has been described by Lane (1962, 1966). It was observed first in the F_2 of a cross between a C3H/HeJ and C57BL/6J. It behaves as completely recessive to wild type and nearly recessive to *s*. Homozygotes are black-eyed whites with occasional small patches of pigmented hairs on the head and/or on the rump (Figure 9-2). Some s^l homozygotes die as early as 1 or 2 days after birth while others survive for more than a year. The usual age at death is about 15 days but regardless of how long they live, they all eventually suffer from and usually succumb to megacolon.[5]

2. Megacolon

According to Lane (1966) megacolon can readily be distinguished in weaning-age or older animals at autopsy by a markedly distended colon that

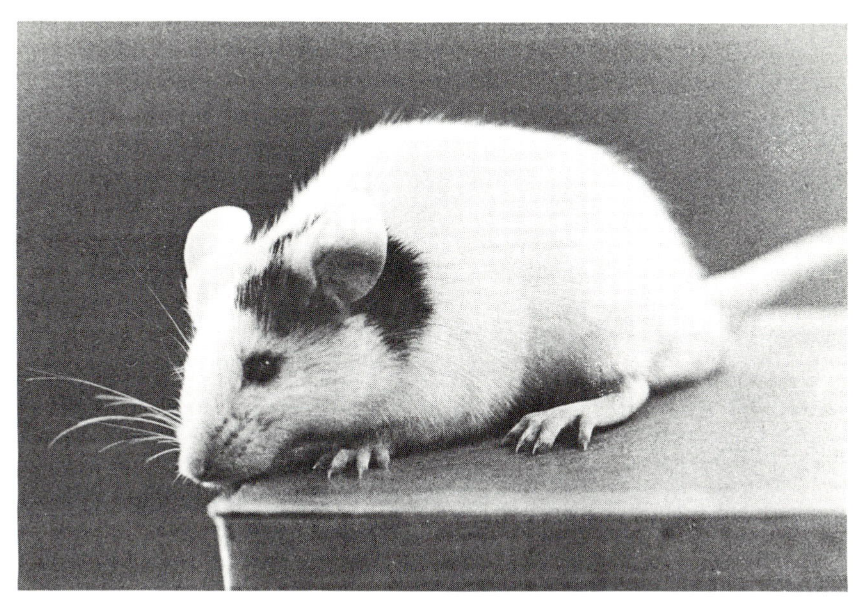

Figure 9-2. A typical piebald lethal (s^l/s^l) mouse. Courtesy of George C. McKay, Jr. and Priscilla W. Lane.

is filled with hard fecal matter, usually extending from the caecum to varying distances above the rectum. Very young (2- to 15-day-old) mice may or may not show obvious signs of the condition but when they do "their colons are distended and full of soft gelatinous fecal material in which the beaded appearance of early pellet formation is not present as it is in normal mice of this age."

Since Bielschowsky and Schofield (1960, 1962) found that megacolon in *s*/*s* mice was associated with a marked reduction of myenteric ganglion cells, Lane determined whether these cells also were deficient in the colons of young s^l homozygotes. She found that whereas in 2- to 12-day-old normal mice groups of myenteric ganglion cells of varying size were present approximately every 50–100 μm from the rectum to the caecum, in s^l/s^l genotypes these cells were lacking at the very distal end of the colon but increased gradually and were present in normal numbers at the proximal end.

3. Inner Ear Abnormalities

Deol (1967, 1970b) has reported that the inner ear of s^l homozygotes lacks melanocytes and displays severe abnormalities; the cochlear duct is never normal.

C. Other Alleles

Besides *s* and s^l one or two other possible piebald alleles have been described. One deviant mentioned by Keeler (1935) was designated "Berlin-blaze." It was exemplified by very little white and was believed to represent the activity of a different allele with slighter effects. However, it is just as likely that this condition was due to the ordinary *s* allele's behavior on a unique genetic background (Grüneberg, 1952).

Another allele, also called s^l, was described by Pullig (1949). It differed from the usual *s* allele in that it produced a spotting pattern which was like "a band around the ears extending in varying degrees down the ventral side and looking much like a white scarf tied around the head with free ends of varying lengths." However, as pointed out by Grüneberg, since this type appeared as a mutation in an experiment in which *s* also was present, and since *s* and its characteristic pattern were eliminated by "rigid selection," it is not easy to determine if the new spotting pattern was wholly a property of s^l, or whether it was to some extent conditioned by the selected genetic background.[6]

D. The "*k*" Complex

As already noted stocks of *s*/*s* mice may differ widely with regard to the amount of white areas in their coats. Thus by selection and inbreeding it is possible to establish reasonably stable lines of piebald animals which are either essentially all white with very few pigmented areas (hereafter referred to as the "all-white" line), or, have very little spotting, amounting to 10% or less of the coat (Dunn and Charles, 1937). Lines selected for intermediate amounts of spotting are much more variable. This situation appears to reflect the fact that whereas most of the variability in the all-white and predominantly pigmented *s*/*s* stocks is due to nongenetic causes, *both* genetic and nongenetic factors influence the variability in the intermediately spotted stocks.

To determine whether these different lines had different piebald alleles or possessed the same "main gene," *s*, with different sets of "modifiers," Dunn and Charles raised an F_2 population after crossing an all-white line with a dark line.[7] It follows that if distinct *s* alleles were responsible for the different amount of spotting in the two parental stocks, this F_2 population should display a trimodal distribution and the extreme parental types should be recovered. In actual fact, however, these types were not present as the F_2, like the F_1, displayed intermediate amounts of spotting with a *unimodal* distribution, though it was more variable than the F_1. It thus appears that the all-white and mostly pigmented piebald lines did not possess different "*s*" genes but differed by a number of "modifying" genes which recombined freely with each other and had cumulative effects. This conclusion was substantiated further by the results of backcrossing F_1 animals back to the all-white line. Indeed, a rough estimate from this backcross indicated that at least three modifying genes were involved (Grüneberg, 1952).

To study the nature of these "modifiers," Dunn and Charles outcrossed their all-white *s/s* mice with mice which putatively did not possess any spotting genes at all. Since the F_1s from such crosses were either fully pigmented or had very small spots on the belly or forehead, it seemed that the all-white type behaved very nearly as a complete recessive when *s* and all the "modifying" factors were heterozygous. When these F_1 animals were backcrossed to animals of the all-white stock two groups of spotted mice, corresponding to the genotypes *s*/+ and *s/s*, were evident. Although the *s/s* group possessed much more spotting than the *s*/+ group, nevertheless, animals of the latter group displayed much more dorsal white spotting than the F_1, presumably because some of the "modifiers" were really not modifiers in the strict sense but were spotting genes in their own right. Dunn and Charles assigned the name of "*k*" genes to these modifiers and, starting from this first backcross generation, repeatedly backcrossed the lighter *s*/+ animals to the all-white stock. In this way a population was established in which all of the "*k*" genes were homozygous in both *s*/+ and *s/s* animals, a situation which was evidently complete, or nearly so, from the fourth backcross onward.

To separate *s* from the "*k*" genes, *s*/+ animals from the second and later backcross generations were mated *inter se*. Such matings produced two kinds of phenotypes: those which were all-white, and presumably *s/s*, constituted one-quarter of the population while the remaining, presumably *s*/+ ($\frac{1}{2}$) and +/+ ($\frac{1}{4}$), genotypes were dark spotted animals. By individual tests the +/+ animals were identified; they ranged from 10 to 35% dorsal white, with a mean of 18.9%. On the other hand the amount of dorsal white spotting in the *s*/+ animals ranged from 15 to 60% with a mean of 32.3%. Since *s* had been eliminated in the +/+ animals of this population their spotting must have been caused by the "*k*" genes introduced from the all-white line. Moreover, the fact that the *s*/+ animals of this same population possessed, on the average, more spotting than the +/+ mice indicated that *s*, in the presence of homozygous "*k*" genes, is not completely recessive. On the background of these genes, the three genotypes +/+, *s*/+, and *s/s* displayed about 20, 35, and 100% dorsal white and, as Grüneberg points out, "the difference between 20 and 35 on the one hand, and between 20 and 100 on the other (actually a ratio 15:80) gives a rough measure for the degree of dominance of *s* when acting on a background of homozygous "*k*" genes."[8]

It seems evident that in the absence of "*k*" genes (or in the presence of their normal allelomorphs), *s* is almost completely recessive to + and in the absence of *s* the "*k*" genes too are almost completely recessive. On the other hand, these "*k*" genes act as semidominants in the presence of *s/s*. In other words "the dominance of *s* is influenced by the presence or absence of the '*k*' genes, and the dominance of the '*k*' genes is influenced by the presence or absence of *s*" (Grüneberg, 1952).

The "*k*" complex has been analyzed further by Dunn (1942) and it seems to consist of a large number of genes which, individually, have very small effects. At least one of these genes is dominant since when spotted mice from a "*k*" stock were crossed with DBAs, a strain devoid of spotting and putatively devoid of "*k*" genes, all F_1 animals were spotted either on the tail, belly, or both. The exact number of *dominant* "*k*" genes is not known, but the frequency of spotting in F_2 and backcross populations of this outcross indicated that there was more than one and probably two.

Dunn also established several sublines from the F_2 generation of this outcross which he subsequently inbred. Of these, two showed a type of spotting like the original "*k*" stock, though of a lesser degree, one showed tail spotting only, and one displayed no spotting. Nevertheless, this last stock proved to be different from the unspotted DBA parental strain in that when crossed with various spotted lines, the resulting F_1 exhibited much higher degrees of spotting than when the same spotted lines were crossed to DBA mice. It therefore seems evident that the extracted "self" line differed from a real "self" line by the possession of "*k*" genes which by themselves could not produce spotting but which, when combined with other spotting genes, could exert a noticeable effect. As emphasized by Grüneberg such "subthreshold" spotting genes are, in a manner of speaking, "specific modifiers" in that they are not phenotypically expressed in the absence of other genes. On the other hand, they seem to differ from "minor" spotting genes which do manifest themselves only by the fact that their influence is not as great.

E. The Pattern of Piebald Spotting

When mice of a given spotted (pied) strain are compared, one finds that certain areas are always unpigmented while others are invariably pigmented. The irregularities in the pattern of the spotting are therefore confined to those regions which may or may not be white. Moreover, when the frequency of pigmentation has been determined for different regions of the coat, it is possible to plot contour lines enclosing areas which display different frequencies of pigmentation. Such areas have been mapped out by Charles (1938) and the reader is referred to his paper as well as to Grüneberg's book for a detailed consideration of this aspect of piebald spotting.

F. Other Minor Spotting Genes

There are a number of so-called minor spotting genes which behave as "*k*" genes even though it is not known whether any of them were included in the studies of Dunn and Charles. All of these genes, with the possible exception of the one described by Pincus (1931), exert their influence in the absence of *s*/*s* and, indeed, in most cases their effect in the presence of *s*/*s* has not been investigated. Nor have the relationships of these determinants with each other been resolved.

One of these determinants occurs in the C57BR strain where it produces a small white ventral patch varying in size from total absence to about 10% of the ventral surface. In a few mice the unpigmented area sometimes extends up onto the side of the animal, and some animals exhibit some white on the tail (with or without ventral spotting). According to Murray and C. Green (1933) more than 50% of C57BR mice display some degree of spotting with the frequency and the size of the belly spots being somewhat greater in males than in females. The amount of spotting among littermates also seems to correlate better with the amount of spotting displayed by their sire than by their dam. The basis for these unexpected observations is not known.

Another minor spotting gene, expressed almost exclusively on the tail, occurs in the C57BL strain (Little, 1924; Grüneberg, 1936b). This factor has been analyzed to some extent by Grüneberg who found that it behaves as a simple recessive which segregates independently of agouti.[9]

Barrows (1939) investigated tail spotting in mice by selecting for an increased amount of it in a genetically heterogeneous stock which did not possess *s*. Although his study did not include any attempts to sort out the responsible factor or factors, he believes the increased spotting he observed was due to the cumulative action of several minor genes.

There are a number of minor spotting genes which appear to affect the head. One of these has been described by Pincus (1931) who assumed that it exerted its effects only in the presence of piebald spotting since when combined with *s*/*s* it makes most of the head and face white. However, as Grüneberg points out, his data are not inconsistent with the possibility that the gene in question (which was assigned the symbol *l*) may produce some spotting, though possibly very little, in the absence of *s*.

A gene producing a similar pattern has been described by Fisher and Mather (1936; see also Grüneberg, 1952) and assigned the symbol *te*. Grüneberg refers to this gene as "light head" and it is conceivably the same gene as the one described above. However, in this case there is no doubt that the gene is a spotting factor in its own right as *te*/*te* mice, in the absence of other spotting genes, have a few white hairs (blaze) on their foreheads. The effect of this determinant, both in the presence or absence of *s*, appears to be confined to the head. Moreover, unlike the factor described by Pincus, in segregating populations there appears to be a deficiency of "light head" animals, a deficiency which Fisher and Mather ascribe to reduced viability rather than to misclassification.

Little (1914, 1917c) also described a recessive gene which produced a small blaze of white hairs on the forehead, and occasionally a little ventral or tail spotting, in the absence of *s*. Sometimes this determinant, which was found in a wild population, had no phenotypic effect at all when homozygous and Little (1926) demonstrated that its expression was influenced by a number of modifying genes which were widely spread in various solid colored races. The relationship of this determinant to the one described by Pincus, and to "light head," is not known and it is possible that they all are genetically related if not identical.

Another, possibly different, gene which produces a head spot (referred to as "headdot") has been described and studied by Keeler and Goodale (Keeler, 1935; Keeler and Goodale, 1936; Goodale, 1937a,b, 1942, 1943, 1948). It also behaves, in general, as a recessive though some heterozygotes display a few whites hairs on the forehead, and some homozygotes fail to display any spotting at all. This spotting factor seems to differ from the others in that in animals displaying spotting of the lowest grade a white spot occurs on the forehead only, and this is the most common type. As the amount of head spotting increases, however, white appears successively on the tip of the tail, lower lip, belly, and feet. This character too is influenced by modifiers.[10]

G. Etiology of Piebald Spotting

Since the pigmentation of the coat results from the synthesis and deposition of melanin by neural crest-derived *pigment cells* in the *hair follicle*, the behavior of transplants which include these components is one means of assessing their relative roles in the etiology of spotting. Accordingly Mayer (1965, 1967a,b) investigated the basis of piebald spotting by combining very small sections of neural crest-containing neural tube with small pieces of embryonic skin of the same or different *s* genotype. The relative importance of these components in the development of pigment was judged by the patterns of pigmentation they produced when allowed to differentiate in the coelom of White Leghorn chick embryos. For these experiments Mayer employed a stock of *a/a;s/s* mice which displayed large areas of white spotting on the belly, sides, and back that were predictable in location from one litter to the next (Figure 9-3). The neural crest-containing pieces of neural tube always were taken from 9-day-old embryos at the level of the hind limb bud (or posterior to it), a region which was potentially pigmented in the *s/s* strain employed. The skin was derived from the ventrolateral flank region between the limb buds of embryos of varying ages, a region that was destined to be unpigmented in the piebald stock.

In his first series of experiments Mayer (1965) combined *s/s* neural tube with 11-day-old neural crest-free +/+ (C57BL/6J) skin. He also combined +/+ neural tube with either *s/s* or with +/+ neural crest-free skin. He found that whereas grafts composed of +/+ neural tube and *s/s* or +/+ skin always produced pigmented hairs, as well as melanocyte populations in the dermis of the graft and in the chick tissues surrounding the graft, grafts of *s/s* neural tube and +/+ skin produced only pigmented hairs. Mayer also noted that some tissues, e.g., choroid, harderian gland, leg musculature, were pig-

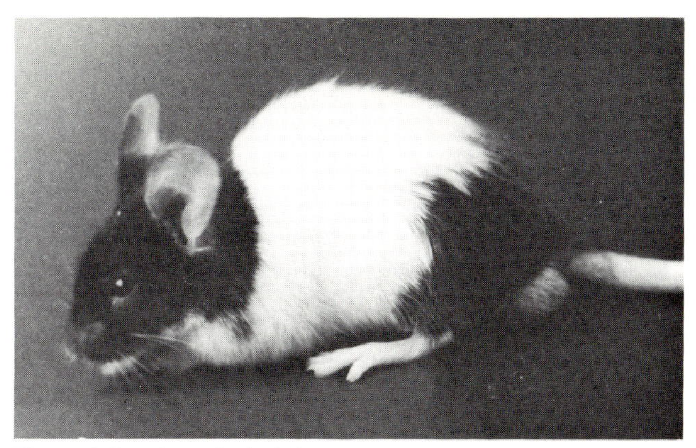

Figure 9-3. A typical piebald mouse (*a*/*a*;*s*/*s*) of the stock employed by Mayer in his studies. Courtesy of T.C. Mayer.

mented in the +/+ but not in the *s*/*s* mice he employed (see Table 9-1). On the basis of these observations he concluded that the primary action of *s* is in the melanoblast but that its effect can be discerned only in certain tissues, i.e., that different tissues, or different regions of the same tissue, such as skin, vary in their capacity to promote the differentiation of *s*/*s* melanoblasts.

To substantiate this hypothesis further Mayer initiated a second series of experiments (Mayer, 1967a) in which he combined *s*/*s* neural tube with neural crest-free *s*/*s* skin. He found that although, as in the case of *s*/*s* neural tube and +/+ skin, melanocytes never occurred in the intrafollicular dermis or in the host tissues surrounding the grafts, unlike the former combination, they also rarely occurred in hair follicles. Thus in contrast to the behavior of grafts composed of *s*/*s* neural crest and +/+ skin, composites which led to

Table 9-1. Estimation of the Pigment Content in Five Tissues of the C57BL, Heterozygous, and Piebald Mice Employed by Mayer[a,b]

Tissues	C57BL (+/+)	Heterozygote (*s*/+)	Piebald (*s*/*s*)
Choroid	+ + +	+ + +	0
Harderian gland	+ + +	+ + +	0
Membranous labyrinth	+ + +	+ + +	+ + +
Leg musculature	+ + +	+	0
Ankle skin	+ + +	+ +	+

[a]Based on Mayer (1965).
[b]+ + +, normal; + +, reduced; +, severely reduced; 0, absent.

pigmented hairs in 21/31 cases, combinations of *s*/*s* neural crest and *s*/*s* skin produced pigmented hairs in only 4/27 transplants.

According to Mayer there are two possible explanations[11] for these results *both* of which involve the participation of the skin in the etiology of piebald spotting. One possibility is that *s* acts not only in the neural crest but also in the skin. The other possibility, which Mayer views as more likely, is that the neural crest is affected by the piebald gene, and that other genetic factors, e.g., members of the "*k*" complex, operating within the piebald stock but absent in the +/+ (C57BL/6J) strain, are responsible for the deleterious influence which the *s*/*s* skin has on melanoblast differentiation.[12]

In a final series of experiments Mayer (1967b) combined 9-day-old *s*/*s* neural tube with potentially white areas of *s*/*s* skin of different ages and found that as the age of the skin increased so did the amount of pigment. Whereas 11-day-old *s*/*s* skin produced pigmented hairs in only 16% of the cases (and no pigment cells were observed in the intrafollicular epidermis), and 13-day-old skin produced pigmented hairs in 79% of the grafts (and again no pigment cells occurred in the epidermis), grafts recovered from implanting embryonic skin 16 days and older produced pigment regularly both in the hair follicles and in the skin. Mayer believes these experiments indicate that there is an inhibitor of melanoblast differentiation or survival in 11-day-old *s*/*s* skin and that this inhibitor gradually disappears.

Although Mayer's interpretation of his findings contrasts with Mintz's hypothesis which proposes that piebald spotting results solely from the demise and partial replacement of inviable melanoblast clones with viable cells, and does *not* involve any direct effect on the skin (see Chapter 7, Section VII), it should be borne in mind that his interpretation also is based upon certain assumptions which are not in accord with her hypothesis and therefore are debatable. For example, while Mayer assumes that his *s*/*s* neural tube grafts were comprised of a uniform population of (viable) melanoblasts because they were derived from a potentially pigmented area, it could be argued that cells of both viable and inviable clones of melanoblasts, and in some cases of only inviable clones, were included in these transplants (especially since the *s*/*s* mice employed were predominantly unpigmented). This possibility is based upon Mintz's contention that spotted phenotypes are a consequence of viable melanoblasts partially replacing inviable cells, a supposition which implies that there may be little or no relationship between the location of viable cells in a 9-day-old embryo and the location of pigmented areas in a mature, white spotted, mouse.[13] Thus, while Mayer's results indicate that piebald spotting cannot be attributed to a defect in the migration of *s*/*s* melanoblasts, and that many of these cells are inviable, they are not as convincing in demonstrating that the skin plays a *direct* role in determining their survival.

Further evidence that the primary effect of *s* and s^l is on the neural crest stems from the megacolon condition associated with these alleles. As already noted this abnormality appears to result from a marked reduction of the myenteric plexus and this plexus is known to originate in the neural crest (Yntema and Hammond, 1954).[14]

III. Lethal Spotting (*ls*)

This recessive determinant on chromosome 2 (R.J.S. Phillips, 1966b) is similar to piebald, especially piebald-lethal. It arose in a subline of C57BL in which a mutation to a^t (which is linked to *ls*) previously had occurred (R.J.S. Phillips, 1958, 1959).

A. Spotting Patterns and Survival of Homozygotes

While the coats of lethal spotting heterozygotes are fully pigmented, homozygotes show variable degrees of white spotting on the back, and their belly is usually white (Figure 9-4). Phenotypically these animals therefore resemble *s*/*s* mice except that their ears and tail are less pigmented than is usually the case with piebald (Lane, 1966). Although on some genetic backgrounds *ls* homozygotes usually die with megacolon at between 2 and 3 weeks of age, and even the few which have lived to breed eventually succumb to this condition (Lane, 1966), on other backgrounds these homozygotes survive quite well. Thus R.J.S. Phillips informs me that *ls* is maintained as a homozygous stock at Harwell with only about 6–7% actual deaths before weaning and with no indication that the incidence of megacolon is any higher than in *s*/*s* animals.

Histological preparations of the colons of two 12-day-old *ls* homozygotes from a "lethal" stock, when compared with the colons of two normal (*ls*/+ or +/+) siblings of the same age, revealed an identical situation with that in

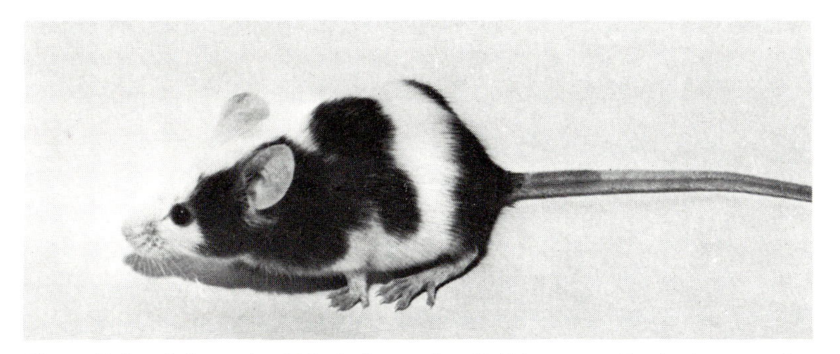

Figure 9-4. A 3-week-old lethal spotting (*ls*/*ls*) mouse displaying spotting typical of this mutant. From T.C. Mayer and E.L. Maltby (1964). Reproduced with permission of the authors and Academic Press, Inc.

Table 9-2. Estimation of the Melanocyte Content in Various Tissues of Lethal Spotting Homozygotes, Heterozygotes, and Normal Mice at 4 Weeks of Age[a,b]

Tissue	*ls/ls*	*ls/+*	+/+
Harderian gland	0	+ +	+ + +
Membranous labyrinth	+	+ + +	+ + +
Retina	+ + +	+ + +	+ + +
Choroid	0	+ +	+ + +
Leg musculature	0	+	+ + +
Ankle skin	0	+ +	+ + +
Hair	0, + + +	+ + +	+ + +

[a]Based on Mayer and Maltby (1964).
[b]+ + +, normal; + +, slightly reduced; +, severely reduced; 0, absent.

piebald-lethal. The mutants displayed a deficiency of myenteric ganglion cells with the distal section of the colon aganglionic and the most proximal part more or less normal (Lane, 1966).

B. Influence on the Occurrence of Melanocytes

A comparison of the pigmentation of various tissues of *ls/ls* and *ls/+* mice with those of +/+ animals disclosed a number of differences (see Table 9-2). The harderian gland, leg musculature, ankle skin, and the choroid of *ls* homozygotes all lacked melanocyte populations, and the number of cells in the membranous labyrinth of these animals was considerably reduced. This reduction was not related to the degree of spotting as not only did animals displaying less spotting show the same extent of nonepidermal pigment reduction, but *ls/+* mice, which have normally pigmented coats, also exhibited less pigment in their harderian glands, ankle skin, and especially their leg musculature. Four-week-old *ls/+* mice were found to possess a total of only 73 ± 30 melanocytes in the leg musculature as compared to 308 ± 68 cells in +/+ animals of similar age. Nevertheless, only the coat pigment of *ls/ls* mice showed the mosaic or pattern effect, i.e., the other regions displayed a dilution and finally a complete loss of pigment (Mayer and Maltby, 1964).

C. Etiology

In an effort to determine the basis of spotting in *ls/ls* genotypes, Mayer and Maltby grafted skin from midway between the two limb buds (a region most likely to show spotting in *ls/ls* individuals) of 12.5- and 13.5-day-old embryos to the coelom of the chick. However, because in the stock employed *ls*/*ls* homozygotes usually die before weaning, these grafts had to originate from embryos derived from *ls*/+ matings; consequently only one-quarter of them would be expected to be *ls*/*ls*. Accordingly only 14 grafts representing 10 of 45 donors were recovered which indicated their genotype to be *ls*/*ls*. Nine of these grafts were mosaics possessing both pigmented and pigment-free hairs and the other five contained only unpigmented hairs. Because no pigmentation was associated with this latter group of transplants, i.e., melanocytes were not found in either their dermis or in the coelomic lining of the host surrounding them, and because the mosaic grafts contained both pigmented and pigment-free hair located in separate regions such that white hair follicles were always surrounded by a dermis that lacked melanocytes, Mayer and Maltby believe that the white areas of *ls*/*ls* mice totally lack pigment cells as a result of some neural crest deficiency. Further support for such a defect stems from their megacolon condition.

IV. Belted

This recessive mutation (*bt*; chromosome 15) has been reported twice: the first time as a spontaneous mutation in the DBA stock (Murray and Snell, 1945) and the second time as a spontaneous mutation in strain CBA/J (Mayer and Maltby, 1964).[15]

A. Description

bt/*bt* mice are characterized by a white patch of hair located transversely across the back posterior to the midline (Figure 9-5). This patch, which may vary from 1 to 20% of the dorsal surface, often joins a ventral white patch to form a complete belt around the body. The amount of ventral spotting in *bt*/*bt* animals, unlike that in piebald (*s*/*s*) mice, is usually less than the amount of dorsal spotting. Although *bt* behaves as a recessive, very occasionally some heterozygotes display a small belly spot (Murray and Snell, 1945).

As is usually, if not always, the case the effect which *bt*/*bt* has on the amount of white spotting is greatly augmented when it is combined with other spotting genes.[16]

B. Etiology

There are two investigations (Mayer and Maltby, 1964; Schaible, 1972) concerned with the etiology of white spotting in *bt*/*bt* mice, and, in spite of the fact that they employed very similar protocols, their results are quite dif-

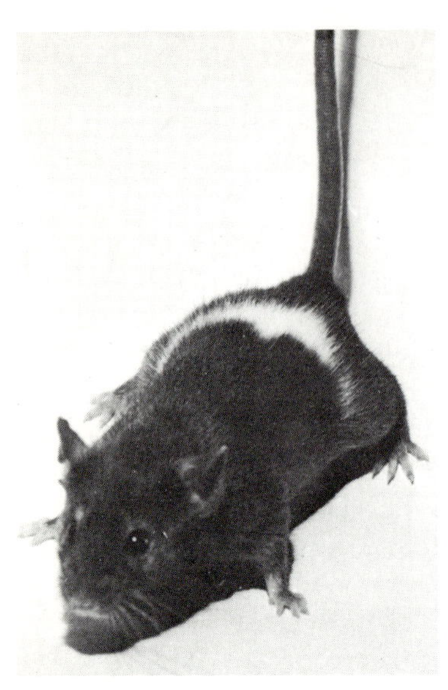

Figure 9-5. Belted (*bt*/*bt*) mouse illustrating the "belt" across the back posterior to the midline. From T.C. Mayer and E.L. Maltby (1964). Reproduced with permission of the authors and Academic Press, Inc.

ferent. In Mayer and Maltby's experiments small explants of ectoderm and underlying mesoderm from potentially pigmented and spotted (belted) areas of 12- to 12.5-day-old *bt*/*bt* embryos were transplanted to the coelom of the chick. Whereas grafts from potentially pigmented regions *always* developed pigmented hair, and in most cases melanocytes were found in the dermis of the graft between the hair follicles, as well as in the lining of the coelom of the host, grafts from the region of the belt usually displayed *no hair pigment*, even though numerous melanocytes occurred in the dermis (between the white hair follicles) and in the coelom of the host. Moreover, some grafts which Mayer and Maltby believe by chance had overlapped both belted and nonbelted areas, formed pigmented hairs in one region and white hairs in another. Since these observations indicate that melanoblasts migrate freely throughout all areas of the skin of *bt*/*bt* mice, even in those areas destined to form white hairs, Mayer and Maltby contend that the unpigmented hairs must result either from "a failure of melanoblasts to gain entrance into the developing follicles, or to their failure to differentiate in this environment." They therefore conclude that *bt* produces a specific genetic block at the level of the hair follicle. To further support this contention they cite their observation that, unlike most other spotting genes, *bt* does not affect the melanocyte populations of other tissues. Thus they noted that the number of melanocytes in the harderian gland, the membranous labyrinth, the choroid, the leg muscles, and the ankle skin of *bt*/*bt* mice was the same as in +/+ animals.[17]

On the other hand, Schaible's transplantation experiments yielded almost diametrically opposite results, and from donors (a^e/a^e;*bt*/*bt*) which had been selected from a wide-belted stock. Thus, using a slightly different grafting procedure he transplanted pieces of 12- to 14-day-old embryonic *bt/bt* skin to the chick coelom. Some of these grafts were proportional to the full width and one-third of the dorsoventral length of the belts (assuming that the belt region of embryos had the same location and was proportional in size to that of the adult), while others (controls) were taken from potentially pigmented regions. Although, in accord with Mayer and Maltby, Schaible found that all grafts derived from potentially pigmented areas always displayed pigmented skin and hair, and all grafts regardless of origin displayed pigment in the skin, unlike Mayer and Maltby, he also observed *pigmented* hairs in grafts which had originated from the belted region. Thus all seven grafts which had originated from the lumbar area of 12- and 13-day-old wide-belt donors, and two of six of the 14-day-old wide belt transplants were *completely* pigmented.

To explain these different results Schaible suggests that perhaps the different ages of the *hosts* (because of the different grafting techniques, Schaible's host embryos were a day older at the start of the experiment than Mayer and Maltby's) as well as the unique genetic backgrounds of the donors were responsible.[18] He also suggests that the failure of some of his 14-day-old embryonic grafts to develop pigmented hair could have been due to the fact that the grafting procedure was not carried out until the guard hair follicles were already formed [the first hair follicles develop on the fourteenth day of gestation (Schumann, 1960)].

Although it may seem surprising that it was the grafts from the wide-belted region of Schaible's mice rather than those from the narrower belted area of Mayer and Maltby's animals that formed pigmented hairs, the different sizes of these transplants could be responsible. Thus if one assumes that the larger lumbar region grafts which Schaible introduced into the coelom did not expand as rapidly as they would have if left *in situ* (or at as rapid a rate as Mayer and Maltby's smaller grafts) this could explain why melanoblasts were able to reach hair follicles in time to become incorporated into the bulb. Indeed, both Mayer and Maltby's and Schaible's observations are most readily explained in terms of Mintz's hypothesis (see Chapter 7, Section VII). Thus the pigment cells which occurred in the skin and sometimes in the hair of their recovered "belted-region" grafts could represent a secondary population which had migrated into the explanted region after the primary population (of inviable cells) had died (but before the transplants were made), a secondary population which migrated in too late to become incorporated into the hair follicles of Mayer and Maltby's transplants, but not too late to become established in the hairs of the less rapidly expanding Schaible grafts.[19] Mintz's hypothesis also readily explains why under normal conditions the belt usually occurs just posterior to the midline. Thus she would argue that *bt/bt* genotypes undoubtedly possess a number of inviable melanoblast clones but that most of these are replaced after they die by melanoblasts from neighboring viable clones. However, because the region just posterior to the midline of the trunk is a rapidly growing area, viable melanoblasts are unable to repopulate it completely and hence a belt is formed.[20]

Notes

[1]Although one might conclude that certain mutations affect only certain pigment centers, e.g., the lumbar centers by *bt*, the coronal center by W^a, the sacral centers by *Rw* (rump-white) etc., Schaible believes his selection experiments support Wright's (1942) gradient concept of pigment formation in spotted mammals. Thus, not only does *bt* appear to restrict the proliferation of melanoblasts in the lumbar region but, to a lesser extent, it also appears to restrict proliferation in pigment areas located adjacently and farther away. Moreover, while Schaible contends that white spotting results from a failure of pigment cells to migrate out from one or more of his proposed "pigment centers," he also believes that the primary mechanism responsible for this migratory failure differs among the various mutants.

[2]*s*/+ mice as a rule are pigmented normally but may have a white tail tip and digits.

[3]Previously Gates (1926) and Dunn and Mohr (1952) had reported that piebald spotting affected the pigmentation of the choroid. Gates found that all the *s*/*s* mice he examined microscopically lacked choroidal pigment, including the pigment in the outer layer of the iris; retinal pigment was unaffected. Dunn and Mohr noted the choroidal pigment of their piebald mice was generally defective. In some of these mice (which were only partially inbred) one or both eyes appeared light ruby or pink. When these eyes were examined with an ophthalmoscope the commonest finding was an enlarged pupil (aniscoria), varying all the way from a slight enlargement to what appeared to be complete absence of the iris (aniridia). This enlargement of the pupil was usually accompanied by a reduction of the amount of pigment in the iris. A smaller proportion of lighter colored eyes had pupils of normal or nearly normal size but lacked a part or all of the dark iris pigment. In the stock examined, about 80% of the mice and 63% of the eyes had iris defects.

[4]In most cases Deol observed that the choroid had one large spot and several small ones, and in some areas there was an intermingling of very small heavily pigmented and lightly pigmented regions. The borders of the spots were generally indented but very occasionally they ran along blood vessels and so were reasonably straight. The intensity of pigmentation in most pigmented regions seemed to be normal. Also, there was more pigment in the ventral half of the choroid, and no regular pattern or symmetry was evident.

[5]The incidence of this condition is very much less in s/s^l animals which, at least on the genetic background studied by Lane, were heavily spotted and indistinguishable from *s*/*s* mice. Lane found that of 12 s/s^l heterozygotes maintained for breeding, three died of megacolon at 11, 13, and 23 months, respectively, while the others were free of the condition when autopsied between 12 and 23 months of age.

[6]There is another simple autosomal recessive piebald mutation which provisionally has been designated *pb*. This mutation, which occurred at Albert Einstein in the o^t12 stock, is of special interest because when homozygous, it causes a progressive dilution of pigmentation so that "some *pb*/*pb* animals become completely white, and

are distinguishable from albino mice only by their pigmented eyes" (Gluecksohn-Waelsch, 1965). Even more remarkable is the fact that this progressive loss of pigment starts early in life and is rapid (Gluecksohn-Waelsch, unpublished). Moreover, during the period of pigment loss many *pb/pb* mice bear a striking resemblance to *c/c* ↔ +/+ allophenics. Indeed, because these mice display conspicuous striping patterns while undergoing pigment loss, it is tempting to speculate that they possess not only some inviable clones of melanoblasts, which are preprogrammed to die prenatally (see Chapter 7, Section VII), but also some which are preprogrammed to die postnatally after they have differentiated and produced pigment. This mutation also induces a high incidence of megacolon (Gluecksohn-Waelsch, 1964a). Symptoms observed in homozygotes include abnormalities of fecal excretion, abdominal distention, and stunting of growth and are reminiscent of those described for the human syndrome known as Hirschsprung's disease (Gluecksohn-Waelsch, 1964b). As far as the author is aware, the relationship of this mutation to either *s* or s^1 is not known.

[7]The all-white line employed possessed from 85 to 100% dorsal white (mean 99.3%) and the dark line had 5 to 25% dorsal white (mean 9.9%).

[8]K. Zimmerman (1941) mentions a wild population from Berlin-Buch in which approximately 70% of the individuals were homozygous for *s* in the absence of any "*k*" genes. These mice possessed either a few white hairs on their belly or one or two central spots with an occasional white tail tip (see Keeler, 1933).

[9]Grüneberg (1936b) states that "genes affecting the pigmentation of the tail tip are very common in tame mice and are also not rarely found in free living mouse populations." From a breading study involving an agouti line, in which most of the animals had dark tail tips, and a nonagouti stock, in which about 75% of the animals had light tail tips, Grüneberg concluded that dark tail tip was dominant, though possibly not completely so, and that a single gene pair was involved.

[10]As pointed out by Grüneberg, the gene flexed-tailed (see Chapter 11, Section II) also fits into this catagory of "minor spotting genes."

[11]To rule out a third possibility, namely that *s/s* melanoblasts do not reach all areas of the integument because they are unable to migrate, Mayer (1967a) studied the migration of +/+ and *s/s* neural tube cells by labeling them with tritiated thymidine before combining them with unlabeled +/+ or *s/s* skin. Although the maximum incubation time of 96 hours precluded the development of melanocytes from these labeled grafts, Mayer found that there was a good correlation between the presence of mouse melanocytes in the host tissues surrounding the +/+ grafts after 15 days incubation, and the presence of labeled +/+ or *s/s* labeled neural tube cells in this same area after an incubation of 96 hours, the only difference was that at 96 hours the area covered by the labeled cells was not as great as the pigment spread after 15 days. It therefore seemed most likely that the labeled cells were in fact melanoblasts and that their migratory capacities were the same regardless of their genotype.

[12]Mayer points out that because *s/s* produces rather small patches of white hair largely restricted to the belly on a C57BL/6J background, the +/+ skin sample used in grafting would actually have been pigmented in a C57BL/6J-*s/s* mouse.

[13]According to Mintz one would also have to assume that the older explants of *s/s* skin were more likely to possess a higher proportion of moribund (inviable) melanoblasts than the younger explants, and this too could affect the ability of viable cells to migrate into and become established in these transplants.

[14]Since both melanoblasts and myenteric ganglion cells are of neural crest origin, it appears that *s* and s^l influence the neural crest *before* it differentiates into these components. Moreover, because the inner ear of s^l homozygotes also displays severe abnormalities, with the neural epithelium always affected, Deol (1967) contends that it is reasonable to assume that there is a deficiency in the acoustic ganglion and that part of this ganglion too is derived from the neural crest.

[15]The latest mutation has been designated bt^J; however we will refer to both as *bt*.

[16]Schaible (1969) contends that the synergistic effect which occurs when two spotting genes, e.g., *bt*/*bt*; Mi^{wh}/+, are incorporated into the same genome may be more apparent than real and he cites an experiment by Reams (1967), as well as some observations by Mayer and his colleagues (Mayer and Maltby, 1964; Mayer, 1965; Mayer and M.C. Green, 1968), to support this contention. Reams inoculated leg buds of White Leghorn embryos with different numbers of chick melanoblasts. He found that when the number of cells introduced was large not only did the skin and feathers of the entire leg become heavily pigmented, but the underlying muscles likewise were pigmented. On the other hand, when the number of cells inoculated was small, the melanocytes remained in the skin without emigrating into either the feather follicles or into the muscle. Mayer and his associates compared the occurrence of melanocytes in the leg musculature of mice heterozygous or homozygous for lethal spotting (*ls*), piebald (*s*), viable dominant spotting (W^v), and steel (*Sl*). They found that whereas melanocytes always occurred in the leg musculature of normal mice they were absent in all mutant homozygotes and reduced in numbers in all heterozygotes. From these findings Schaible argues that while each spotting gene by itself causes a reduction in the number of pigment cells so that few or no cells emigrate to the muscle, the skin population is only slightly reduced. As a result, the white regions of the coat are small. However, when two spotting genes are combined, a much greater proportion of the reduction in the cell population occurs in the skin. As a consequence, "the visible effect on the amount of white regions in the coat is much greater than the additive effects of the two mutant genes even though the effects on the total pigment cell populations probably were additive." While this explanation cannot be refuted it is difficult to reconcile with the fact that some spotting genes, such as belted (*bt*), which affect only hair pigment [estimates of the number of melanocytes in the muscles of the leg, as well as in all other regions of the body which normally possess melanocytes, are the same in *bt*/*bt* and +/+ mice (Mayer and Maltby, 1964)] nevertheless have a hyperadditive effect when combined with other spotting factors.

[17]In a systematic study of 11 inbred strains of pigmented mice, Mayer and Maltby (1964) found that in addition to the retina and choroid of the eye, harderian gland, and ankle skin, melanocytes also occur in the leg musculature and the membranous labyrinth of the ear (where they are associated with the membranes of the utriculus and the semicircular canals). For example, in the C3H/J strain they observed that the lateral and medial heads of the gastrocnemius plantaris, soleus, and the peroneal group were pigmented whereas the muscles comprising the anterior compartment of the leg were generally pigment free. They report that "the total number of melanocytes in the leg musculature is 390 ± 47, and these cells are located in the endomyseal and perimyseal connective tissue between the muscle fibers."

[18]Schaible also believes that because in his and Mayer and Maltby's experiments

the grafts were exposed to an embryonic environment much longer than is normally the case (until the equivalent of 6 days postpartum in Mayer and Maltby's experiments and 7 to 9 days postpartum in his), that melanoblast migration is enhanced; i.e., he contends that the stimulus for proliferation and migration of pigment cells is for some unknown reason much stronger in the chick coelom than in the normal environment of the mouse. As far as I am aware there is no evidence to support this assertion.

[19]Actually, when one considers that Mayer and Maltby found pigmented hairs in 50% (9/18) of the grafts which originated from the level of the belt, their findings are not too different from Schaible's. Moreover, the fact that five of these grafts possessed *both* pigmented and white hairs is undoubtedly significant. One could also argue that the reason why some of the 14-day-old embryonic grafts of Schaible were unpigmented was because they had one or two extra days of (intussusceptive) growth *in vivo*. Clearly if differences in growth are responsible for the discrepancy in results, grafts from the lumbar region of "wide-belt" embryos, of the size employed by Mayer and Maltby, should produce many more *unpigmented* hairs.

[20]In 1975 Kelly reported a recessive mutation which produced a narrow white belt. The deviant occurred among the progeny of a male treated with *N*-methyl-*N'*-nitro-*N*-nitrosoguanidine (MNNG). Since this mutation is *not* allelic with either *bt* or *s* it has been designated "belted-2" (*bt*-2).

Chapter 10

Dominant Spotting, Patch, and Rump-White

I. Dominant Spotting (*W*-Locus)

There are probably no coat color determinants which have received more attention than the alleles at the *W*-locus (chromosome 5). Most of this attention, however, has been devoted to the influence which two alleles at this locus, *W* and W^v, have on hematopoiesis for, like the steel (*Sl*) series of alleles (see Chapter 11, Section I), the *W*-series too affect erythropoiesis and gametogenesis. While it is beyond the scope of this chapter to review in detail all the investigations which have been carried out on the nonpigmentary manifestations of these genes, they must nevertheless be given some consideration as they could be related to the influence the locus has on coat color. Consequently, after considering *W* and W^v's influence on pigment formation, a more general account of their other consequences will be reviewed before the effects of the other alleles at the locus are described.

A. The Influence of *W* on Spotting

The influence of *W* on white spotting has been well reviewed by Grüneberg (1952) and much of this section is based upon his treatment of this subject. *W* was originally recognized by Durham (1908) and initially studied by Little (1915) and by Sô and Imai (1920) who demonstrated that it was lethal when homozygous.

W/+ mice generally have a well-defined belly spot with sharp edges (see Figure 11-1) and a very variable amount of white in the dorsal coat which usually is of a "variegated" nature, i.e., white hairs are usually interspersed among pigmented hairs giving a kind of roan or silvering effect, which increases in intensity with the amount of white in the fur.[1]

1. The "m(*W*)" Complex

The amount of white on the dorsum of *W*/+ heterozygotes is subject to selection. Thus Dunn (1937) succeeded in establishing a stock in which the dorsum of *W*/+ animals displayed, on the average, about 90% white with a typical roan pattern. On the other hand the +/+ segregants of this strain either displayed no white at all, had occasional small belly spots, or had a very small solid (not variegated) white patch on the center of the back. Moreover, the spotting in these +/+ animals seemed to be due not to the same modifiers which increased the amount of white in the *W*/+ segregants but rather to some "*k*" genes (see Chapter 9, Section II, D) which were carried in the stocks and simultaneously selected for.

As far as the increase in the amount of white spotting in the *W*/+ animals was concerned, it seemed to be due to the selection of genes which acted in concert only with *W*, i.e., determinants which, in the absence of this mutation, had little, if any, phenotypic effect. Thus, in contrast to most of the "*k*" genes, these genes were modifiers of *W* in the strict sense and accordingly Dunn referred to them collectively as the "m(*W*)" complex. It follows that if selecting for some of these m(*W*) factors results in increasing the amount of white spotting, selecting for others should result in a strain in which *W*/+ mice have very little spotting. This proved to be the case as Dunn produced one strain in which, except for an occasional belly spot, such heterozygotes displayed no spotting at all. In these animals, the normal alleles of the m(*W*) complex had been selected for.

As a consequence of these modifying genes,[2] which seem to behave as almost complete recessives, *W* can behave either as nearly completely dominant in its effect on the fur, or it can act as almost completely recessive.[3] Indeed, because in a stock homozygous for the normal alleles of the m(*W*) complex *W*/+ mice are generally uniformly pigmented, and those *W*/*W* (severely anemic) animals which survive long enough to grow fur are black-eyed whites,[4] the customary designation of "*W*" as "dominant spotting" can be doubly misleading; it may act as a recessive and does not always produce a spotted phenotype! As most mouse stocks contain some mutants of the m(*W*) complex, *W* usually behaves as a semidominant [and, at least in the C57BL/6 strain, rarely produces dorsal spotting (E. Russell et al., 1957)]. Furthermore, even when incorporated into an inbred strain *W*/+ mice show varying degrees of variegation as a consequence of nongenetic factors.

2. Interaction with Piebald (*s*) and the "*k*" Complex

So far, we have considered the expression of *W* and its specific modifiers in the absence of the gene for piebald spotting (*s*) and the "*k*" genes. Since the simultaneous occurrence of *W* and *s*/*s* in the same genome has a synergistic effect on the amount of white spotting, *W*/+;*s*/*s* mice are either entirely white with black-eyes, or have small pigmented areas in the region of the ears and/or the haunches.[5] In fact, even when *s* is heterozygous with its normal allele it slightly augments the amount of white spotting in *W*/+ mice in the presence of the m(*W*) genes.

Insofar as the interactions of m(*W*) with *s* and with the "*k*" genes are concerned the evidence is incomplete. According to Dunn (1937) "the indications are that the effects of m(*W*) do not cumulate with those of *s* but do to a slight extent with those of "*k*" when both "*k*" and m(*W*) are homozygous. The latter two complexes may have some common genes although it is probable that some of the m(*W*) genes are independent of and act differently from those of the "*k*" series, since the latter may themselves initiate reactions without *W*."

B. Viable Dominant Spotting (W^v)

The second mutation at this locus, W^v (for viable dominant spotting), was recognized and reported by Little and Cloudman (1937).[6] When heterozygous with the normal allele W^v produces a mid-ventral spot of variable size on the trunk, and very frequently a small mid-dorsal spot on the head as well (see Grüneberg, 1939). The rest of the coat, unlike in *W*/+ mice, is diluted (Plate 3–D).[7] While Little and Cloudman found that, like *W*, this mutation when homozygous produced black-eyed whites [Plate 3–E; occasionally pigment occurs in the skin of the ear pinna but not in the hair (Figure 10-1)],[8] these homozygotes survived significantly longer than *W*/*W* mice. In contrast to *W*/*W* genotypes, all of which died within 2 weeks after birth, and the majority of which died earlier—many of them *in utero* (deAberle, 1927)—many W^v/W^v mice lived for more than 3 weeks and some lived to be adults. Indeed, Little and Cloudman found a W^v/W^v female and one male to be fertile, although this fertility was temporary and lasted only for a limited time.[9]

C. Influence of *W* and W^v on Pigment Granules

Various attributes of the pigment granules in *W*/+ and W^v/+ mice have been studied in some detail by E. Russell (1949c) who found that whereas one dose of *W* has no effect on hair pigment intensity, one dose of W^v has a slight but significant one. Thus on full-color black (*a*/*a*;*B*/*B*), chinchilla black ($a/a;B/B;c^{ch}/c^{ch}$), full-color brown (*a*/*a*;*b*/*b*), pink-eyed sepia (*a*/*a*;*B*/*B*;*p*/*p*), and full-color yellow (A^y/–) backgrounds, W^v/+ mice have fewer pigment granules than +/+ animals, a difference which is most apparent in the number of medullary granules. In some of these genotypes the size of the granules also are smaller in W^v/+ than in the corresponding +/+ type. As pointed out by Russell these observations are compatible with the concept

Figure 10-1. Mouse homozygous for viable dominant spotting (W^v). Note that the right ear is pigmented. Although not obvious in this photograph this pigmentation is limited to the skin, the hairs arising through the skin being nonpigmented. Courtesy J. Gordon.

that $W^v/+$ substitution causes a reduction in the general level of pigmentation.[10]

D. Etiology of Spotting in W and W^v Genotypes

All available evidence indicates that the inviability of melanoblasts is responsible for the white spotting of the coat of W-series genotypes. Moreover, this lethality appears to be due to these alleles acting within the melanoblasts themselves rather than via the skin. Evidence that this is the case stems from (1) Mintz's (1970) observation that $W/W \leftrightarrow +/+$ allophenics are either completely pigmented or white-spotted;[11] (2) grafting results which indicate that the pigmented and nonpigmented areas of the allophenics which are spotted are composed of both W/W and $+/+$ cells (see Chapter 7, Section VII); and (3) the results of experiments carried out by Mayer and M.C. Green (1968) and Mayer (1970, 1973a).

Mayer and M.C. Green observed that when neural tube, including neural crest cells, from 9-day-old embryos derived from $W^v/+$ matings, was combined with putative neural crest-free $+/+$ skin (originating from the lateral flank between the fore and hind limb buds of 11-day-old embryos) and allowed to differentiate in the coelom of the chick, 61% of the transplants failed to produce pigment. Since this number is significantly greater than the expected number of W^v/W^v combinations, i.e., one would expect 25% of the grafts to include W^v/W^v neural tubes, it seems likely that a considerable number of $W^v/+$ neural tubes must likewise have failed to give rise to melanocytes (see note 16).

Mayer and Green also found that when neural tubes from 9-day-old +/+ embryos were combined with skin from 11-day-old embryos produced by W^v/+ matings, *all* the grafts were pigmented. Thus it seems apparent that there are no differences in the ability of skin from +/+, W^v/+, and W^v/W^v embryos to support the differentiation of +/+ melanoblasts, again an observation completely in accord with Mintz's hypothesis.

In further support of these conclusions Mayer (1970) obtained similar results when he combined small pieces of 9-day-old +/+ neural crest-containing neural tube with W^v/W^v skin obtained from 13- to 18-day-old embryos, skin whose genotype could at this age be confirmed by the pale color of the donor's liver. When such composite transplants were allowed to incubate for 15 days in the flank of White Leghorn chick embryos, all of them possessed large numbers of pigmented hairs. In fact, in no case were any pigment-free hair follicles observed.

Because Mayer obtained different results when he employed neural crest-containing albino skin, i.e., such skin, presumably because it possessed *c*/*c* melanoblasts, prevented the entrance of +/+ melanoblasts (see Chapter 11, note 11), he notes that one possible conclusion of his findings is that melanoblasts never enter the skin of W^v/W^v embryos. However, as he also points out, an alternative explanation, and one which seems more likely, is that W^v/W^v melanoblasts enter the skin but die soon after.[12]

Mayer (1973a) also determined the fate of reciprocal combinations of 13-day-old embryonic W/W^v[13] and +/+ epidermis and dermis when grafted to the chick coelom, and these results too were completely in accord with the fact that there was nothing wrong with the mutant's skin. Thus melanoblasts which were present in either the +/+ dermis or epidermis could move freely into the corresponding W/W^v component, differentiate, and form pigment.[14]

Despite the fact that all the evidence indicates that *W*-locus alleles act within the melanoblast, Gordon (1977) has reported some observations which he believes are difficult to reconcile with Mintz's notion that *W*/+ mice possess two kinds of melanoblast clones, *completely normal* and *inviable*. He made *W*/+;+/+ ↔ +/+;*c*/*c* allophenics and found that such animals usually were completely white although a few possessed traces of pigment. Their eyes were ruby colored and microscopic examination showed the eye pigment to be located only in the retinal epithelium which was a mosaic of black and white sectors. It thus appears that *W*/+ melanoblasts while capable of pigmenting most of the coat (as well as the eyes) of *W*/+ mice, populated exceedingly few hairs when in competition with +/+;*c*/*c* (albino) melanoblasts. Although this is not surprising in the sense that it is known that in some allophenic combinations the melanoblasts of one genotype often have an inherent ability to dominate over those of the other, due at least in part to their greater proliferative capacity, Gordon also found that +/+ cells of similar genetic origin as the *W*/+ cells frequently pigmented most of the coat of +/+;+/+ ↔ +/+;*c*/*c* allophenics. It therefore appears that *none* of the melanoblasts of *W*/+ mice are as competent as those of coisogenic +/+ animals in competing with albino melanoblasts. Accordingly, Gordon believes that the melanocytes of *W*/+ mice are inherently weaker than those of +/+ animals (or it could be argued that they possess both inviable and weaker than viable +/+ clones.)[15] However, there is an alternative explanation for these findings, one which again is in complete accord with Mintz's hypothesis. Thus if one accepts the fact that to begin with *W*/+;+/+ ↔ +/+;*c*/*c* allophenics possess, on the average, only *half* the number of potentially pigmented melanoblast clones as +/+;+/+ ↔ +/+;*c*/*c* allophenics (and these are outnumbered 2:1 by melanoblast clones of albino origin) it does not seem very surprising that so far all of them have been completely or almost completely unpigmented. This is especially the case since only relatively few of these allophenics have been produced (see note 11).[16]

E. Expression of Extraepidermal Melanocytes in W^v/– Mice

Although all the evidence is consistent with the conclusion that the effect of the *W*-series of alleles on white spotting of the coat is due entirely to a defective population(s) of melanoblasts, and is completely independent of any direct influence of these genes on the skin itself, this does not rule out the possibility that other tissue environments can influence the capacity of *W*-mutant melanoblasts to survive and/or form melanin. Indeed, evidence that such is the case is derived from the study of Markert and Silvers (1956) who, after surveying the occurrence of melanocytes in the nictitans, harderian gland, hair follicle, ear skin, choroid, and retina of 50 different genotypes (see Table 10-1), concluded that "the capacity of a tissue to elicit melanogenesis depends upon the embryonic history of the tissue (i.e., what kind of tissue it has become—nictitans, harderian gland, etc.) and upon the genetic composition of the tissue." This fact has been reemphasized by two ingenious studies of Deol (1971, 1973). In the first of these investigations he analyzed the pigmentation patterns in the choroid, harderian gland, and inner ear in a number of spotted genotypes and concluded that the host tissue plays an important role in determining the pattern of spotting and that all melanoblasts may not be affected to the same degree. For example, he found that in the harderian gland of W^v/+ mice the number of melanocytes was not only slightly reduced from that observed in +/+ glands but they were on average smaller than normal, although cells of normal size were quite common.[17] Deol also observed that although spotting of the choroid was significantly heavier in W^v/+ animals with mid-dorsal head spots than in those without them, the amount of white spotting of the coat was not a reliable guide to internal pigmentation, nor was any general trend evident when the effects of different genes were compared (see Table 12-1).

In his later study, Deol focused his attention on the pigmentation of the eye and compared the pigmentation of the iris with that of the choroid and

Table 10-1. The Influence of Genotype on the Occurrence of Melanocytes in Six Tissues of the House Mouse[a]

Genotype	Retina	Choroid	Hair follicle	Ear skin	Harderian gland	Nictitans
A/A	+[b]	+	+	+	+	+
$A/A;b/b$	+	+	+	+	+	+
A^y/a	+	+	+	−	+	+
$A^y/a;b/b$	+	+	+	−	+	+
A^w/A^w	+	+	+	+	+	+
a^t/a^t	+	NE	+	+	+	+
$a^t/a^t;b/b$	+	NE	+	+	+	+
a/a	+	+	+	+	+	+
$a/a;B^{lt}/B^{lt}$	+	+	+	+	+	+
$a/a;b/b$	+	+	+	+	+	+
$a/a;b/b;c^{ch}/c^{ch}$	+	NE	+	+	+	+
$A/A;c^e/c^e$	+	NE	+	+	+	+
$A/A;b/b;c^e/c^e$	+	NE	+	NE	+	NE
$A^y/a;c^e/c^e$	+	NE	0	0−	+	−
$a/a;c^e/c^e$	+	NE	+	+	+	+
$A/A;c/c$	0	0	0	0	0	0
$A^y/a;d/d$	+	NE	+	−	+	+−
$a/a;d/d$	+	NE	+	+	+	+
$a/a;b/b;d/d$	+	NE	+	+	+	+
$a/a;b/b;d/d;ln/ln$	+	NE	+	+	+	−+
$a/a;b/b;ln/ln$	+	NE	+	+	+	+−
$a/a;p/p$	−	−	+	0	0	0
$a/a;p/p;d/d$	−	NE	+	0−	0	−
$a/a;b/b;p/p;d/d$	−	NE	+	0−	0	0−
$A/A;Sl/+$	+	+	+	+	−	+
$a/a;Sl/+$	+	NE	+	+	−	+
$A/A;Sl/+;W^v/+$	+	+	+	+	0−	0−
$a/a;Sl/+;W^v/+$	+	+	+	+	0−	0−
$a/a;W^v/+$	+	+	+	+	+	+
$a/a;W^v/W^v$	+	0	0	0−	0−	0
$a/a;W^v/W$	+	0	0	0	0	0
$a/a;W/W$	+	0	0	0	0	0
$a/a;W/+$	+	+	+	+	+	+
$A/A;W/+;s/s$	+	0	+	+	0	0
$a/a;W/+;s/s$	+	0	+	+	0	0
$A^y/a;s/s$	+	+	+	−	+	+
$A^y/a;b/b;s/s$	+	+	+	−	+	+
$A^w/A^w;s/s$	+	+	+	+	+	+
$a/a;s/s$	+	+	+	+	+	+
$a/a;b/b;s/s$	+	+	+	+	+	+
$a/a;To/+$	+	NE	+	+	+	+
$a/a;Mi^{wh}/+$	+	0	+	+	0	+−
$a^t/a^t;Mi^{wh}/+$	+	0	+	+	0	+−

Table 10-1. *continued*

Genotype	Retina	Choroid	Hair follicle	Ear skin	Harderian gland	Nictitans
$a/a;Mi^{wh}/Mi^{wh}$	–	0	0	0	0	0
$A^w/A^w;gl/gl$	+	NE	+	+	+	+
$a/a;ru/ru$	–	+	+	0–	–	–
$a/a;tp/tp$	+	NE	+	+	+	–
$a/a;Va/+$	+	NE	+	+	+	0
$a/a;b/b;Va/+$	+	+	+	+	+	0

[a]Based on data of Markert and Silvers (1956).

[b]+, substantially normal occurrence; –, greatly reduced number of melanocytes; 0, no melanocytes; NE, tissue not examined. Spotted tissues are recorded as + when numerous melanocytes are present in the pigmented areas even though the white areas contain no melanocytes. Hair follicles are classified into two types: +, pigmented; 0, white. Although an entry is made for the retina of each genotype, an examination of sectioned eyes was made only for those genotypes in which a positive entry is also recorded for the choroid. Entries for the retinas of unsectioned eyes are based on gross observations alone. The c^e/c^e genotypes have numerous melanocytes but their granules are lighter than normal.

the retina in the same eye of different white spotted genotypes. This was particularly appropriate since the outer layer of the iris derives its melanocytes from the choroid while the melanocytes of the inner layer of the iris are derived from the retina. Thus, as Deol points out, any pigmentation differences between the choroid and outer layer of the iris or between the retina and inner layer of the iris would constitute evidence that different tissue environments can affect the survival, differentiation, and/or melanogenic capacity of pigment cells. Such differences were found (see Table 12-2). In $W^v/+$ mice, the choroid was partially pigmented (spotted) while the outer layer of the iris was fully pigmented suggesting that identical melanocytes (from the choroid) were responding independently to these different environments. In W^v/W^v mice, however, both the choroid and outer layer of the iris were devoid of pigment, presumably because the melanoblasts of these animals were so abnormal that they could not differentiate (or survive) in either environment.

F. Influences of W^v on the Inner Ear

The inner ears of all W^v/W^v mice have marked abnormalities in the cochlea (the most striking abnormalities occur in the organ of Corti and the stria vascularis) and many have severe abnormalities in the saccule as well.[18] These anomalies also accur, though in a more benign form, in a small part of the cochlea of a few old $W^v/+$ animals (Deol, 1970c). Since there is good evidence that the neural crest contributes to the formation of the acoustic ganglion, it seems most likely that it is via some neural crest defect that these pathological changes in the inner ear are produced (Deol, 1970c). These observations, therefore, provide further support for the thesis that the effect which W^v has on coat color likewise results from a direct influence on the neural crest.[19]

G. Influence of *W* and W^v on Hematopoiesis

There are three coat-color determinants in mice, flexed-tailed, steel, and dominant spotting, all of which produce white spotting and all of which are associated with congenital anemias. The most thoroughly investigated of these anemias is the one associated with the *W*-series of alleles, especially *W* and W^v. Indeed, the anemic condition caused by these alleles has been studied so extensively that this aspect of *W*-gene action alone could easily form the basis of an impressive monograph. The early efforts on this subject are well reviewed by Grüneberg (1952) and by E. Russell (1954). A more recent review is included in the second edition of the *Biology of the Laboratory Mouse* (E. Russell and Bernstein, 1966) and much of what follows is based on this reference (see also E. Russell, 1970).

As we have already noted animals homozygous for *W* are characterized by a very severe macrocytic anemia (Attfield, 1951; E. Russell and Fondal, 1951), so severe in fact that they usually die within a few days after birth. W/W^v and W^v/W^v mice likewise suffer from this condition and although it is not as deleterious in these genotypes it can be lethal in some, especially between birth and weaning when they are growing most rapidly. Animals which survive this period usually live more than a year (E. Russell and Bernstein, 1966).

While *W* has no affect on erythropoiesis when heterozygous with the normal allele, the number of erythrocytes is slightly reduced and their mean cell volume slightly increased in $W^v/+$ genotypes (Grüneberg, 1942; E. Russell, 1949a). Thus the anemic conditions caused by the alleles *W* and W^v can be ranked as follows in terms of the number of erythrocytes: $+/+$ (normal) $= W/+ > W^v/+ > W^v/W^v > W/W^v > W/W$.

In the severely affected genotypes the anemia already is apparent when the liver succeeds the yolk sac blood islands as the principle site of hematopoiesis at 12.5 days gestation (Borghese, 1954) and it persists for as long as the animal lives. On a heterogeneous background the mean erythrocyte counts of newborn W/W, W/W^v, and W^v/W^v mice are 0.8×10^6, 1.4×10^6, and 2.2×10^6 RBC/mm^3, respectively (E. Russell and Fondal, 1951), counts which differ significantly from the 4.8×10^6 RBC/mm^3 for normals of the same age (E. Russell, 1954).[20] Although on specially selected genetic backgrounds the postnatal survival of W/W anemics can be prolonged to a mean of 10 days, with an occasional animal surviving to adulthood, under these conditions as well the number of red cells/mm^3 ($1.5-1.8 \times 10^6$) at birth is very low (E. Russell and Lawson, 1959).

Nevertheless, the anemia in these severely affected genotypes is not aplastic but *hypoplastic*. The absolute number of erythrocytes shows the same relative increase from the sixteenth day of gestation to birth in W/W mice as it does in +/+ genotypes (E. Russell and Fondal, 1951), and the proportion of reticulocytes in the blood of W^v/W^v adults (Niece et al., 1963) and of W/W newborns (deAberle, 1927) is higher than in their +/+ littermates. The fetal liver and neonatal marrow of W/W, W/W^v, and W^v/W^v animals is hypoplastic (E. Russell, 1954; Borghese, 1959), while the marrow cellularity of adult W/W^v and W^v/W^v mice is nearly normal (E. Russell et al., 1953). All the evidence indicates also that there is nothing wrong with the life span of the erythrocyte (Niece et al., 1963).[21]

There is good evidence that the anemia produced by the W-alleles results from a delay in the maturation of erythrocytes (E. Russell et al., 1953; Borghese, 1959; Benestad et al., 1975; Shaklai and Tavassoli, 1978). When isotopically labeled heme-precursors were inoculated into W/W^v mice and their +/+ littermates, erythrocytes with labeled protoporphyrin appeared in the circulation of the normal mice after 3 days but not until after 7–14 days in the anemics (Altman and E. Russell, 1964). This delay in the capacity of W/W^v mice to form erythrocytes, a delay which seems to be caused by a deficiency in the number or differentiating capacity of erythropoietic stem cells (McCulloch et al., 1964; M. Bennett et al., 1968a), is responsible also for their great sensitivity to X-irradiation (Bernstein, 1962; E. Russell et al., 1963).

That this defect in the maturation of erythrocytes results from W-alleles acting directly upon the blood-forming tissues, and *not*, as in the case of steel (see Chapter 11, Section I, E), indirectly by affecting the microenvironment in which hemopoiesis occurs, is demonstrated by the fact that W/W, W/W^v, and W^v/W^v anemic animals can be cured completely and permanently by implantation of histocompatible (syngeneic) +/+ blood-forming tissue from adult marrow or from fetal livers (E. Russell et al., 1956b, 1959; Bernstein and E. Russell, 1959; Bernstein, 1963; E. Russell and Bernstein, 1968), or even from the placenta (Dancis et al., 1977). Indeed, this anemia can also be cured by the intravenous administration of histoincompatible (allogeneic) hemopoietic cells from either very compatible donors (E. Russell and Bernstein, 1967; Harrison, 1972a; Harrison and Cherry, 1975) or into immunologically tolerant or immunosuppressed hosts (Seller, 1966, 1967, 1968, 1970, 1973; Seller and Polani, 1966, 1969).[22]

H. Influence of W and W^v on Gametogenesis

In addition to their effect on coat color and hematopoiesis the alleles of the W-series influence the development of the primordial germ cells. W/W^v mice always are sterile and, on most genetic backgrounds, W^v/W^v mice are too (Grüneberg, 1952; Veneroni and Bianchi, 1957).[23]

A histological study of the gonads of W/W, W/W^v, and W^v/W^v mice, and their normal counterparts, from 0 to 28 days postpartum revealed a drastic deficiency in the number of germ cells in the potentially sterile gonads of both sexes (Coulombre and E. Russell, 1954). The severity of this defect seemed to parallel that of the anemia, i.e., it was most severe in W/W animals and more severe in W/W^v than in W^v/W^v genotypes. This impairment seemed to reflect both a reduction in the number of definitive germ cells and a retardation of their maturation. Thus, in contrast to the situation in the normal (+/+) ovary which contains many large follicles at 1 month, corpora lutea by 6 weeks, and continues to produce large numbers of ova which develop normally long after 1 year (E. Russell, 1954), the ovaries of W^v/W^v females possess very few ovarian follicles at any time and these cease to grow and develop after the age of 2 months (E. Russell and Fekete, 1958).[24] The adult W^v/W^v testis too is abnormal, being almost devoid of spermatogenesis. Most of the tubules contain only Sertoli cells, and the few germ cells which occur are almost all spermatogonia (Coulombre and E. Russell, 1954; E. Russell, 1954).

1. Embryological Basis of Germ Cell Defect

It follows that if the influence which these *W*-alleles have on gametogenesis is a consequence of their effect on erythropoiesis then this latter effect must *precede* the germ cell anomaly. To determine if this was the case Borghese (1956) explanted gonads from 12-day-old W/W and +/+ fetuses into a favorable tissue culture environment. Because these gonads continued to develop as they would have if left *in situ*, Borghese concluded that the sterility was independent of the anemia. Similar observations were made by E. Russell and her associates (1956a) who transplanted W/W and W^v/W^v gonads of the same age to the spleens of histocompatible +/+ hosts. Nevertheless, while these observations indicated that the germ cell defect had reached full expression at the earliest stage (12.5 days) at which evidence of the erythropoietic defect had been identified, they did not prove that the germ cell defect *preceded* the anemia. These experiments could just as easily be interpreted as a simple failure to revive germ cell formation in an already deficient gonad.

The best evidence that the germ cell defect is independent of the anemia stems from the elegant analysis of Mintz and E. Russell (1957). By elective staining of primordial germ cells of embryos derived from $W/+$ and $W^v/+$ matings with the azo dye coupling technique for alkaline phosphatase, they were able to follow the migration of germ cells from their place of origin, the yolk sac splanchnopleure, to their definitive positions in the paired germinal ridges. They found, in agreement with the observations of Chiquoine (1954), that in normal embryos the migration of these cells occurs between the eighth and twelfth days of embryonic life and that this migration is accompanied by their continued multiplication.[25] Thus during this 4-day interval the number of germ cells in +/+ mice was found to increase from a maximum of 76 to a maximum of 5711 (Mintz and E. Russell, 1957). On the other hand, while W/W, W/W^v and W^v/W^v embryos possessed normal numbers of primordial germ cells at 8 days of age, the number of these cells failed to increase thereafter and were retarded in their migration to the germinal ridges. By day 12 these affected genotypes possessed as few as 18 and a maximum of 72 germ cells.[26]

It is therefore apparent that the defect in gametogenesis in W/W, W/W^v, and W^v/W^v mice is expressed as a mitotic failure, evident at 9 days, and it is very likely that the cause of this failure is present *before* this time. Consequently it is very unlikely that the sterility is a secondary result of the anemia. Indeed, even if an undetected defect in yolk sac hematopoiesis should occur at 8 days, i.e., earlier than the observed germ cell anomaly, it is most unlikely that it could physiologically affect the activities of the germ cells because of the absence of a functional embryonic circulation at this early stage of development (Mintz, 1957a). It therefore seems reasonable to conclude that the pleiotropic effects of W-series alleles "stem from a single gene-mediated alteration to which certain kinds of cells are peculiarly vulnerable because of their own special activities" (Mintz, 1957a).[27]

I. Other Alleles

In addition to W and W^v the W-locus is represented by the following alleles:

1. Ames Dominant Spotting (W^a)

There is not much information on this allele. It was found among the offspring of an X-rayed male of the Z strain and resembles W except that $W^a/+$ heterozygotes have a prominant head blaze. Homozygotes are anemic and die within a few days after birth. W/W^a young are also anemic (Hollander, 1956; Schaible, 1963a). Schaible (1969) employed this W-allele in his studies on white spotting and notes that when heterozygous (with +) W^a not only causes spotting but dilutes black pigment to some shade of grey. He also states that some animals "may show variegation in that one or more patches will be of a color different from the expected shade of grey."[28]

2. Ballantyne's Spotting (W^b)

This W-series allele occurred in the C57/St strain at Roswell Park. Its characteristics have been described by Ballantyne (from whom it gets its name) and his associates (1961) and what follows is based on their description. The original mutant was a female who showed "an extensive diffuse-white-spotting of the back and sides with a small frontal blaze." The belly was predominantly white and the pigmented areas on the dorsum were significantly lighter in color than the intense black characteristic of C57/St mice. When hairs were plucked from these various regions and examined microscopically it was found that they could be classified as follows: (1) completely white; (2) pigmented at the tip but nonpigmented for variable lengths basally; (3) pigmented at the tip and irregularly pigmented thereafter; and (4) relatively uniformly pigmented from tip to base. The "basal dilution" of pigmentation observed in many hairs appeared to be related to a striking reduction in medullary granules.

Breeding studies revealed that, like W, W^v, and W^a, W^b is inherited as a semidominant. W^b/W^b genotypes are black-eyed white and presumably anemic since they are pale at birth with a median life span of 8 days. All W^b homozygotes which survive are sterile. Test matings with W^v, and with another putatively unique W-allele called W^s, have confirmed its W-locus assignment.[29]

Although $W^b/+$ mice more closely resemble $W^v/+$ than $W/+$ animals, they usually are more extensively spotted than $W^v/+$ and microscopic examination of pigmented hairs from $W^v/+$ mice failed to reveal the extensive terminal deficiency of medullary pigmentation regularly found in $W^b/+$ hairs (Ballantyne et al., 1961).[30]

3. *W*-Fertile (W^f)

This mutation has recently been described by Guénet and his associates (1979) and the following is based entirely on their account (see also Guénet and Mercier-Balaz, 1975).

The first animals carrying this mutation were discovered when in the course of producing a $W^v/+$ C3H/He congenic line, one C3H/He female produced two black-eyed white males (presumably W^v/W^f heterozygotes) with greyish patches at the roots of the ears. Both of these mice looked healthy, fared well in competition with their littermates, and subsequently proved fertile.

$W^f/+$ animals on a C3H/He background are characterized by sharply demarcated white spots on the forehead and belly. They also have a white tail tip (Figure 10-2). The spots vary in size, sometimes consisting of only a few white hairs. The coat itself is not diluted as in $W^v/+$ heterozygotes. When transferred onto other isogenic backgrounds, or when expressed in F_1 hybrids (e.g., with C57BL/6 or 129Sv) the phenotypic expression of W^f may completely vanish so that $W^f/+$ heterozygotes are indistinguishable from +/+ animals (Figure 10-2), a situation which suggests that modifier genes, perhaps similar to those of the m(W) complex, are responsible for the expression of this allele when heterozygous with +.

W^f/W^f homozygotes display extensive white areas (Figure 10-2) lacking any well-defined pattern. According to Guénet and his colleagues this pattern is similar to that displayed by varitint-waddler heterozygotes ($Va/+$) (see Chapter 11, Section IV) although in this case the phenotype does not change with age. Unlike the situation with $W^f/+$ heterozygotes, the expression of W^f/W^f seems to be independent of the genetic background.

Both $W^f/+$ and W^f/W^f males and females are fertile. Between 1 and 6 weeks of age W^f homozygotes are less viable than their $W^f/+$ or +/+ litter-

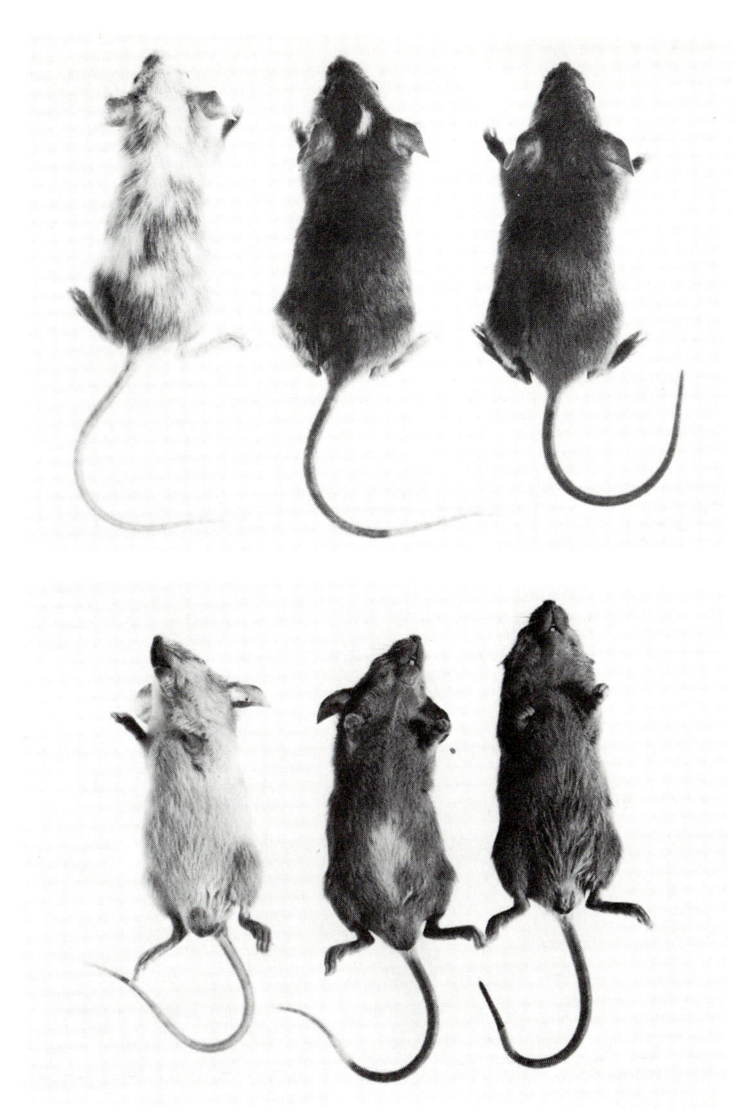

Figure 10-2. Dorsal and ventral views of W^f/W^f (left), $W^f/+$ (middle), and $+/+$ (right) mice. Note that while the W^f homozygote exhibits extensively depigmented areas lacking any well-defined pattern, the $W^f/+$ heterozygote (on the C3H/He background) is characterized by sharply limited white spots on the forehead and on the belly as well as a white tail tip. These spots are variable in size and sometimes reduced to a few white hairs. Courtesy of J.L. Guénet.

mates and some become "runted" and die after 1 or 2 weeks of age. Deaths during adulthood however are rare.

As is the case with other W-alleles, W^f interacts with several other spotting genes. Thus, compounds with rump-white (Rw) (see Section II, C), i.e., $Rw+/+W^f$, exhibit spotting patterns analogous to those of $Rw/+$ mice except that the white area extends up to the belt. When combined with steel-Dickie (Sl^d) (see Chapter 11, Section I, C,1) ($Sl^d/+;W^f/+$) a phenotype is produced which looks very much like $Sl^d/+;W^v/+$. Pigmented hairs are scattered among the unpigmented ones over most of the body without a definite pattern and the ventrum is almost invariably white. Finally, $W^f/+;s^l/s^l$ (piebald lethal; see Chapter 9, Section II, B) compounds are almost completely white with the exception of some relatively intensely pigmented areas on the shoulders or haunches.

In addition to its influence on pigmentation, W^f also produces a chronic macrocytic anemia. Thus Guénet and his associates found adult +/+, $W^f/+$, and W^f/W^f mice to have red cell counts ($\times 10^6/\text{mm}^3$) of 8.00 ± 0.10, 7.29 ± 0.12, and 6.29 ± 0.19, respectively. The mean cell volumes of their erythrocytes (μm^3) were 46.55 ± 0.73 (+/+), 49.25 ± 0.53 ($W^f/+$), and 54.47 ± 1.13 (W^f/W^f). These results, therefore, are consistent with data concerning other *W*-alleles. Moreover, these investigators have also found that the bone marrows of $W^f/+$ and W^f/W^f mice do not possess a normal number of cells capable of forming macroscopic colonies in spleens of irradiated coisogenic recipients, a finding also in accord with those reported for other *W*-locus genotypes (E. Russell, 1970).

Clearly the most interesting feature of this allele is that, unlike the others, when homozygous it neither produces an all-white phenotype nor does it appear to have any demonstrable affect on gametogenesis. This occurs despite the fact that such homozygotes are anemic. Although it is conceivable that further study will reveal some influence on gametogenesis, nevertheless, taken together, this mutation provides further evidence that there is no *direct* relationship between the triad of effects associated with *W*-series alleles.

4. Jay's Dominant Spotting (W^j)

This *W*-locus allele was found by George E. Jay in his C3H colony at the National Institutes of Health. It has been analyzed by E. Russell and her colleagues (1957) and the following is based largely on their findings. When heterozygous with the normal allele W^j produces extensive white spotting on the ventrum with some white on the back especially on the crown of the head. There is no apparent diminution of pigment intensity in the colored areas of the coat in $W^j/+$ heterozygotes nor are they anemic. When homozygous W^j produces black-eyed whites with an average survival of about 8 days (0–18). These animals are severely anemic with blood counts averaging 1.04 (± 0.04) $\times 10^6$ RBC/mm^3 at 0–1 day of age [as compared with a mean of 4.58 (± 0.24) $\times 10^6$ RBC/mm^3 for their normal littermates]. They also are deficient in germ cells, a deficiency which has been analyzed embryologically by Mintz (1957a) and which is indistinguishable from her observations for *W* and W^v (Mintz and E. Russell, 1957).

Test crosses with both *W*/+ and $W^v/+$, as well as with *Sl*/+, have confirmed that W^j is a member of the W-series, a member which appears to resemble W in all respects except for the increased amount of white spotting, especially on the ventrum.[31]

5. Panda-White (W^{pw})

This spontaneous mutation occurred at Oak Ridge. In heterozygotes ($W^{pw}/+$) the pigmentation of the coat is restricted mainly to the head (the snout, near the eyes, and at the base of the ears) and the base of the tail, but sometimes small pigmented patches occur on the hips and shoulders as well. Homozygotes are severely anemic and die within 3 days postpartum. W^{pw}/W^v heterozygous are black-eyed whites and sterile (Steele, 1974).

6. Sash (W^{sh})

This putative W-allele was recently reported by Lyon and Glenister (1978). It occurred spontaneously at Harwell in a pair set up to provide (C3H × 101)F_1 hybrid stock. The original mutant had a broad sash of white around its body in the lumbar region and produced offspring like itself when bred to a normal animal. When crossed to mice carrying patch (*Ph*) or rump-white (*Rw*), the double heterozygotes showed an additive interaction in their spotting pattern, but heterozygotes with W^v were black-eyed whites, with small patches of pigment around the ears and eyes. All three types of heterozygotes proved fertile, and no crossing-over between sash and *Ph*, *Rw*, or W has yet been observed in several hundred offspring. Since sash displays a nonadditive interaction with W, it is considered to be an allele of W and has been given the symbol W^{sh}.

Originally, sash appeared to be lethal when homozygous. However, in crosses of $W^{sh}/+ \times W^{sh}/+$ about 1% of the offspring were black-eyed whites. These black-eyed whites proved to be heterozygous for the original lethal sash and a new viable type. Homozygotes for the viable sash are black-eyed whites, viable and fertile. The interpretation is that the original sash carried a linked recessive lethal, and that in the viable type this lethal was lost by crossing-over. The data indicate that the lethal occurs about 1 cM from W, but which side is not yet known.

No evidence of anemia has been found in $W^{sh}/+$, W^{sh}/W^{sh}, or W^{sh}/W^v genotypes. Thus W^{sh} is an unusual allele in that it shows the full effects of the W-locus on spotting, but appears to have no effect on erythropoiesis or gametogenesis.

7. Extreme Dominant Spotting (W^e)

This W-allele also has only recently been described (Cattanach, 1978). It occurred at Harwell in a Rb(5.15)3Bnr stock maintained on a C3H/H-101/H genetic background. It closely resembles Ballantyne's spotting (W^b) in that when heterozygous (with +) it produces a white belly spot with extensive white markings on the head and body, and a general lightening of the coat. Homozygotes are anemic black-eyed whites which, so far, have not survived beyond 10 days. Compounds with W^v are also black-eyed whites but survive to maturity and are sterile. Compounds with rump-white (*Rw*) are viable and fertile black-eyed whites with some very limited pigmentation in the ear skin. This new *W*-mutation has been given the provisional name of extreme dominant spotting (W^e).

8. Other Alleles

In addition to the above, Edwin Geissler and E.S. Russell (1978) currently are analyzing 10 putative *W*-alleles all of which occurred spontaneously in the C57BL/6J strain at the Jackson Laboratory (see also E. Russell and Bernstein, 1974). Since these mutations are on the same genetic background they are especially advantageous for comparing their pleiotropic effects and in many cases the severity of these effects in different tissues does not correlate at all well. For example, while seven of these alleles produce severe anemic conditions when homozygous—so severe that only 0–11% of animals homozygous for these alleles are viable at birth and all succumb shortly thereafter—the amount of spotting associated with these same alleles when heterozygous (with +) varies from as little as 4% (limited to the ventrum) to as much as 95%. Moreover, some of these heterozygotes display a severe pigmentary dilution while in others this manifestation may not be present at all. These observations are especially interesting because they provide further evidence that the influence which the *W*-series of alleles have on hematopoiesis, gametogenesis, and melanoblast survival are unrelated. Indeed, one wonders if the *W*-locus is after all a functional unit, or a complex of separate regions for different functions (E. Russell, personal communication)?

The effects of some of the *W*-locus genotypes considered above are summarized in Table 10-2.

Table 10-2. Descriptions of Various *W*-Locus Genotypes[a]

Genotype	Description
$W/+$	Generally a well-defined belly spot and a very variable amount of white on dorsum (usually "variegated"); normal viability and fertility
W/W	White with black eyes; severe macrocytic anemia; usually lethal soon after birth; severe germ cell deficiency
$W^x/+$	Like $W/+$
W^x/W^x	Like W/W
$W^v/+$	Diluted coat; belly spot of variable size; frequently small mid-dorsal head spot; slight macrocytic anemia; fully viable and fertile

Table 10-2. *continued*

Genotype	Description
W^v/W^v	White with black eyes (occasionally pigment in skin of ear pinna); macrocytic anemia (not as severe as in W/W); many survive to maturity; usually sterile but on some backgrounds males considerably more fertile than females; develop ovarian tumors; moderate to severe inner ear abnormalities
W/W^v	White with black eyes; severe macrocytic anemia (about one-half survive to adulthood); severe germ cell deficiency
W^x/W^v	Like W/W^v; develop ovarian tumors
$W^a/+$	Dilutes black pigment to some shade of grey; belly spot and prominent head blaze; some may show variegation in that one or more patches may be a different shade of grey
W^a/W^a	Severely anemic and die within a few days after birth
W/W^a	Severely anemic
$W^b/+$	Dilutes black pigment (more than $W^v/+$); extensive diffuse white spotting of back and sides (more than $W^v/+$); small frontal blaze; belly predominantly white
W^b/W^b	White with black eyes; presumably anemic (median life span 8 days); survivors sterile
$W^f/+$	Spotting variable and on some backgrounds not expressed; mild macrocytic anemia
W^f/W^f	Extensive white areas lacking any pattern; fully viable and fertile; macrocytic anemia
W^f/W^v	White with greyish patches at the roots of the ears; black eyes; fertile
$W^j/+$	Extensive white spotting on ventrum; some white on dorsum especially on crown of head (otherwise intensely pigmented); normal viability and fertility
W^j/W^j	White with black eyes; severely anemic (average survival about 8 days); severe germ cell deficiency
W^j/W^v	White with black eyes; anemic; germ cell deficiency; develop ovarian tumors
$W^{pw}/+$	Pigmentation of coat restricted mainly to head (snout, near eyes, base of ears) and base of tail; sometimes pigmented patches on hips and shoulders
W^{pw}/W^{pw}	Severely anemic (die within 3 days after birth)
W^{pw}/W^v	White with black eyes; sterile
$W^{sh}/+$	Broad sash of white around body in lumbar region; normal viability and fertility
W^{sh}/W^{sh}	White with black eyes; viable and fertile
W^{sh}/W^v	White with black eyes but with small patches of pigment around ears and eyes; viable and fertile
$W^e/+$	Resembles $W^b/+$; white belly spot with extensive white markings on head and body; diluted coat; viable and fertile
W^e/W^e	White with black eyes; anemic; lethal soon after birth
W^e/W^v	White with black eyes; viable but sterile

[a]Only effects noted in literature are included.

II. Patch (*Ph*) and Rump-White (*Rw*)

Although patch (*Ph*) and rump-white (*Rw*) will for the most part be considered separately, they are introduced together because as noted above they are very closely linked to each other as well as to the *W*-locus on chromosome 5. In fact, these three loci are so closely associated that it seems logical at present to regard them "as a gene triplet which has arisen by repeated duplication of the original chromosome segment" (Searle and Truslove, 1970). The upper fiducial limits (P=0.05) for map distances are 0.08 cM for *Ph* and W^v, 1.4 cM for *Ph* and *Rw*, and 0.4 cM for *Rw* and W^v (Grüneberg and Truslove, 1960; Searle and Truslove, 1970).[32]

A. Patch (*Ph*)

Almost all the information on this mutant stems from a detailed analysis by Grüneberg and Truslove (1960) and what follows is essentially a summary of their findings.

Although patch arose as a spontaneous mutation in the C57BL strain, Grüneberg and Truslove's study was conducted on an outcross to CBA/Gr.

1. Spotting Patterns of Heterozygotes

White spotting in *Ph*/+ animals is like the spotting in piebald (*s*/*s*) and belted (*bt*/*bt*) mice in that the areas of white fur usually are sharply demarcated from the pigmented areas and only occasionally does roan or variegated spotting, i.e., the intermixture of white and pigmented hairs as is characteristic of dominant spotting (*W*/+ and W^v/+), occur. While *Ph*/+ mice always display at least a large belly-spot as well as extensive tail spotting and white digits, and there is frequently a large white patch in the middle of the trunk, the amount of white spotting is highly variable (Figures 10-3 and 10-4a). Moreover, as with other spotting genes, this variability depends upon the genetic background. Thus when Grüneberg and Truslove outcrossed their patch stock to three different inbred strains (A, C57, and CBA) the F_1 *Ph*/+ mice displayed widely different amounts of dorsal white. The spotting pattern itself seems to result from "the gradual spread of a belly-spot to form a belt (sometimes with a secondary belt in the shoulder region) which then expands first posteriorly and then anteriorly."

2. Interactions with Other Spotting Determinants

When tests were carried out to establish the relationship between patch and some other spotting genes it was found that they interacted synergistically. Thus when *Ph*/+;*s*/+ mice were backcrossed to +/+;*s*/*s* animals the *Ph*/+;*s*/*s* offspring resembled the black-eyed white (*W*/+;*s*/*s*) of the Mouse Fancy, being either completely white, or with small pigmented patches round the eyes, ears, or on the haunches (Grüneberg and Truslove, 1960). Patch likewise interacts with belted (*bt*), white (Mi^{wh}), and viable dominant spotting

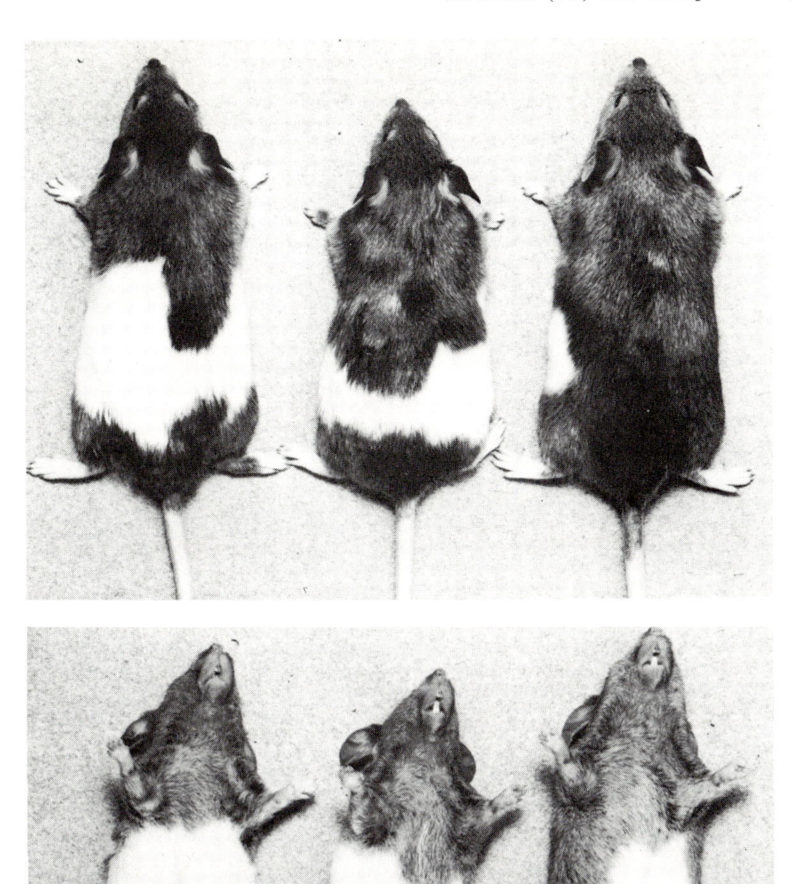

Figure 10-3. Dorsal and ventral views of three patch (*Ph*/+) mice. From H. Grüneberg and G.M. Truslove (1960). Reproduced with permission of the authors and Cambridge University Press.

(W^v) (Grüneberg and Truslove, 1960) as well as with dominant spotting (*W*), steel (*Sl*) (Wolfe and Coleman, 1966), steel-Dickie (Sl^d), and rump-white (*Rw*) (Deol, 1970b; Searle and Truslove, 1970). Patch, belted (*Ph*/+;*bt*/*bt*) mice usually have pigmented heads and shoulders while the rest of the body is either completely white or has a little pigment on the haunches. Patch,

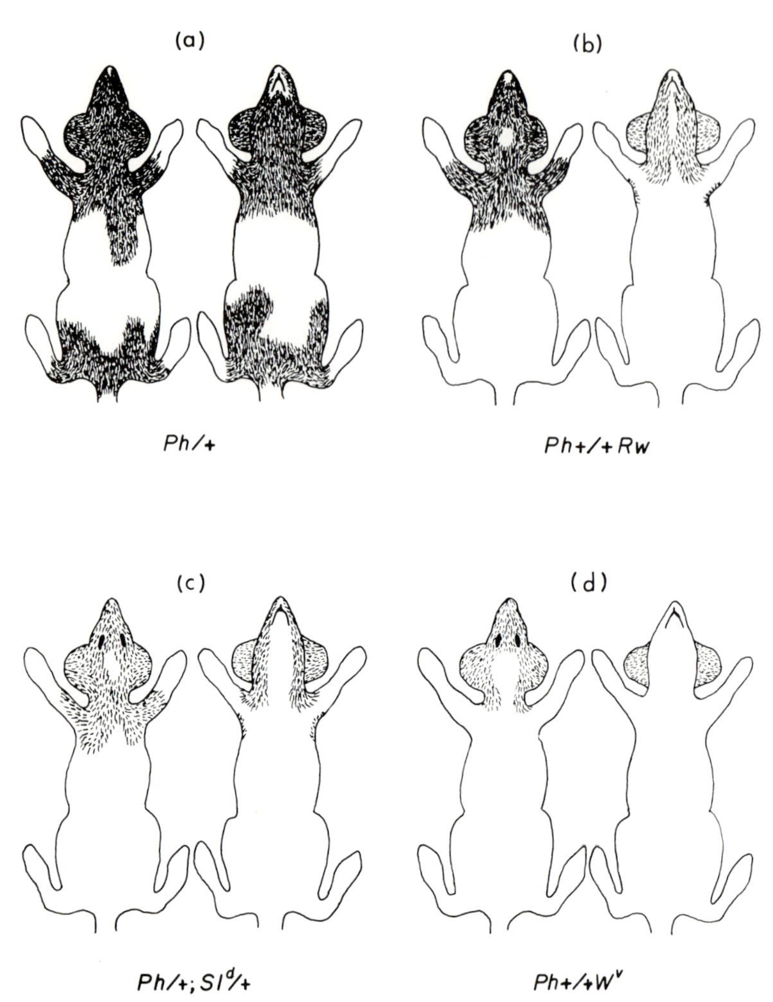

Figure 10-4. Typical spotting patterns of (a) *Ph*/+ alone and in combination with (b) *Rw*/+, (c) Sl^d/+, and (d) W^x/+. Dorsal view on the left and ventral view on the right. Based on Deol (1970b). Modified with permission of the author.

white (*Ph*/+;Mi^{wh}/+) mice resemble *Ph*/+;*bt*/*bt* animals in that pigment is confined to the head, shoulders, and forelegs, with sometimes a little on the haunches. However, unlike *Ph*/+;*bt*/*bt* animals, the pigmented areas of these mice are dilute as a consequence of the effect of Mi^{wh}/+ (see Chapter 12, Section I, B, 1). In the double heterozygote of patch and rump-white (*Ph*+/+*Rw*) the head (except for a small head spot) and the shoulder region are almost fully pigmented, while the remainder of the body is white although occasional pigment patches occur (Figure 10-4b). The same description applies to the patch and steel-Dickie ($Ph/+;Sl^{d}/+$) double heterozygote, except that in these the pigmented fur is diluted (Figure 10-4c).

The most profound effect on pigmentation is found when patch is combined with W^{v} (Figure 10-5). $Ph+/+W^{v}$ mice are completely unpigmented except for their head which is usually pigmented as in Figure 10-4d but is sometimes pigmented only on the cheeks and ears. Moreover, even the pigmented areas are diluted to a greater extent than can be accounted for solely by the influence of $W^{v}/+$. These animals also have less pigment in the inner ear than either heterozygote alone, a deficiency associated with more severe inner ear abnormalities (see Deol, 1970b).

3. Other Effects

The only other anomaly found in $Ph/+$ mice is that their skull is a little wider and shorter than normal and has a large interfrontal bone (Grüneberg and Truslove, 1960).

4. Viability of Homozygotes

Inasmuch as all patch mice tested have proved to be heterozygous and since matings between such animals produce a 2:1 rather than a 3:1 ratio of spotted to self mice, it is evident that the Ph/Ph homozygote dies sometime before birth. To determine when, Grüneberg and Truslove examined embryos from both F_2 ($Ph/+ \times Ph/+$) and backcross ($Ph/+ \times +/+$) matings at various stages in their development. They found in the F_2 matings (but *not* in the backcrosses) that some late embryos (12–17 days old) were severely disfigured by a condition they designated as "cleft face." Never-

Figure 10-5. Typical $a/a;Ph +/+ W^{v}$ mouse. Courtesy of George C. McKay, Jr.

theless, the number of these "monsters," which obviously were *Ph/Ph* homozygotes, only accounted for about a third of their expected number. Clearly, about two-thirds of the *Ph/Ph* embryos had died before they reached this stage.

Further studies revealed that the primary abnormality of *Ph* homozygotes is an increase in their water content (hydrops) which seems to develop between the 8- and 9-day stage.[33] Abnormal accumulations of liquid are observed first flanking the notochord posteriorly and later anteriorly. To a varying extent, excess fluid also occurs in the circulation, pericardium, under the epidermis, and within the tissues. The more severely affected embryos die around the tenth day but about one-third survive longer. These subsequently develop a subepidermal bleb in the middle of the face which interferes mechanically with the formative movements of the nose and palate and hence is responsible for the "cleft-face."

It appears obvious from the more extreme manifestations of the *Ph* homozygous condition that the total water content of the embryo is greatly increased and that this excess fluid must be of external origin. Whether it enters via the allantois, or whether it is derived from the amniotic cavity or from the extraembryonic coelom is, however, not known.

5. Pigmentation of Homozygotes

The pigmentation of *Ph* homozygotes has recently been investigated by Mayer (unpublished) who transplanted skin from various regions of two 14-day-old *Ph/Ph* embryos, identified by "cleft-face," to the testes of compatible hosts. He found that whereas skin from the vibrissae region always produced pigmented hairs, and skin from the shoulder and hip regions gave rise to either pigmented, unpigmented, or a mixture of pigmented and unpigmented hairs, all grafts from the flank region were unpigmented. It thus appears that, unlike splotch (*Sp*) homozygotes (see Chapter 11, Section III, A, 2), *Ph/Ph* animals have some viable melanoblast clones. They also possess pigmented eyes.

6. Influence on Erythropoiesis and Gametogenesis

Because the *Ph*-locus is so closely linked to the *W*-locus, Grüneberg and Truslove determined whether *Ph* influenced erythropoiesis or germ cell development. They found that 13- to 14-day-old *Ph*/+ heterozygotes were not anemic (see also Searle and Truslove, 1970) and that *Ph* homozygotes appeared to have normal numbers of primordial germ cells at 10 days gestation.[34] Whether these homozygotes are, or would be, anemic was not possible to determine but they did find that $Ph+/+W^v$ animals suffered from a slightly more severe macrocytic anemia than that associated with W^v ($W^v/+$) alone. Thus while these data do not provide a strong case for the existence of a specific (allelic) interaction between the *Ph* and *W* loci, neither do they rule it out (Grüneberg and Truslove, 1960).

B. Patch-Extended (Ph^e)

Recently a second allele has been reported at the patch locus (Truslove, 1977). It has been given the name of patch-extended (Ph^e) because only the head and shoulder region of $Ph^e/+$ animals are pigmented (Figure 10-6a).[35]

When $Ph^e/+$ mice are crossed with $W^v/+$ animals the double heterozygote ($Ph^e +/+ W^v$) is a black-eyed white (Figure 10-6b), phenotypically indistinguishable from W^v/W^v. Moreover, these animals look slightly anemic at birth and, unlike $Ph +/+ W^v$ and $Rw +/+ W^v$, may be sterile as so far none of them (three males and three females) has bred (Truslove, 1977).

$Ph^e +/+ Rw$ mice look like $Ph +/+ W^v$ animals as they have diluted fur restricted to the sides of the face and ears (Figure 10-6c).

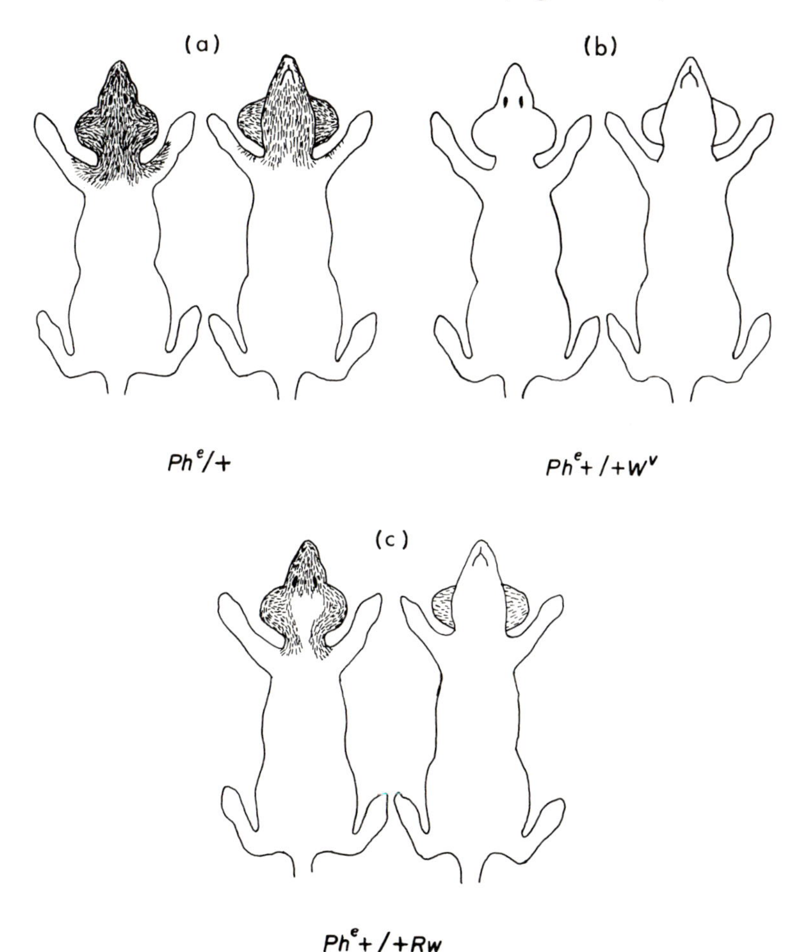

Figure 10-6. Typical spotting patterns of (a) $Ph^e/+$ alone and in combination with (b) $W^v/+$ and (c) $Rw/+$. Dorsal view on the left and ventral view on the right. Based on G.M. Truslove (1977). Modified with permission of the author.

Although, as in the case of *Ph* homozygotes, both Ph^e homozygotes and Ph/Ph^e heterozygotes usually die early in gestation, one presumed Ph/Ph^e and one presumed Ph^e/Ph^e embryo with "cleft-face" have been observed. The latter homozygote survived to birth (Truslove, 1977).

C. Rump-White (*Rw*)

This mutation originated in the course of an experiment in which hybrid mice were exposed to fast neutron irradiation (Batchelor et al., 1966). It has been analyzed by Searle and Truslove (1970) and the following is based on their observations.

1. Spotting Patterns of Heterozygotes

Mice heterozygous for rump-white (*Rw*/+) usually "have white tails, apart from frequent distal pigmentation,[36] white hind legs and a variable area of depigmentation in the sacral and lumbar regions. This tends to be rather more extensive ventrally than dorsally. Occasionally there are islands of pigmented hair surrounded by white. Digits of forefeet also tend to be white." Three typical *Rw*/+ mice are shown in Figure 10-7 (see also Figure 10-8a).

As in *Ph*/+ mice, the pigmented areas of the coat of *Rw*/+ animals show no signs of dilution. In fact, rump-white shows a definite phenotypic resemblance to patch, except that the main area which is unpigmented is in the lumbosacral instead of the thoracolumbar region. The average amount of tail pigmentation also is greater in patch.[37]

2. Homozygotes

Rump-white like patch behaves like a fully penetrant dominant mutation which is lethal when homozygous. Although the exact cause of death is not known, dissections of pregnant females from *Rw*/+ × *Rw*/+ matings revealed an excess of dead embryos in mid-pregnancy, at about the frequency expected if they comprised the *Rw*/*Rw* class (Searle and Truslove, 1970)

3. Interaction with Other Spotting Determinants

Rump-white also interacts synergistically with other white spotting genes. We have already described the $Rw+/+W^f$ and the *Rw*+/+*Ph* (Figure 10-4b) phenotypes. In $Rw+/+W^v$ mice (Figures 10-8b and 10-9) pigmented hairs are scattered among unpigmented ones in thoracic and lumbar regions as well as on the head. Moreover, there is always a large mid-dorsal head spot and very little pigmentation ventrally. The ears are always pigmented but, as in $Ph+/+W^v$ animals, there is less pigment in the inner ear (and more inner ear abnormalities) than in either *Rw*/+ or $W^v/+$ mice (Deol, 1970b).

In general, the interaction of *Rw* with other spotting genes results in the posterior part of the coat being affected more than the anterior (Searle and

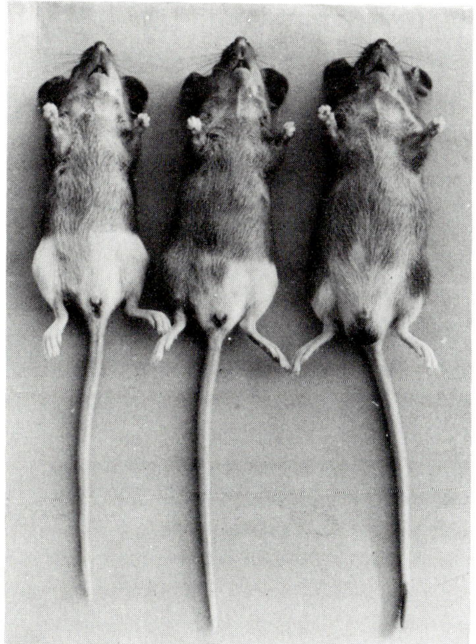

Figure 10-7. Dorsal and ventral views of three rump-white (*Rw*/+) mice. Note that the scrotal skin of the male is pigmented but that the hairs arising through it are not. This male also has a pigmented tail tip. From A.G. Searle and G.M. Truslove (1970). Reproduced with permission of the authors and Cambridge University Press.

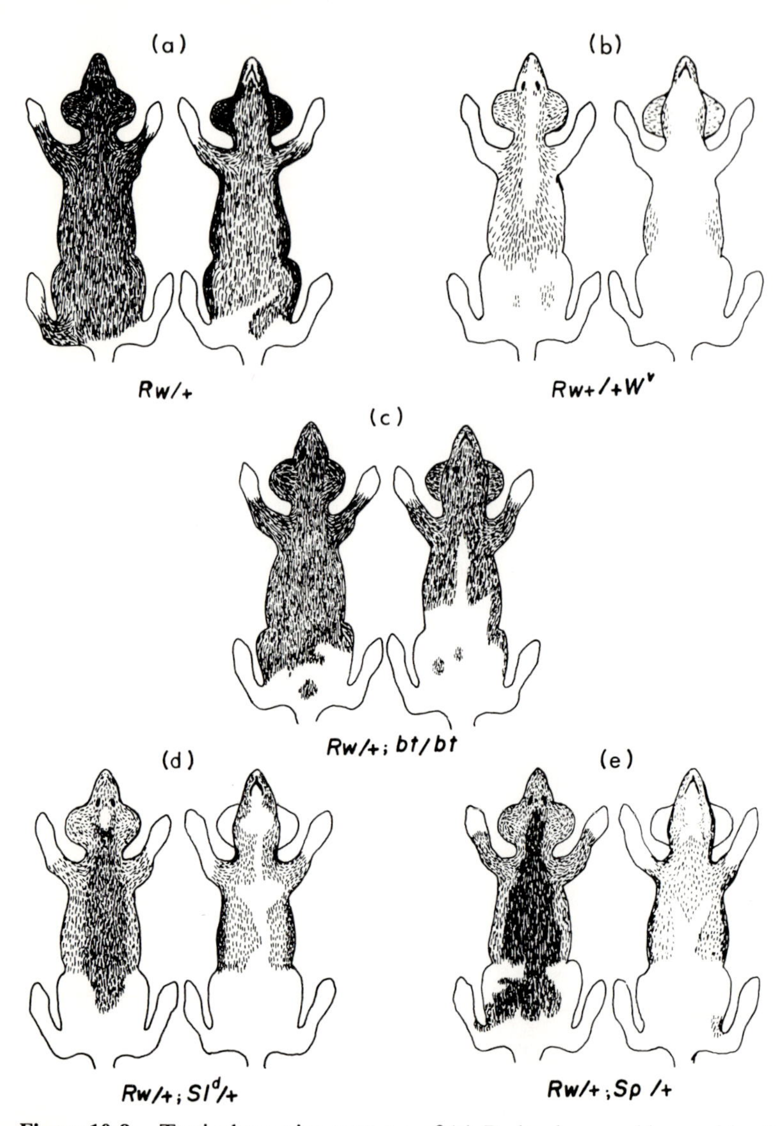

Figure 10-8. Typical spotting patterns of (a) *Rw*/+ alone and in combination with (b) W^v/+, (c) *bt*/*bt*, (d) Sl^d/+, and (e) *Sp*/+. Dorsal view on the left and ventral view on the right. Based on A.G. Searle and G.M. Truslove (1970). Modified with permission of the authors.

Truslove, 1970). Thus rump-white, piebald (*Rw*/+;*s*/*s*) mice have pigmented hairs on their head and shoulders, but little elsewhere. With lethal spotting (*ls*), *Rw* displays a greater reduction of pigmentation than with *s*, so that only one or very few colored patches, mainly dorsal and anterior to the sacrocaudal region, remain (Searle and Truslove, 1970). Typical examples

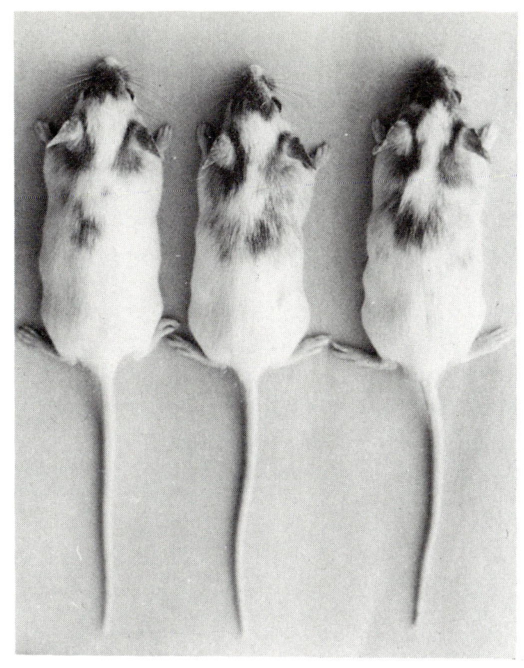

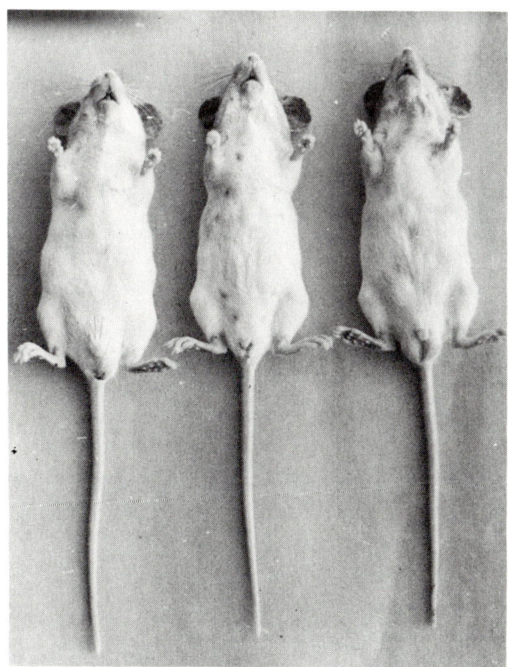

Figure 10-9. Dorsal and ventral views of three *Rw* +/+ W^v mice. Courtesy of G.M. Truslove.

of the spotting patterns displayed by rump-white, belted (*Rw*/+;*bt*/*bt*), rump-white, steel-Dickie (*Rw*/+;Sl^{d}/+), and rump-white, splotch (*Rw*/+;*Sp*/+) mice are shown in Figure 10-8c, d, and e. With *bt*, as with *s*, the area without pigmentation seems less than Grüneberg and Trusolve (1960) reported for *Ph*. Furthermore, a similar situation seems to prevail when *Rw*/+;Mi^{wh}/+ mice are compared with *Ph*/+;Mi^{wh}/+ animals; although both of these genotypes possess pigmented heads, shoulders, and forelegs, only in the former is there considerable pigment in the thoracic region.

4. Influence on Erythropoiesis

Because of the close linkage between *Rw* and *W* the blood picture of *Rw*/+ mice and of *Rw*+/+W^{v} animals has also received attention. As in the case of *Ph*/+, *Rw*/+ mice have a normal blood picture while the double heterozygote of *Rw* and W^{v} is slightly more affected than W^{v}/+ alone both with respect to the number of erythrocytes, and their mean corpuscular volume (Searle and Truslove, 1970).

D. Etiology of White Spotting in *Rw*/+ and *Ph*/+ Mice

Although, as the above testifies, there are obvious differences in the pigmentation patterns of *Rw*/+ and *Ph*/+ mice, the most interesting difference was found when thin slices of 1- to 3-day-old skin from various pigmented and nonpigmented regions of these genotypes were cleared, prepared for microscopic examination, and compared. Whereas it was found that in 1-day-old *Rw*/+ mice dendritic melanocytes were scattered throughout the epidermis, even in areas which were devoid of pigmented hair bulbs, in *Ph*/+ animals very few pigmented epidermal melanocytes were seen either in regions of pigmented or of white hair. Thus, in the pigmented rump region of *Ph*/+ mice there were decidedly fewer melanocytes in the epidermis than were found in normal littermates or in the white rump area of *Rw*/+ mice (Searle and Truslove, 1970).

The presence of many extrafollicular melanocytes in the white rump region of *Rw*/+ infant mice indicates that the absence of hair pigment is not due to a failure of melanoblasts to migrate into or survive in this region. Moreover, as pointed out by Searle and Truslove, this is reinforced by the fact that the white spotting of the scrotum and of the tail of these mice includes only the hairs, i.e., the skin through which these hairs emerge is pigmented (see Figure 10-7). While these observations could be taken as evidence that the white spotting of *Rw*, unlike that of *Ph*, is due to some genetic defect which prevents melanoblasts from either entering or maturing in the hairs of the affected region, as has been claimed for belted (Mayer and Maltby, 1964) (Chapter 9, Section IV, B), there is an alternative explanation. If one assumes that the melanocyte population in the white rump area of *Rw*/+ mice represents a *secondary* population which migrated in after the original population failed to survive because it originated from a defective clone(s) (Mintz, 1969a), the unpigmented hairs in this region could result from the fact that this "new" population entered the region too late to become incorporated into the developing hair bulbs. Similarly, the white spotting in *Ph*/+ mice also could be due to the preprogrammed death of specific clones of melanoblasts but at a time too late for them to be replaced by viable cells from adjacent regions. In accord with this interpretation is the observation that whereas in *Ph*/+ mice there is a sharp demarcation line between the areas of pigmented and nonpigmented hair bulbs (Grüneberg and Truslove, 1960), in *Rw*/+ animals unpigmented hair bulbs are sometimes surrounded by pigmented ones and vice versa (Searle and Truslove, 1970).[38]

Notes

[1]An independent mutation to a *W*-allele occurred in the C3H/J colony in 1952. This mutation is in all respects indistinguishable from *W* and has been called W^x. In C3H/J, W^x/+ individuals have ventral spots of moderate size and no dorsal spots (E. Russell et al., 1957).

[2]According to Grüneberg (1952), the number of m(*W*) genes involved in Dunn's experiments was probably few (about three).

[3]In a stock homozygous for the whole m(*W*) complex *W* is nearly completely dominant in its effects on the fur and *W*/+ × *W*/+ matings produce one-quarter *W*/*W* anemics (potentially white), one-half *W*/+ roan (nearly white), and one-quarter +/+ solidly colored animals. On the other hand, in a stock homozygous for the normal alleles of the m(*W*) complex, *W* is completely recessive and so heterozygous matings produce one-quarter *W*/*W* anemics (potentially white) and three-quarters solidly colored animals (two-thirds *W*/+ and one-third +/+).

[4]The black-eyes in these animals is due entirely to the fact that the retinal layer of the eye is pigmented normally (see Chapter 1, note 5). It should also be noted that the m(*W*) complex has no affect on *W*/*W* melanoblasts nor has it any apparent effect in *W*/*W* (or *W*/+) mice on erythropoiesis or gametogenesis.

[5]Thus the genotype *W*/+;*s*/*s*;*p*/*p* may be phenotypically indistinguishable from an albino (*c*/*c*).

[6]The original W^v mutation occurred in a male of a strain of silvered black self mice who, unlike his sibs, had a number of small irregular white spots or streaks on his dorsal surface and a white streak on the mid-ventral line. The colored areas of his coat were also markedly paler than the typical coat color of the black silvered stock from which he originated. An independent mutation, indistinguishable from W^v, also occurred in the C57BR colony at the Jackson Lab in 1945 (E. Russell et al., 1957).

[7]In 1952 Carter described a female mosaic mouse heterozygous for W^v which had two large patches of wild type hair, as if a somatic mutation had occurred from W^v to +. Nevertheless, when this animal was bred she produced a significant excess of W^v offspring indicating that a mutation in the *opposite* direction, i.e., from + to W^v, had occurred in her gonads. Carter discusses the possible explanations for this paradoxical situation and concludes that the most likely ones are somatic crossing-over (see

Chapter 3, note 12) or somatic reduction at an early cleavage stage. If somatic crossing-over between W^v and the centromere had occurred both W^v/W^v and +/+ daughter cells would be formed; if the former cells had given rise to one ovary while the latter produced the wild type colored patches, this would explain the mosaicism. This seems a more likely explanation than somatic reduction at an early cleavage division, leading to haploid W^v and + cells, since it is most unlikely that a haploid ovary or haploid melanocytes could function successfully. Still another possibility, and one also discussed by Carter, is nondisjunction leading to aneuploid cells of genotype $W^vW^v/+$ (going to the ovary) and + alone. However, the observed segregation of W^v and + in the offspring of this mosaic female does not agree as well with this explanation as with the other two.

[8]The harderian glands of these mice likewise sometimes possess melanocytes (see Table 10-1) (Markert and Silvers, 1956).

[9]Another characteristic of W^v/W^v genotypes is that, unlike normal mice which are able to utilize both the D and L form of certain amino acids, they are unable to utilize the D form and this form is excreted. $W^v/+$ mice are more or less intermediate in this regard and excrete less of the D form (Goodman, 1955, 1956, 1958). The basis of this change in amino acid metabolism, and its significance, is not known.

[10]$W^v/+$ also was found to produce more white spotting than $W/+$ on four different genetic backgrounds. Furthermore, the amount of white spotting associated with each of these heterozygotes is greatly augmented when combined with $Mi^{wh}/+$ (E. Russell et al., 1952).

[11]Recently Heath (1978) has produced allophenics between mice carrying W^v and normal (+/+) animals of the AG/Cam strain. Six such allophenics have so far survived to adulthood and one of these is of the class $W^v/W^v \leftrightarrow +/+$. The coat of this mouse displays the belly spotting characteristic of $W^v/+$ mice, whereas its blood is of the +/+ type and there is no evidence for transmission of mutant gametes. Two other mice are putatively $W^v/+ \leftrightarrow +/+$ allophenics. These mice too have belly spots but also show evidence of contributions from both partners in their blood and germ line.

[12]Schumann (1960) has presented evidence which she believes indicates that white spotting in mice is caused by a delay in the migration of melanoblasts from the neural crest. She contends that "as a result of this delay, the pigment cells do not reach some parts of the integument until after the skin has already reached such a stage of development that it is impervious to the melanoblasts" [see, however, Mayer and Maltby (1964) and Chapter 9, note 11].

[13]W/W^v mice, like W/W and W^v/W^v animals, are severely anemic black-eyed whites. On most genetic backgrounds about one-half of these heterozygotes survive to adulthood (E. Russell et al., 1957).

[14]One interesting result of this study was that whereas the combination of +/+ epidermis and W/W^v dermis *always* resulted in pigmented grafts, in only 50% of reciprocally combined grafts, i.e., W/W^v epidermis and +/+ dermis, did pigment occur. This variability in the occurrence of melanocytes in grafts in which the mesoderm is the +/+ melanoblast carrier undoubtedly reflects the migratory pattern of melanoblasts into the skin. As pointed out by Mayer (1973a), and in accord with other findings (Mayer, 1973b), these results are interpreted best by assuming that melanoblasts initially migrate through the dermal mesoderm and, secondarily, gain access to the

ectoderm. Thus, by 13 days gestation the epidermis of +/+ embryos would be expected to be heavily populated with melanoblasts, which a few days previously occurred only in the dermis but which by now had, for the most part, left this layer (see Chapter 11, note 12).

[15]Gordon draws attention to the fact that the trace of pigment which occurred in one alleged *W*/+;+/+ ↔ +/+;*c*/*c* allophenic was located at the tip of the right hind foot, an area which is consistently unpigmented in *W*/+ mice. He believes this suggests that the absence of pigment cells in this area of *W*/+ mice may not be due to a migratory insufficiency, but rather to the fact that *W*/+ pigment cells are marginally viable. He raises the possibility that "as the cells migrate dorsolaterally in *W*/+ mice, cells more proximal to the dorsal midline may die, leaving empty spaces which are filled by backward migration of more distally located melanoblasts." In the *W*/+;+/+ ↔ +/+;*c*/*c* mouse, however, these empty spaces may be filled by laterally migrating albino cells from a neighboring clone, thus permitting the *W*/+;+/+ cells to proceed to the tip of the foot.

[16]Mayer's observation that 61% of neural tube grafts (instead of the expected 25%) from W^v/+ matings did not produce any pigment when combined with neural crest-free +/+ skin may likewise be interpreted in terms of some inherent weakness of W^v/+ melanoblasts. On the other hand this observation also is in complete accord with Mintz's notion that some of the pieces of W^v/+ neural tubes, by chance, possessed only inviable melanoblast clones.

[17]Deol also found the choroid of W^v/+ mice frequently to have "one or more moderately large unpigmented patches of an extremely irregular shape, the total unpigmented area being always less than half. The borders of these patches were sharp but heavily indented, somewhat like the skull sutures in old mice. The pigment, where present, seemed to be of normal density, and there was generally more of it in the ventral half than the dorsal. There was no tendency towards a regular pattern of symmetry."

[18]It is interesting to note that although the coat of W^v/W^v mice is invariably white, Deol (1970c) frequently found the inner ear of these animals to be pigmented. According to him in normal mice the distribution of pigment in the labyrinth is not uniform, but follows a characteristic pattern, occurring largely in the vestibular part. In the cochlea of +/+ mice pigment is confined to the stria vascularis, and the saccule is on the whole free of it. The free wall of the utricle of +/+ animals is heavily pigmented, as are certain well-defined areas in the semicircular ducts; pigment likewise occurs around the cristae, and among the nerve fibers, particularly near their external endings. In contrast, in the 33 W^v/W^v mice examined by Deol, 7 had no pigment in the inner ear on either side, 7 displayed it only on one side, and in 19 animals both sides were pigmented. Nevertheless, even in the 45 ears which possessed pigment its distribution was never normal. Thus, "the stria was always unpigmented except for a small region in the basal part in those few cases in which the basal stria was normal."

[19]Although Deol (1970b) notes that there is a good correlation between the severity of the effect which different spotting genes have on the inner ear with their effect on the pigmentation of this part of the ear, this does not always apply to animals of the *same* genotype. Thus, he found no clear correlation between the pigmentation of the inner ear and its abnormalities in W^v/W^v mice (Deol, 1970c). Because of this

Deol deems it possible that "the region of the neural crest which produces melanoblasts for the inner ear is different from that which produces the primordium of the acoustic ganglion."

[20]The mean erythrocyte count in the normal young adult mouse is about 9–10 × $10^6/mm^3$ whereas in the viable W^v/W^v it is about 5–5.5 × $10^6/mm^3$ (E. Russell, 1954).

[21]W/W^v mice also have been reported to have normal numbers of platelets and granulocytes in their peripheral blood (J. Lewis et al., 1967) but decreased numbers of megakaryocytes (Chervenick and Boggs, 1969; Ebbe et al., 1973a; Ebbe and Phalen, 1978) and neutrophils (Chervenick and Boggs, 1969) in their marrow. Moreover, although these mice respond poorly to erythropoietin (Keighley et al., 1966), they respond as well as +/+ animals to lowered oxygen tension by developing elevated hematocrits, reticulocytosis, and increased blood volume (Keighley et al., 1962; Bernstein et al., 1968). In fact during prolonged periods of hypoxia such mice produce large amounts of erythropoietin but, because of their genetic defect in RBC precursors, they remain anemic (E. Russell and Keighley, 1972; see also Fried et al., 1967). Erythropoiesis is also stimulated in W/W^v mice by bleeding (Grüneberg, 1939; Harrison and E. Russell, 1972).

[22]Recently Flaherty and her associates (1977) have demonstrated that the cell surface antigens of normal erythrocytes change during their maturation and that the appearance and disappearance of these antigens differ in +/+ and *W*/*W* cells. Employing both a rat anti-mouse erythroblast serum and a rat anti-mouse adult RBC serum they found that the former antiserum recognized antigen(s) present on erythroid cells early in development, while the latter recognized antigen(s) present on mature erythroid cells only. Relative to normal cells, the erythroid cells of *W*/*W* mice were out of phase; the developing cells prematurely lost the "early" antigen(s) recognized by the antierythroblast serum and prematurely gained the "late" antigen(s) recognized by the anti-RBC serum. Flaherty and her colleagues suggest "that the *W*-locus may affect the erythroid cell surface in such a way as to preclude normal recognition of the mitogenic signals needed for continued proliferation." Such a cell surface defect could, of course, be either a primary one or one which results from some abnormal differentiative process in these cells which then leads to an altered makeup of the cell surface. For other hematological investigations on W/W^v mice see Till et al.(1967), Boggs et al. (1973) and Harrison (1972b, 1975a,b).

[23]In addition to the original W^v/W^v deviants of Little and Cloudman which occasionally were fertile, a stock displaying fertility of some homozygous W^v mice was produced by transferring this allele into a stock of *W*/+ animals selected for prolonged survival of *W*/*W* anemics (E. Russell and Lawson, 1959). Following this procedure the incidence of fertile W^v/W^v males was much higher than the incidence of fertile W^v/W^v females. Indeed only 2 females of 14 were fertile and they produced only one small litter each (Mintz, 1960). This sex difference is not unexpected since, in contrast to the situation in males, any increase in the number of germ cells in the female can occur only *before* meiotic prophase sets in. Thus, as emphasized by Mintz (1960), if conditions favoring germinal proliferation are slow in operating, the female is at a marked disadvantage as compared with the male. It should also be noted that in fertile W^v/W^v adult males the testis does not present a normal histological picture. Although normal-looking regions are evident, spermatogenic elements may be lack-

ing in some cross-sections of tubules and even where spermatogenesis has clearly been occurring, it may be arrested or abnormal (Mintz, 1960).

[24]E. Russell and Fekete (1958) also observed that these W^v/W^v females displayed diminishing numbers of young and atretic follicles until 5 months of age. This diminution was accompanied by an invagination of the germinal epithelium, culminating by the seventh month in the formation of tubular adenomas of all ovaries. Similar findings have been reported for the ovaries of other W-mutant heterozygotes (W^x/W^v and W^j/W^v) which progressed through stages of tumorigenesis classified as tubular adenomas, complex tubular adenomas, and finally granulosa-cell tumors and luteomas (E. Murphy and E. Russell, 1963). It is thought that tumors develop in these females as a consequence of the lack of developing ovarian follicles, i.e., due to the deficiency of germ cells there is an underproduction of estrogen and an overproduction of pituitary gonadotrophin with excess stimulation of the gonad. That nothing is wrong with the pituitary of W^v/W^v mice is demonstrated by the fact that when ovaries from normal (+/+) C57BL females are implanted into the ovarian capsules of weaning-age W^v/W^v females, they function normally. Such recipients mate and rear healthy normal litters. This indicates that the W^v/W^v pituitary can react normally to the estrogens from the +/+ gonad, and that there is also a normal response of the W^v/W^v uterus and mammary gland to this pituitary stimulation (W. Russell and E. Russell, 1948).

[25]According to Mintz (1957a), primordial germ cells are first observed in the normal mouse embryo at 8 days "when they are seen in the yolk sac splanchnopleure, caudal end of the primitive streak, and root of the allantois. Migration through neighboring tissues begins at 9 days, when the cells occur in the gut splanchnopleure and may proceed up the dorsal mesentery to the gut. At 10 days they are found in the mesentery and at its root, near the dorsal aorta, around the coelomic angles, in the mesonephric regions, and in the paired germinal ridges." Migration is largely completed at 12 days (see also Chiquoine, 1954; Mintz and E. Russell, 1957).

[26]Inasmuch as the experimental matings in this study were between heterozygotes, they should theoretically yield 25% defective offspring. The actual frequency of embryos with a severe paucity of germ cells was 28–29%. It should also be noted that there were no apparent differences in the number of germ cells in W/W, W/W^v, and W^v/W^v embryos. This is interesting because, as shown by Coulombre and E. Russell (1954), at 0–28 days postpartum the gonads of surviving "defective genotypes" can be ranked with W/W gonads exhibiting the most severe defect and W^v/W^v the least. Thus it appears that the different effects which W and W^v have on germ cell development occurs between the end of germ cell migration and parturition (Mintz, 1957a).

[27]W/W^v mice are also very deficient of mast cells; their skin possesses less than 1% the normal number, and they do not occur in other tissues (Kitamura et al., 1978). After transplantation of bone marrow cells from normal (+/+) donors, however, the number of mast cells in the skin, stomach, caecum, and mesentery of these mice increases to normal levels (Kitamura et al., 1978). Although the fact that W/W^v genotypes lack both melanocytes and mast cells seems to be consistent with Okun's (1976) claim that these cells have a common precursor, as pointed out by Kitamura and his associates, this contention is inconsistent with the observation that the white spots of $W^v/+$ and $W/+$ mice have as many mast cells as their pigmented dorsal skin.

W/W^v mice also appear to be deficient of a thymus derived cell. Thus when they are injected with normal (+/+) bone marrow cells that have been treated with antiserum to the thymus cell antigen theta (Thy-1) and complement (C′), their anemia is not cured. The addition of +/+ thymocytes to these cells, however, restores their capacity to cure the anemia (Wiktor-Jedrzejczak et al., 1977). These findings suggest that a theta-sensitive cell is required for the promotion of mouse hematopoietic stem cells into erythrocytes, and that W/W^v mice are deficient of such a cell (Wiktor-Jedrzejczak et al., 1977).

[28]Schaible (1969) reported that $Mi^{wh}/+$ and $Va/+$ heterozygotes also display a similar type of variegation with no evidence of germinal mosaicism and that "reversion to the non-mutant color occurred in 5.3, 8.0 and 90.5%, respectively, of 4083 $Mi^{wh}/+$, 1738 $W^a/+$ and 317 $Va/+$ individuals from the stocks having minimum white."

[29]In 1953 Strong and Hollander published a paper entitled "Two Non-Allelic Mutants Resembling "*W*" in the House Mouse." One of these mutants, designated as "type-1," originated in the N-strain which is dilute, brown, nonagouti, and piebald ($d/d;b/b;a/a;s/s$). This mutant displayed more dilute pigmentation and extensive white areas, and breeding tests indicated it "was of the *W*-type, the homozygote being anemic and dying young." In general, however, the heterozygotes were more like $W^v/+$ in phenotype. The second mutant, designated as "type-2," appeared in a strain called 3CAMG. This mutant "showed a large frontal blaze, a large patch on the belly, and several spots on the shoulders and back." When these "type-2" animals were mated to each other it became evident that homozygotes for the mutation did not survive until birth. A subsequent examination of fetuses from these matings revealed that at 15 days some were anemic and some were already dying. Crosses between "type-1" and "type-2" indicated that they were *not* alleles, and crosses between "type-2" and $W/+$ mice produced no anemic young but the phenotypes of the F_1 and the 2:1 ratio obtained indicated intrauterine lethality. Unfortunately "type-1" was never crossed with a known *W*-allele. Nevertheless, "type-2" evidently was the consequence of a *W*-locus mutation and accordingly has been designated W^s for Strong's dominant spotting (Ballantyne et al., 1961).

[30]A spontaneous mutation in the CBA/H inbred strain has been reported (Searle et al., 1974) which could very well be a remutation to W^b. On an agouti background, animals heterozygous for this mutant have a large white frontal blaze, much white spotting on the body, a general lightening of the coat, and a white belly. They are easily distinguishable from $W^v/+$ littermates. Compounds with W^v are black-eyed whites which survive to maturity. Animals homozygous for this mutation are very anemic black-eyed whites which do not survive for more than about 10 days.

[31]In addition, a separate analysis revealed that the incidence and the size of the white spots on the crown of the head of $W^j/+$ mice was significantly greater than in $W/+$ animals, when both heterozygous genotypes were on the *same* genetic background (E. Russell et al., 1957).

[32]This "triplet" is very likely a "quartet" as more recently Southard and M.C. Green (1971) have reported that another white spotting determinant, recessive spotting (*rs*), also appears to be very closely linked to W^v. *rs* was originally described by Dickie (1966b). It was first noticed when two animals with large head blazes and large belly spots and diluted bellies were found in a C3H/HeJ litter. Although breed-

ing tests indicate that *rs* is a viable recessive mutation, nevertheless, 9–21% of *rs*/+ animals have a long thin white spot on the ventrum (but no head blaze and no dilution). Both W^v+/+*rs* and *Sl*/+;*rs*/+ genotypes are black-eyed whites, fertile, and nonanemic.

[33]At 9 days of gestation the *Ph*/*Ph* embryos can be distinguished from their normal sibs by external inspection. They all display wavy neural tubes. The more severely affected embryos have irregularities of the somites and the most abnormal individuals are inflated with enormous hearts, or hearts of about normal size inside a very large paricardium. Depending on the degree of abnormality, *Ph* homozygotes are retarded to a greater or lesser degree as compared with their normal littermates (Grüneberg and Truslove, 1960).

[34]Nevertheless there is some evidence that when the amount of white spotting is increased in *Ph*/+ mice by selection, there is a concomitant decrease in fertility (Truslove, 1977).

[35]In contrast to *Ph*, where the correlation in the amount of white spotting between parents and offspring is incomplete (Grüneberg and Truslove, 1960), in Ph^e mice it appears to be complete (Truslove, 1977).

[36]When the tail is partly pigmented the tip frequently has black hairs as well as pigmented skin while the proximal portion may have pigmented skin with unpigmented hairs.

[37]Also, the scrotal skin of *Rw*/+ males remains dark in spite of the fact that the hair arising through it as well as the surrounding skin and hair are unpigmented. In females the perianal skin likewise is pigmented. In contrast, in *Ph*/+ males when the ventral white area includes a portion of the scrotal region the scrotum itself lacks pigment. Moreover, unlike *Ph*/+ the size of the interfrontal bone of *Rw*/+ mice is normal.

[38]*Ph*/+ mice may also have fewer *viable* melanoblast clones than *Rw*/+ animals and this too could contribute to the deficiency of epidermal melanocytes in the unpigmented (and pigmented) regions of their skin.

Chapter 11

Steel, Flexed-Tailed, Splotch, and Varitint-Waddler

I. Steel (*Sl*)

The alleles at the *Sl* locus (chromosome 10)[1] are of special interest because, like those of the *W*-series, they too adversely affect melanogenesis, erythropoiesis, and germ cell development. The first mutation at this locus, *Sl*, arose spontaneously in a C3H line at Oak Ridge (Sarvella and L. Russell, 1956).

A. Heterozygotes

Steel receives its name from the fact that when heterozygous it produces an overall dilution of hair pigment (Plate 3–F) which is more severe on the ventrum. On a C3H background *Sl*/+ mice also have a white snout tip and, almost invariably, a small white spot in the middle of the forehead as well as one on the belly (see note 27). Very occasionally *Sl*/+ mice may also have a white blaze between the eyes or a few white hairs on the back. They also display an almost complete lack of pigmentation in the skin of the feet and frequently an unpigmented tail tip (D. Bennett, 1956).

On some genetic backgrounds the fur dilution of *Sl*/+ heterozygotes is very slight, and is not invariably accompanied by the pattern of small white spots present on the C3H background, the head spot being absent most often, the belly spot the least (Sarvella and L. Russell, 1956).

Although *Sl*/+ mice have a moderate macrocytic anemia (D. Bennett, 1956), their life span is normal. At 14 days gestation they possess about 75% of the normal number of erythrocytes and from 2 weeks after birth onward about 80%.[2] When heterozygous *Sl* also produces a reduction in gonad size which, according to Bennett, but not Mintz (1957b, 1960),[3] may be associated with a slight deficiency of primordial germ cells.

The macrocytosis associated with *Sl* is barely detectable in the heterozygote. At birth the mean cell diameter of *Sl*/+ vs +/+ erythrocytes is on the borderline of statistical significance but the maximum diameter in the heterozygote is considerably greater than the maximum in the +/+ group (and, conversely, the minimum diameter in the wild type is much smaller than in the heterozygote). At 15 days gestation the distribution of cell diameters in *Sl*/+ and +/+ animals is also very similar although the cell diameters of the former group may still be slightly larger. No differences are discernible in 14-day-old embryos (D. Bennett, 1956).

Although the average weight of the testis of *Sl*/+ males is about 75% that of normals (+/+) it looks normal histologically and the fertility of heterozygotes is not diminished.

B. Homozygotes

In contrast to the relatively slight effects which *Sl* has when heterozygous, it is lethal when homozygous, *Sl*/*Sl* mice rarely surviving to birth. The anemia already is severe at 14 days gestation when *Sl*/*Sl* embryos have only 28% of the normal number of erythrocytes and are readily distinguished by their pale color. At 15 days gestation the anemia is even more pronounced with *Sl* homozygotes having less than 16% of the normal number of red cells. The very severe anemia at this stage of development marks the beginning of the lethal period for these embryos, which die on or after the fifteenth day (Sarvella and L. Russell, 1956). Very occasionally some *Sl*/*Sl* homozygotes survive to birth with an erythrocyte count about 26% of normal, a value that is undoubtedly higher than it really is since such homozygotes never nurse and are probably dehydrated when assayed (D. Bennett, 1956).

Paralleling the more severe anemia suffered by *Sl* homozygotes over heterozygotes is a more marked macrocytosis. Although this macrocytosis is not evident in 14-day-old embryos it is very obvious at 15 days. Moreover, the percentage of nucleated cells, i.e., cells of the primitive generation, is much higher in 14- and 15-day-old *Sl* homozygotes than in +/+ mice. Since at this stage of development most of the blood cells are being produced by the liver this organ is normally (and in *Sl*/+ mice) densely packed with hematopoietic foci of intermediate cells. In *Sl*/*Sl* embryos, however, large areas of the liver are totally devoid of erythropoietic activity, and the foci which are present are smaller and much less dense (D. Bennett, 1956).

Sl homozygotes are also completely deficient in germ cells. As previously noted (Chapter 10, Section I, H, 1) in normal mice germ cells can first be detected in the yolk sac splanchnopleure, the allantoic mesoderm, and the caudal primitive streak area of 8-day-old embryos, and at 9 days they can be found in increasing numbers migrating anteriorly to take up their definitive positions in the genital ridge (Chiquoine, 1954). While *Sl*/*Sl* embryos cannot be distinguished from their +/+ or *Sl*/+ littermates at 8 days, a paucity of germ cells is evident at 9 days and by 14 days gestation they are completely deficient in these cells. Bennett suggests that this progressively increasing deficiency of germ cells in *Sl* homozygotes may be due to the fact that the cells which are formed fail to divide or to migrate to their proper location, and so degenerate.

Because *Sl* is lethal when homozygous it was not possible to evaluate directly how two doses of the mutation affected pigmentation. Histological examination of the few homozygotes which survived till birth revealed a total absence of pigment granules in the follicles of the vibrissae, but this could hardly be taken as evidence that the entire body lacked pigment. That such was the case, however, was indicated from the results of inserting skin from the mid-dorsal area at the level of the forelimbs of 14- and 15-day-old embryos and from newborn animals, all of which had been classified as *Sl*/*Sl* on the basis of blood counts, under the skin of the head of histocompatible adult hosts. All of these grafts, when maintained in these hosts for from 10 to 20 days, produced normal but *unpigmented* hairs, whereas similar transplants from *Sl*/+ and +/+ littermates always gave rise to pigmented hairs (D. Bennett, 1956). It thus appears that *Sl*/*Sl*, like *W*/*W* and W/W^v (as well as like other *W* and *Mi*-series (Chapter 12, Section I) genotypes, and like *Sp*/*Sp* (Section III)), results in a complete absence of neural crest-derived melanocytes. This was confirmed when other, more viable, *Sl* allelic combinations became available.

C. Other Alleles

The *Sl* locus seems to be highly mutable. In populations exposed to X-irradiation at Oak Ridge and at Harwell more than 30 mutations have been encountered which behave as *Sl* alleles. Moreover, in nonirradiated populations of C57BL/6J and DBA/2J mice maintained at the Jackson Laboratory, at least five *Sl* mutations were reported over a 3-year period (E. Russell and Bernstein, 1966). Induced and spontaneous mutations include the following:

1. Steel-Dickie (Sl^d)

Next to *Sl*, this allele has been the most widely studied. Indeed, most of the experimental efforts on the effects of steel have involved the Sl/Sl^d heterozygote since these heterozygotes display all the pleiotropic effects of *Sl*/*Sl*, i.e., they are anemic, sterile, and, except for their retinas, lack pigment, but are viable and may live for as long as a year.

The Sl^d mutation appeared spontaneously in the DBA/2J strain (Bernstein, 1960). Sl^d/+ mice have a similar phenotype as *Sl*/+ animals; they have a slight overall dilution, which is more pronounced on the belly, an occasional belly spot, and a mild macrocytic anemia (Deol, 1970b). Although $Sl^d/+$ mice have no inner ear abnormalities, such abnormalities are produced when $Sl^d/+$ is combined with recessive spotting (*rs*/*rs*) (see Chapter 10, note 32) another mutation which, by itself, is *not* associated with inner ear lesions. Thus Deol (1970b) has reported that $Sl^d/+;rs/rs$ genotypes are completely white and suffer from moderate to severe abnormalities of the inner ear with every ear affected. Moreover, although such genotypes usually (in 91% of the cases) have some pigment in their inner ear, they never have a full complement of melanocytes.

Sl^d homozygotes, like Sl/Sl^d heterozygotes, are black-eyed whites, sterile, and severely anemic. These genotypes also have similar life spans which vary with their genetic background. For example, on a C57BL/6J background 20% of Sl^d/Sl^d mice survive until they are a month old, and their mean survival is 79 days, whereas when incorporated into a (C57BL/6J × DBA/2J)F_1 background these values increase to 25% and 113 days, respectively (E. Russell and Bernstein, 1966).

Erythropoiesis has been studied extensively in Sl/Sl^d fetuses and all the observations are consistent with the fact that *Sl* genes do not affect the primitive erythroid cell lineage derived from the yolk sac blood islands, but interfere seriously with the development of the definitive erythroid cell lineage of fetal liver origin (Chui and Loyer, 1975a).[4] Thus whereas the number and sizes of yolk sac-derived nucleated red blood cells are similar in Sl/Sl^d and +/+ fetuses from days 13–17 of gestation, Sl/Sl^d fetuses have significantly fewer and much larger fetal liver derived nonnucleated cells (Chui and E. Russell, 1974; Chui and Loyer, 1975a). Because the number of hemoglobin-containing mature erythroblasts in Sl/Sl^d fetal livers is markedly reduced from that in +/+ livers, while the number of immature erythroid precursors per unit area of fetal liver is not, it has been suggested that "the mutant *Sl* gene product(s) interferes with or fails to support the differentiation of immature erythroid precursors into hemoglobin synthesizing cells" (Chui and E. Russell, 1974).

Adult Sl/Sl^d mice display a reduction in the absolute number of nucleated marrow cells (Bernstein et al., 1968; Travassoli et al., 1973; Wilson and O'Grady, 1976), have normal numbers of blood platelets, reduced numbers of megakaryocytes (Ebbe et al., 1973, 1977) and neutrophils (Ruscetti et al., 1976), and their granulocytopoiesis is not totally normal (Sutherland et al., 1970; Knospe et al., 1976).[5] Furthermore, it has been shown that although these animals, like W/W^v anemics, are very unresponsive to erythropoietin (Bernstein et al., 1968; Harrison and E. Russell, 1972; E. Russell and Keighley, 1972),[6] they have much more difficulty than the *W*-anemics in adapting to sudden exposure to constant hypoxia (Bernstein et al., 1968).[7]

In contrast to the situation in *Sl*/*Sl* mice where no primordial germ cells reach the genital ridges, in Sl/Sl^d embryos the germ cell defect does not seem to be as severe and some cells reach these ridges. In fact McCoshen and McCallion (1975) found that while there is a great paucity of primordial germ cells in Sl/Sl^d mice, these cells appear in the same locations on about the same days as in normal genotypes. In normal mice they found that less than 2% of the germ cells had reached the gonadal ridges by day 10 and more than 70% after day 11. On the other hand, in Sl/Sl^d embryos, 23% of the germ cells reached the ridges by day 11 and an additional 12% were found in the adjacent mesenteric root and coelomic angles. Indeed, these investigators point out that considering the fact that the mutant's germ cells do not proliferate (or, if they do, have a high death rate), their rate of migration seems to be comparable to those of normal animals.[8]

2. Cloud-Gray (Sl^{cg})

This mutation occurred in an F_1 of a neutron-irradiated male. $Sl^{cg}/+$ mice are barely lighter than wild type whereas Sl^{cg} homozygotes are a very light grey with white blotches (Owens, 1972). When heterozygous with *Sl*, Sl^{cg} produces an off-white coat, black eyes, and dark ears (Kelly, 1974). Sl^{cg}/Sl^{cg} mice are not very fertile (Owens, 1972) and about one-third of the females have an imperforate vagina (Kelly, 1974).

3. Contrasted (Sl^{con})

This mutation occurred in a neutron irradiation experiment (Searle, 1968c). $Sl^{con}/+$ mice can be classified at or soon after birth by the presence of dark pigmentation on the genital papilla. The adult coat, however, tends to be a little lighter than normal. Animals homozygous for this allele also have dark external genitalia but a markedly diluted coat.[9] Both eumelanin and phaeomelanin are affected. The underfur is more severely diluted than the overfur, especially in nonagouti (a/a) mice. While Sl^{con} homozygous males are fertile (Beechey and Searle, 1971), many Sl^{con}/Sl^{con} females are sterile and none has borne more than one litter (Searle and Beechey, 1974). Vaginal smears indicate that they generally do not come into oestrus (Searle, 1968c), and histological studies show that there is a gradual degeneration of oocytes in Graafian follicles so that practically all have gone by 2 months (Beechey and Searle, 1971). $Sl^{con}Sl^d$ heterozygotes display markedly increased white-spotting and dilution (Beechey and Searle, 1972). Moreoever, these heterozygous females are completely sterile with very small ovaries. This allele also has hematological consequences. Thus Sl^{con} homozygotes, and $Sl^{con}Sl^d$ and $Sl^{con}/+$ heterozygotes, have fewer erythrocytes than $Sl^{con}/+$, $Sl^d/+$, and $+/+$ mice, respectively. There is also evidence that this anemic condition involves a macrocytosis (Searle and Beechey, 1974).

4. Grizzle-belly (Sl^{gb})

This *Sl* allele arose spontaneously in a white belted female (Schaible, 1960). $Sl^{gb}/+$ heterozygotes have light bellies while Sl^{gb} homozygotes are anemic and die during the first week of life (Schaible, 1961). Sl^{gb}/Sl^{d} and Sl^{gb}/Sl genotypes are black-eyed white anemics (Nash, 1963); Schaible, 1963c). Sl^{gb} interacts with $Mi^{wh}/+$ to give a white coronal area, as well as a grizzled belly (Schaible, 1960). It also interacts with $W^{a}/+$ to produce a roan phenotype with white coronal area (Schaible, 1961).

5. Sooty (Sl^{so})

Sooty arose in the C57BL/6 strain (D.S. Miller, 1963). $Sl^{so}/+$ heterozygotes have a dilute coat and light tail whereas Sl^{so} homozygotes are white with black eyes and anemic (Hollander, 1964). On a C57BL background Sl^{so} homozygotes usually die before maturity. On other backgrounds, however, they may survive longer but are sterile (Hollander, personal communication).

6. Steel-Miller (Sl^{m})

This mutation occurred in the same colony as Sl^{so} (D.S. Miller, 1963). The coat of $Sl^{m}/+$ mice is diluted but displays no white spotting. Sl^{m} homozygotes are black-eyed whites and somewhat anemic but viable. All Sl^{m}/Sl^{m} males are sterile but a few females have borne one litter (D.S. Miller, 1963; Hollander, 1964).

The anemia of C57BL/6-Sl^{m}/Sl^{m} mice has been studied in some detail by Kales and his associates (1966). It is a macrocytic condition characterized by an increased number of reticulocytes which they attribute to a gastrointestinal bleeding defect of unknown etiology. These investigators also report that Sl^{m}/Sl^{m} homozygotes, like Sl/Sl^{d} heterozygotes, do not respond normally to erythropoietin or to hypoxia. Although they also found this erythropoietic defect in $Sl^{m}/+$ heterozygotes, in these animals it evidently remains latent as they are not anemic.

7. Dusty (sl^{du})

Dusty, which appears to act as a recessive, occurred as a spontaneous mutation in the C3H/HeJ strain and has been described briefly by M.C. Green and H. Sweet (1973). Homozygotes have a slightly diluted coat and small white patches which vary in number from none to extensive white speckling. They are not noticeably anemic. Occasionally $sl^{du}/+$ heterozygotes can be recognized by the fact that they possess whitish toes and the ventral side of their lower jaw is lighter than normal. Very infrequently sl^{du} homozygotes cannot be distinguished from wild type as Green and Sweet found that matings of $Sl/sl^{du} \times sl^{du}/sl^{du}$ produced six apparently wild type offspring (out of a total of 334), four of which (the other two died) proved to be sl^{du}/sl^{du}. Sl/sl^{du} heterozygotes are moderately anemic at birth and, as a consequence, paler than normal. These heterozygotes also usually have more white spotting than $Sl/+$ mice.

The effects of the *Sl*-locus genotypes considered above are summarized in Table 11-1.

Table 11-1. Descriptions of Various *Sl*-Locus Genotypes[a]

Genotype	Description
$Sl/+$	Slight dilution of coat (more severe on ventrum); may have (occurrence and incidence depends on genetic background) white snout tip, tail tip, and white feet; usually small white spot in middle of forehead and/or belly; occasionally white blaze between eyes or few white hairs on dorsum; moderate macrocytic anemia; normal viability and fertility
Sl/Sl	Severe macrocytic anemia and rarely survive to birth; completely deficient in germ cells and in neural crest-derived melanocytes
$Sl^d/+$	Similar to $Sl/+$
Sl^d/Sl^d	White with black eyes, severe anemia; survival depends on genetic background but may survive up to a year; sterile and may develop ovarian tumors
Sl/Sl^d	Similar to Sl^d/Sl^d
$Sl^{cg}/+$	Barely lighter than wild type
Sl^{cg}/Sl^{cg}	Very light grey with white blotches; not very fertile; one-third females have imperforate vagina
Sl/Sl^{cg}	Off-white with black eyes; dark ears
$Sl^{con}/+$	Little lighter than wild type (can be classified at or soon after birth by dark pigmentation of genital papilla); slight macrocytic anemia
Sl^{con}/Sl^{con}	Markedly diluted coat (underfur more severly affected); dark external genitalia; males fertile but females frequently sterile; macrocytic anemia but viable
Sl^{con}/Sl^d	Markedly increased white spotting and dilution; females sterile with very small ovaries; macrocytic anemia but viable
$Sl^{gb}/+$	Light belly
Sl^{gb}/Sl^{gb}	Severe anemia and die during first week of life
Sl^{gb}/Sl^d	White with black eyes; anemic
Sl^{gb}/Sl	White with black eyes; anemic
$Sl^{so}/+$	Diluted coat; light tail
Sl^{so}/Sl^{so}	White with black eyes; anemic; variable survival; sterile
$Sl^m/+$	Diluted coat
Sl^m/Sl^m	White with black eyes; macrocytic anemia but viable; males sterile but some females have borne one litter
$sl^{du}/+$	Usually like wild type; occasionally some have whitish toes and ventral side of lower jaw lighter than normal
sl^{du}/sl^{du}	Slightly diluted coat and small white patches which vary from none to extensive white speckling (few, however, indistinguishable from wild type); no noticeable anemia
Sl/sl^{du}	Usually have more white spotting than $Sl/+$; moderate anemia

[a]Only effects noted in literature are included.

D. Influence on Pigmentation

Steel's influence on pigmentation has been studied by Mayer (1970, 1973a) and by Mayer and M.C. Green (1968) who on the basis of their findings concluded that *Sl* acts solely via the developing skin to block the survival or differentiation of melanoblasts. In support of this contention they offer the following observations:

1. When grafts of neural tubes, including neural crest cells, from 9-day-old embryos from matings segregating for *Sl* and Sl^d, were combined with 11-day-old neural crest-free embryonic +/+ skin, and grafted to the chick coelom, melanocytes always were found in either the hairs of recovered grafts or in the tissues of the host in the operated region. However, when +/+ neural tubes were combined with embryonic skin derived from *Sl*/+ × Sl^d/+ matings pigment was absent in 33% of the cases even though all grafts produced normal hairs. Moreover, these pigment-free grafts could be subdivided into two groups with respect to pigment development in the tissues of the host embryo. In one group the grafts were surrounded by host tissues possessing melanocytes of graft origin, whereas in the other group the coelomic lining and skin of the host chick, like the graft itself, was completely devoid of melanocytes. From these observations it was concluded that the 33% of the cases in which pigment failed to be produced represented combinations in which (by chance) the skin was *Sl*/Sl^d, and that the presence of this skin adversely affected the differentiation of +/+ melanoblasts which migrated out into the surrounding tissues of the host (Mayer and M.C. Green, 1968).
2. When the occurrence of melanocytes was surveyed in a number of tissues of +/+, *Sl*/+, Sl^d/+, and *Sl*/Sl^d 5-day-old mice it was found that *Sl*/Sl^d mice were devoid of pigment in all locations save the retinal layer of the eyes (see Table 11-2). From this it was concluded either that

Table 11-2. Occurrence of Pigment Cells in Six Tissue Environments of 5-Day-Old +/+, *Sl*/+, Sl^d/+, and *Sl*/Sl^d Mice[a]

	+/+	*Sl*/+ and Sl^d/+	*Sl*/Sl^d
Hair follicle	+[b]	Spotted	0
Leg	+	–	0
Harderian gland	+	+	0
Membranous labyrinth	+	+	0
Choroid layer	+	+	0
Retinal layer	+	+	+

[a]This table is based on information from Mayer and Green (1968).
[b]+, normal number; –, number reduced; 0, melanocytes absent.

Sl/Sl^d skin affected the differentiation of nonepidermal melanoblasts or that other tissues too were affected by the Sl/Sl^d genotype (Mayer and M.C. Green, 1968).

3. In a later study Mayer (1970) extended his earlier efforts by combining neural tubes (including neural crest) from 9-day-old +/+ embryos with skin from 13- to 18-day-old Sl^d/Sl^d embryos. By employing skin of this age Mayer was able to assess its genotype accurately.[10] The results were consistent with his and Green's earlier observations, i.e., the large majority of the Sl^d/Sl^d skin grafts failed to display pigmented hairs when combined with +/+ neural tubes even though such neural tubes always produced pigmented hairs when combined with W^v/W^v skin or with 11-day-old (neural crest-free) albino skin.[11] Nevertheless, this absence of pigment in Sl^d/Sl^d hair follicles was not absolute since in a few grafts (3 of 63) some pigmented hairs were found along with those lacking pigment. In these experiments, too, the Sl^d/Sl^d skin appeared to inhibit the differentiation of melanoblasts which migrated into the surrounding tissues of the chick host for in only five cases were such melanocytes found, the area of the host immediately adjacent to the remaining 58 grafts being completely melanocyte free.
4. Finally, in an attempt to analyze more critically this block of pigmentation by the skin, small pieces of embryonic skin from 13-day-old +/+ and Sl/Sl^d embryos were separated into their dermal and epidermal components, recombined, and transplanted to the chick coelom (Mayer, 1973a).[12] The results indicated that *both* components of steel skin adversely affected the development of pigment. Thus, when +/+ epidermis was combined with Sl/Sl^d dermis, all of the grafts failed to display pigment in the dermis as well as in the host tissues surrounding the graft. Nevertheless, in all but 2 of 24 grafts pigmented hairs were produced. Conversely, when Sl/Sl^d epidermis was combined with +/+ dermis none of the recombinants had pigmented hairs (and in only 2 of 18 cases were melanocytes found in the host tissues surrounding the graft) even though in 10 of 18 cases melanocytes were observed between the hair follicles in the dermis. On the basis of these results, and those involving recombinations of W/W^v and +/+ dermis and epidermis (see Chapter 10, Section I, D), Mayer concluded that both the dermis and epidermis of steel skin adversely affects the development of pigment, and that there is no observable or consistent evidence of a transfer of an inhibitory effect from a Sl/Sl^d skin component to its associated normal component.[13]

The results of these experiments certainly indicate that Sl acts solely via the environment to block the survival or differentiation of melanoblasts.[14] The most convincing experiment in this regard is the one in which neural tubes from 9-day-old embryos produced from $Sl/+ \times Sl^d/+$ matings were implanted with normal, 11-day-old neural crest-free skin. The fact that all 33 such combinations in which donor skin grafts were recovered were characterized by pigmented hairs, along with the observation that the 12 others which did not form skin nevertheless produced large numbers of melanocytes (which occurred in the chick tissues extending out from the operated region), implies that the expected 10 or so Sl/Sl^d neural tube-containing composites in this population of transplants must have possessed *viable* melanocyte clones. The only other possibility is that the putatively neural crest-free +/+ skin was not actually free of melanoblasts. This, however, does not seem likely since skin of the same genotype, age, and orgin usually did *not* produce pigment when combined with neural tubes derived from $W^v/+$ matings (see Chapter 10, Section, I D). It is therefore hard to escape the conclusion that *Sl* has no *direct* effect on melanoblast survival,[15] a conclusion which, at least to the author, is a little surprising not because it implies that there may be more than one mechanism involved for producing white spotting, but because it is somewhat out of line with Mintz's attractive model.[16] Thus, if Mayer and Green are correct, and Mintz is also, not only might one expect *Sl*/+ mice to look differently, but one would also anticipate $Sl/Sl \leftrightarrow +/+$ allophenics to display pigment patterns different from $W/W \leftrightarrow +/+$ animals. Indeed, on the basis of Mayer and Green's observations one might expect such allophenics to resemble more closely the pigment patterns produced by hair follicle phenoclones [such as perhaps the phenotypes displayed by silver (*si*/*si*) mice] than those produced by viable and inviable melanoblast clones. Such allophenics are eagerly awaited.

1. Behavior of +/+ Melanocytes in Adult Sl/Sl^d Skin

While the results noted above indicate that *embryonic* steel skin adversely affects the survival and/or differentiation of +/+ melanoblasts, this may *not* be the case with *adult Sl* skin. Thus when small histocompatible black (*a*/*a*;*B*/*B*) ear skin grafts are placed in the center of well-established adult Sl/Sl^d trunk skin grafts, some of the (Sl/Sl^d) hairs immediately surrounding these implants become pigmented (Poole and Silvers, upublished). It therefore appears either that the adverse influence *Sl* skin has on pigment cells is a transitory one limited to embryogenesis and early life, or, alternatively, that the melanocyte is susceptible to the "steel environment" only during a stage in its differentiation, i.e., pigment cells which have passed this "critical" stage can continue to survive and function in Sl/Sl^d hair follicles.[17]

E. Influence on Erythropoiesis

The influence which steel has on melanogenesis appears to parallel its affect on erythropoiesis as there are now a considerable number of studies which indicate that the anemia of Sl/Sl^d mice results from a derangement in the hemopoietic microenvironment in which erythropoiesis takes place (McCulloch et al., 1965; Bernstein, 1970; M. Bennett, et al., 1968b; Altus et al., 1971; Fried et al., 1973; McCuskey and Meineke, 1973; Wolf, 1974). Thus when marrow cells from Sl/Sl^d or Sl^d/Sl^d animals were implanted into heavily irradiated normal (+/+) mice, they were capable of forming macroscopic spleen colonies with approximately the same frequency as cells from +/+ mice. Indeed, Sl/Sl^d marrow cells were found to be as capable as +/+ cells in curing the anemia of histocompatible W/W^v mice. On the other hand, when +/+ marrow cells were transplanted into irradiated Sl/Sl^d animals they failed to proliferate and differentiate normally (McCulloch et al., 1965; see also Bernstein et al., 1968).[18]

Further evidence that steel adversely affects the hemopoietic microenvironment is provided by the fact that the anemic condition of Sl/Sl^d mice can be ameliorated if they are grafted with an intact spleen from a genetically normal (and histocompatible) donor. Moreover, the anemic condition of these mice also improves following receipt of a W/W^v histocompatible spleen (Bernstein, 1970).[19]

II. Flexed-Tailed (*f*)

Flexed-tailed (*f*), a recessive spotting determinant on the thirteenth chromosome, appeared in a stock maintained by H. Hunt (1932; H. Hunt et al., 1933). It receives its name from the fact that *f*/*f* mice have flexed tails, as well as occasional fusions of other vertebrae (H. Hunt et al., 1933; Kamenoff, 1935). This mutation also produces a transitory siderocytic anemia. Indeed, it is this condition which has received the most attention. While it is beyond the scope of this review to present a detailed analysis of the nonpigmentary effects of this mutation (see Grüneberg, 1952; E. Russell and Bernstein, 1966; E. Russell et al., 1968),[20] they nevertheless deserve some consideration for the bearing they may have on the etiology of the white spotting. This is especially the case since we are again dealing with a spotting gene which has a major influence on erythropoiesis. Thus before considering flexed-tailed's influence on pigmentation, its morphological and hematological consequences are briefly described.

A. Influence on Tail

The most obvious effect of *f* when homozygous is the flexures of the tail it produces. These are highly variable (H. Hunt et al., 1933). As a rule there are one or more permanent angles in the tail, though occasionally there may be as many as five. These may be acute, obtuse, or right angles, occurring most frequently in the proximal half of the tail, though sometimes one occurs near the tip. In addition to these flexures the tail may be significantly shortened. The tail is usually very stiff where the flexures occur and even when there is no visible flexure palpation may reveal rigid areas. These stiff regions in tails which appear straight may be so small, and approach the normal so closely in flexibility, that it is often difficult to classify some *f*/*f* animals solely on the basis of their tails (H. Hunt et al., 1933). In fact, in some populations the tails of as many as 45–50% of *f*/*f* mice appear normal (Clark, 1934).

The development of the tail anomalies of *f*/*f* mice has been investigated by Kamenoff (1935) who found that there was an ankylosis of vertebrae which could be accounted for by an anomaly of the intervertebral disks. This disk consists of the nucleus pulposes, which stems from the notochord and is surrounded by a hyaline sheath, and the annulus fibrosus, which consists of felted connective tissue fibers which form a ring round the nucleus pulposes. In normal embryos the early cartilage, which later forms the annulus fibrosus, differentiates into fibers on the fourteenth to fifteenth day of gestation. However, in flexed embryos, this differentiation of fibers is disturbed, and in places does not take place at all. As a consequence as many as 10 or more successive tail vertebrae may be joined by cartilage bridges which subsequently ossify. If this happens on one side of the intervertebral disc, flexing occurs, and if it occurs on both sides, a stiff fusion without flexure results.

B. Anemic Condition

In addition to *f*'s effect on the tail, Hunt noted early in his experiments that flexed-tailed young were considerably lighter in color than their normal tailed littermates. He suspected that this was because they were anemic and this suspicion was confirmed when their blood was compared with their normal tailed siblings (Mixter and H. Hunt, 1933).[21] The anemia is a microcytic one, characterized by a high frequency of siderocytes[22] (Grüneberg, 1942b,c), which seems almost to disappear after the first week of postnatal life.[23]

Because initial studies suggested that this anemia could be detected only *after* erythropoiesis was initiated in the liver on the twelfth day of gestation, and because it is most severe during that period of gestation (thirteenth to sixteenth day) when the liver is the only erythropoietic organ (after the sixteenth day the burden of erythropoiesis is gradually assumed by the bone marrow), the anemia originally was believed to be due to a disturbance of the hematopoietic function of this organ (Kamenoff, 1942; Grüneberg, 1942b), A more recent study (E. Russell et al., 1968), however, has revealed that the anemia is sufficiently severe on the twelfth day of gestation (when only primitive nucleated erythrocytes of yolk sac origin are present) to indicate that it is not transitory and that all generations of erythrocytes are affected.

This conclusion is further supported by the observation that adult flexed-tailed mice also display abnormalities of hemopoiesis when placed under conditions of physiological stress (Margolis and E. Russell, 1965; Thompson et al., 1966; Fowler et al., 1967; Coleman et al., 1969; see also Gregory et al., 1975). Thus, during recovery from phenylhydrazine-induced anemia there is a delay in reticulocyte production in *f/f* animals (Coleman et al., 1969), and hemopoietic stem cells from these mice do not proliferate normally in lethally irradiated hosts (Thompson et al., 1966).

Although the specific cause of the hemopoietic defect produced by *f* when homozygous is not known, a recent investigation by Chui and his associates (1977) indicates that there is a relative excess of free heme in *f/f* cells. For this reason they suggest "that one effect of the mutant *f* gene product is to interfere with the normal expression of globin genes during fetal erythropoiesis, leading to decreased production of globin chains, hypochromic microcytic anemia, and a relative excess of intracellular free heme pool due to decreased utilization." [For other hematological investigations of flexed-tailed see A.E. Bateman and Cole (1972) and A.E. Bateman et al. (1972)].

C. White Spotting

The white spotting associated with the *f/f* genotype, like the tail flexures, is variable in its expression. Although many flexed-tailed mice have a white spot about 2 cm in diameter on their abdomen, in segregating populations the size of the spot varies from no spot at all (normal overlap) to an unpigmented area covering most of the ventral surface. In rare cases the spot is even larger, extending up the sides to form a partial belt. Mice which display white spotting have white toes and may or may not have a spot on the tip of the tail (Clark, 1934).

As in the case of the tail abnormality, the incidence of white spotting in *f/f* mice depends upon the genetic background. Whereas in some genetically uniform stocks all *f/f* mice show a moderate degree of white spotting (E. Russell and McFarland, 1966), in some heterogeneous populations from 40 to 45% normal overlaps have been reported (Clark, 1934).

D. Combined Influence of Flexed-Tailed and Dominant Spotting

Inasmuch as *f* and *W* combine effects on quality and distribution of hair pigmentation with pathological effects on erythropoiesis, it was of interest to determine how these genes would affect these traits when they occurred together. This was pertinent not only insofar as providing information on the etiology of the anemias, but of the white spotting as well. It follows that if the white spotting produced by these genes is in some way a secondary effect of the anemia, then the combined effect of *W* and *f* should have a similar influence on both conditions.

To investigate this E. Russell and her associates (E. Russell and Mc-

Farland, 1965, 1966; E. Russell et al., 1968) established congenic stocks differing at the *W* and *f* loci so that all possible combinations of *W*- and *f*-locus genotypes could be compared on the *same* uniform genetic background. They found, insofar as the white spotting was concerned, that *W*/+ and *f*/*f* acted synergistically. Whereas the ventrums of *W*/+ and +/+;*f*/*f* mice displayed similar amounts of white spotting, and there was almost no white on their backs, *W*/+;*f*/*f* animals displayed much more than the sum of the amounts of white produced by *W*/+ and *f*/*f* alone (E. Russell and McFarland, 1966). This hyperadditive effect of *W*/+;*f*/*f* was particularly striking on the dorsum where the amount of spotting was increased from 1% or less to more than 20% (see Table 11-3 and Figure 11-1).

In contrast to the hyperadditive influence of *W*/+;*f*/*f* on spotting the effect of this genotype on the blood was exactly what one would anticipate from the combination of effects of *W*/+ versus +/+ and of *f*/*f* versus *f*/+ genic substitutions. Thus, there was no evidence of an association between the amount of white spotting (much higher in *W*/+;*f*/*f* than in +/+;*f*/*f* mice) and the degree of anemia (similar in *W*/+;*f*/*f* and +/+;*f*/*f* mice). The evidence is therefore consistent with the notion that in the case of *W*, and probably in the case of *f*/*f*, white spotting results from an independent *primary* gene effect on pigment-forming cells (E. Russell and McFarland, 1966).

III. Splotch (*Sp*) and Delayed Splotch (Sp^d)

Five mutations, all of which occurred at the Jackson Laboratory, have been noted for this locus on the first chromosome (Snell et al., 1954; Parsons, 1958). Four of these are presumed to be mutations to splotch (*Sp*) and the other to a second allele, delayed splotch (Sp^d) (W. Russell, 1947; Dickie, 1964b). Both alleles are semidominant and lethal when homozygous.

Table 11-3. Proportional Areas of White Spotting When Dominant Spotting (*W*) and Flexed-Tailed (*f*/*f*) Occur Separately and Together[a]

	Females		Males	
Genotype	Dorsal	Ventral	Dorsal	Ventral
W/+;*f*/+	1.0 ± 0.5[b]	22.0 ± 1.8	0.5 ± 0.1	27.5 ± 1.0
+/+;*f*/*f*	0	22.9 ± 2.2	0.7 ± 0.4	23.6 ± 1.4
W/+;*f*/*f*	21.4 ± 2.2	54.7 ± 2.0	23.7 ± 2.3	59.7 ± 2.2

[a]From E. Russell and McFarland (1966).
[b]Mean ± standard error, proportion of white area to total area, expressed as percentage.

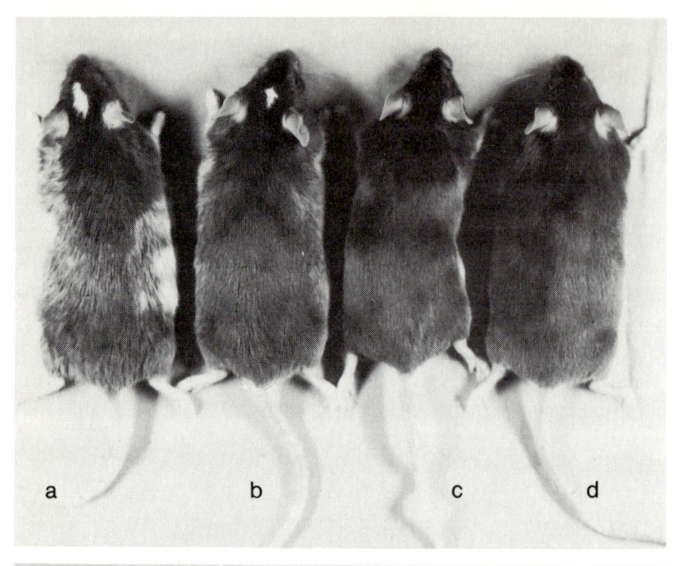

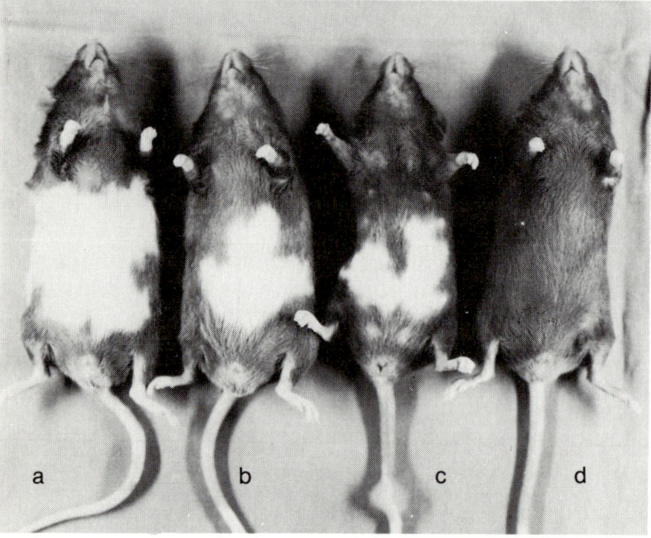

Figure 11-1. Dorsal and ventral views of littermate 6-week old (a) *W*/+;*f*/*f*, (b) *W*/+;*f*/+, (c) +/+;*f*/*f*, and (d) +/+;*f*/+ genotypes. Note the hyperadditive effect on spotting when *W*/+ is combined with *f*/*f*. From E.S. Russell and E.C. McFarland (1966). Reproduced with permission of the authors and *Genetics*.

A. Splotch (*Sp*)

The first mutation to *Sp* occurred in the C57BL strain and was described by W. Russell (1947). Heterozygotes (*Sp*/+) display white spotting on the belly and occasionally on the back, feet, and tail.[24] *Sp*/*Sp* embryos usually die at

13 days of gestation (Auerbach, 1954)[25] with multiple abnormalities.[26] These have been described in detail by Auerbach (1954) and the following is based on his account (see also Hsu and van Dyke, 1948).

1. Abnormalities When Homozygous

Sp/Sp embryos can first be recognized at 9—9.5 days of gestation by the fact that in contrast to the virtually completely closed neural folds of normal embryos, their neural folds are open in the region of the hind limbs and there is an aggregation of neural tissue on either side of the dorsal midline. This development of open neural folds and of neural overgrowth (which is known as *rachischisis*) is variable but usually is confined to a discrete area of the lumbosacral region. The degree of neural outgrowth in the region of the rachischisis may be relatively slight or quite excessive. Indeed, by the twelfth day of gestation the "overgrowth of neural tissue may manifest itself by large flaps of tissue projecting laterally from the embryo."

In addition to this abnormality, just over half of *Sp/Sp* embryos 10 days or older display open neural folds of the head region (*cranioschisis*). The condition is limited to the region of the hind brain (the myelencephalon and rhombencephalon) and is invariably associated with outgrowth of neural tissue. The lumen of the brain of homozygous *Sp* embryos is also "highly distorted throughout, being partially collapsed or obliterated by the irregular and excessive growth of neural tissue."

Although the neural tube abnormalities are most pronounced in the lumbosacral and cranial regions they are not confined to these areas as the neural tube displays some degree of malformation throughout its length. While its pathway is straight in normal embryos, in splotch homozygotes it "weaves more or less erratically and is collapsed at irregular intervals throughout the embryo."

Splotch homozygotes also are characterized by abnormalities of the dorsal root ganglia and, in some cases, the sympathetic ganglia. Tail abnormalities also occur. These may be recognized in 10-day-old embryos and include a wide variety of morphological distortions.

2. Pigmentation of Homozygotes

In view of the fact that the presence of one *Sp* gene produces white spotting, it was of interest to determine how this mutation would affect pigmentation when homozygous. This was achieved by transplanting tissue from 9.5- to 12.5-day old *Sp/Sp* embryos to either the coelom of chick embryos or to the anterior chamber of adult albino mouse eyes. The tissue implanted from 11.5-day-old or older embryos was comprised of ectoderm and underlying mesoderm, whereas neural crest and somite material were also included in grafts from younger embryos. Similar implants from +/+ and *Sp*/+ embryos served as controls (Auerbach, 1954).

Auerbach found that these grafts, which were allowed to develop in their new environments for 16 days, developed pigment only when derived from normal (*Sp*/+ or +/+) embryos. Thus no trace of pigment was found in either the implant or the surrounding tissue when it originated from an *Sp*/*Sp* embryo, even though in many cases numerous (unpigmented) hairs were present. These findings indicate that splotch when homozygous results in a complete failure to produce functioning melanocytes. In this regard the effect of *Sp* parallels the effect of dominant spotting (*W*); one dose of either of these genes is associated with white spotting, whereas two doses eliminates all pigment save that of the retina which is normal. This of course is not surprising since, as noted previously (Chapter 1, note 5), the pigmented cells of the retina are not of neural crest origin.

It thus appears that the splotch syndrome is due to a disturbance of the region of the embryo which includes the neural crest and the dorsal part of the neural tube, and that as a result of this disturbance *Sp* homozygotes do not produce any viable melanoblast clones.

B. Delayed Splotch (Sp^d) and Other Mutations

Three other apparent mutations to splotch have been reported by Dickie (1964b). Two of these (Sp^J and Sp^{3J}) occurred in C57BL/6 males and the other (Sp^{2J}) in a C3H/HeJ female.[27]

Dickie reported two other semidominant mutations which were lethal when homozygous and one of these proved to be another splotch locus allele.[28] This allele also occurred in C57BL/6J and is called "delayed splotch" (Sp^d) because homozygotes are not as severely malformed and their death is delayed in comparison with *Sp* homozygotes. Sp^d/+ mice have a large belly spot. Sp^d/Sp^d homozygotes survive until birth and although they consistently suffer from caudal rachischisis in no case has any gross anomaly in the anterior portions of the fetus been observed. Sp/Sp^d heterozygotes are lethal and similar in appearance to Sp^d homozygotes. Linkage studies have provided further evidence that *Sp* and Sp^d are alleles (Dickie, 1964b).

IV. Varitint-Waddler (*Va*) and Varitint-Waddler-J (Va^J)

There are two semidominant alleles at this locus on the twelfth chromosome. One of these, *Va*, occurred in 1942 at the Jackson Laboratory (Cloudman and Bunker, 1945) and the other, Va^J, in 1967 at the same laboratory (Lane, 1972).

A. Varitint-Waddler (*Va*)

Va was observed first in two offspring of a (C57BL × C57BR)F_1 female (one in each of her first two litters) which was backcrossed to a C57BL male.

Inasmuch as the male produced numerous normal but no abnormal young when mated to other females, it seems likely that the mutation arose in the mother and that she was a germinal mosaic (Cloudman and Bunker, 1945).

1. General Characteristics

The mutation combines a peculiar type of variegated pigment pattern with a modified shaker-syndrome (Grüneberg, 1952). All *Va* mice are deaf and show circling behavior, head tossing, and hyperactivity. These behavioral abnormalities are more marked in homozygotes than in heterozygotes.

While the viability of the *Va*/+ heterozygote is nearly normal, homozygous *Va* mice are greatly reduced in number presumably because a considerable number of them die before birth (Cloudman and Bunker, 1945).

Most *Va* homozygotes and some heterozygous males are sterile, and female heterozygotes often make poor mothers.

2. Influence on Coat Color and Hair Structure: A Complex Locus?

The coats of young *Va* heterozygotes display a combination of spotting with dilution (Plate 3G; Figures 11-2 and 11-3c). There are some islands of fur, some large, others small, which are pigmented intensely. These never cross the midline of the body either dorsally or ventrally and do not seem to change with age. Other regions of the coat show a dilute pigmentation. These areas become fainter as the animal gets older and ultimately end up white or nearly so. Finally, some regions including most of the belly are nonpigmented from the beginning although occasionally fully pigmented areas extend to the mid-ventral line (Cloudman and Bunker, 1945; Grüneberg, 1952; see note 29).

Va heterozygotes are detectable at about 3 days of age when scattered white areas appear on the head and body. These areas have a very broken pattern and sometimes there is an intermingling of white and colored hairs

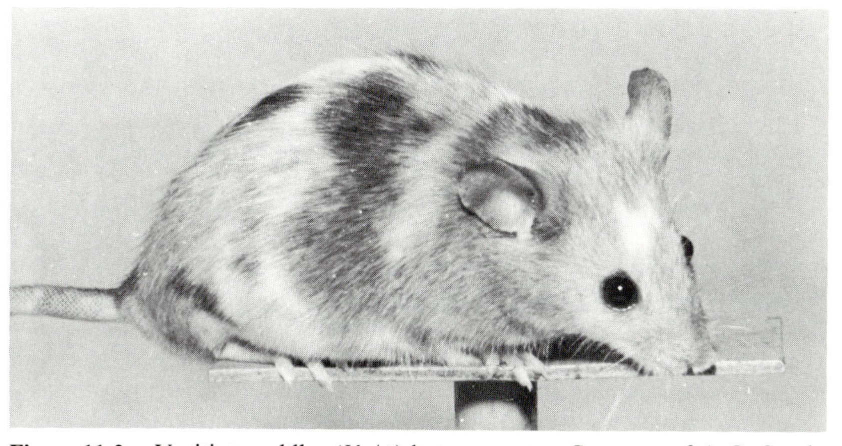

Figure 11-2. Varitint-waddler (*Va*/+) heterozygote. Courtesy of A.G. Searle.

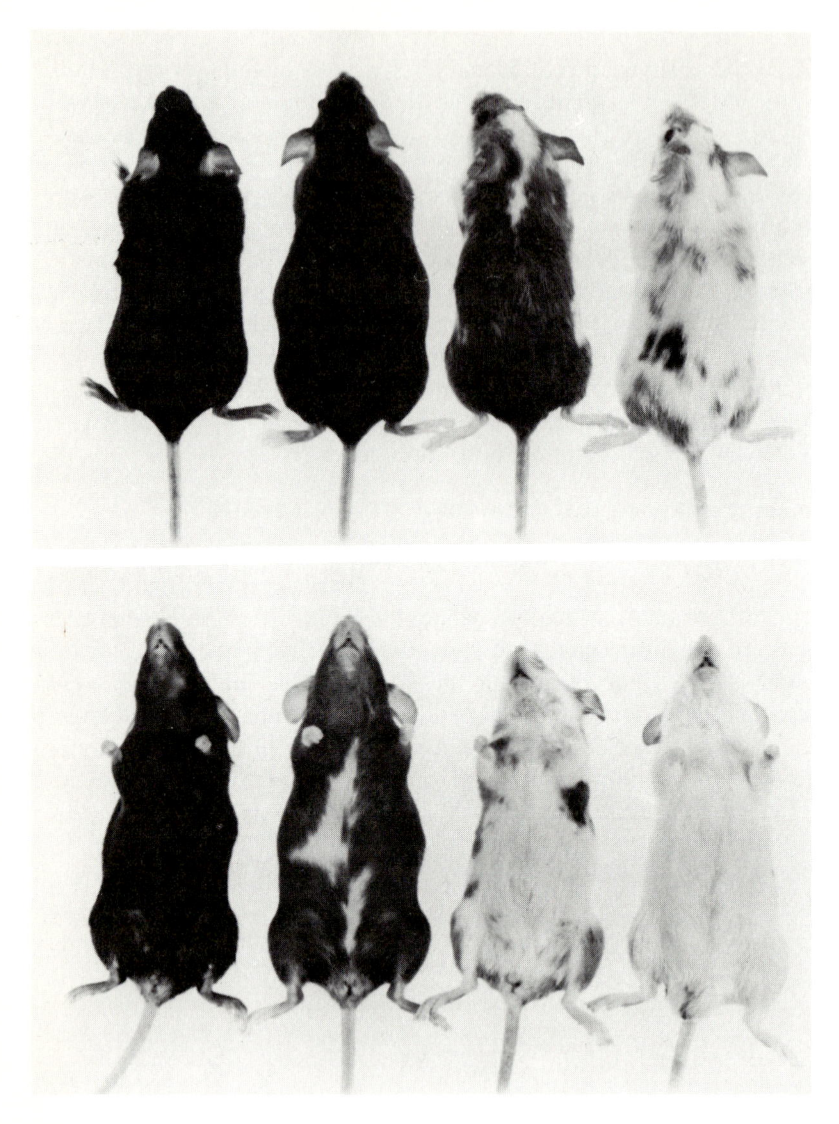

Figure 11-3. Dorsal and ventral views of (a) +/+, (b) $Va^J/+$, (c) $Va/+$, and (d) Va^J/Va^J mice. From P.W. Lane (1972). Reproduced with permission of the author and the American Genetic Association

giving an appearance that might be called roan (Cloudman and Bunker, 1945).

According to Mintz (1971a) the pigment patterns displayed by $Va/+$ mice can be attributed to both melanoblast and hair follicle phenoclones occurring simultaneously (see Chapter 7). Indeed, in her opinion "there are some inviable melanoblast clones and also two phenotypic colors of viable clones, yielding a three-color melanoblast pattern." It thus appears that *Va*, like mottled (*Mo*), may be a complex locus.

Grüneberg (1966c) draws attention to the fact that the pigment patterns of *Va*/+ mice resemble some of the patterns produced by the *Mo* series of alleles (see Chapter 8, Section II). Moreover, he reports that, as in the case of the *Mo*-series, *both* hair structure and melanin formation are affected. "The white hairs are grossly abnormal in structure, the grey ones rather less so, whereas the normally pigmented hairs are structurally nearly, but apparently not quite, normal." It is his contention that the defective pigmentation of *Va*/+ animals is secondary to abnormal hair structure, and that the color changes with age reflect a gradual deterioration of hair structure.[29]

Va/*Va* mice are white except for small patches of unaltered colored hairs which usually occur near the ears and the base of the tail.[30] At birth these animals are very pale but there is no evidence that they are anemic [although Grüneberg (1952) believes this may need verification].

3. Interaction With Other Spotting Determinants

As in the case with many of the other spotting genes there is a striking interaction when *Va*/+ is combined with either dominant spotting (*W*) or viable dominant spotting (W^v). Thus the double heterozygotes *Va*/+;*W*/+ and *Va*/+;W^v/+ are nearly completely white except for some pigmented spots on the head and near the tail (Cloudman and Bunker, 1945; Grüneberg, 1952). This synergism evidently does not occur between *Va*/+ and piebald spotting (*s*/*s*) since preliminary data indicate that *Va*/+;*s*/*s* and *Va*/+;*s*/+ genotypes may be phenotypically difficult to separate (Cloudman and Bunker, 1945).

4. Behavior and Relationship to Inner Ear Abnormalities

The behavior of *Va* homozygotes and heterozygotes has been described by Cloudman and Bunker. At about 2 weeks of age *Va*/*Va* mice are very excitable. They are deaf but react quickly and violently when disturbed. Usually they begin by rolling over and over. This is followed by their jumping wildly into the air and finally by a convulsive stiffening of the body. These convulsions may account for the loss of many *Va* homozygotes under 1 month of age. Quite often young *Va*/*Va* mice do not right themselves when turned upside down gently (Deol, 1954). *Va* heterozygotes are not as excitable and do not display a tendency to roll and leap; otherwise both *Va*/*Va* and *Va*/+ behave very much alike. When awake they are usually constantly in motion, nodding and tossing their heads. When walking they waddle (hence their name) and when excited they run in circles sometimes clockwise and sometimes counterclockwise. Although all *Va*/*Va* mice and elderly *Va*/+ animals, i.e., those over 10 months of age, cannot swim, young heterozygotes swim easily in small circles (Cloudman and Bunker, 1945; Deol, 1954).

The inner ear of varitint-waddlers has been studied in detail by Deol (1954) and while lesions in the labyrinth[31] can explain their deafness it cannot account for all of their behavioral disorders. According to Deol, the shaking of the head may be "due to loss of tone in the neck muscles resulting from the defects of the utricular macula and the ampullary cristae or it may be central in origin."

B. Varitint-Waddler-J (Va^J)

1. Origin and General Characteristics

This mutation, which has been described by Lane (1972), occurred in a male which was heterozygous for *Va*. All indications are that he was both a germinal and a somatic mosaic for *Va*/+ and Va^J/+ (and hence that Va^J arose as a mutation from *Va*). He was originally classified as *Va*/+ but was not as varicolored as these mice should be. Because of his unusual color he was mated to a C57BL/6J (+/+) female. This mating produced not only *Va*/+ and +/+ mice, as expected, but a third phenotype as well. This new phenotype (Va^J/+) is characterized by "a slightly diluted coat color, a large irregular belly spot, white feet, and a white tail tip, closely resembling the viable dominant spotting heterozygote, W^v/+" (Figure 11-3b). Although Va^J/+ mice are deaf they behave normally.

Va^J/Va^J mice have a dilute coat color and are predominantly white (Figure 11-3d). They are deaf but swim and behave normally. Both sexes are fertile.

A considerable number of Va/Va^J mice, like *Va* homozygotes, evidently die *in utero* as only a dozen such mice (four of which died young) occurred in 231 progeny of *Va*/+ × Va^J/+ matings. Such heterozygotes are somewhat similar to the Va^J homozygote but are smaller with more white spotting. They are also deaf, cannot swim, and circle vigorously.

Clearly the *Va*-locus is among the more interesting coat color determinants and as such merits further attention.

Notes

[1]Beechey and Searle (1976) report that Sl^{con} is about 16 cM from grizzled (*gr*). This is in reasonable accord with the estimate of 6–14 cM made by Cacheiro and L. Russell (1975).

[2]The most severe effect on erythropoiesis seems to occur between 15 days gestation and 1 week after birth when *Sl*/+ mice possess only about 65% of the normal number of erythrocytes (D. Bennett, 1956).

[3]Mintz (1957b, 1960) investigated the effect which *Sl* has on embryonic gametogenesis and found no apparent influence of this mutation on primordial germ cell number when present in a single dose. Thus *Sl*/+ embryos displayed similar numbers of primordial germ cells as *W*/+, W^v/+, and +/+ embryos. Moreover, she found no evidence that a simultaneous substitution of one mutant gene at the *W* and *Sl* loci (*e.g.*, *W*/+;*Sl*/+) created any special difficulty in germ cell proliferation or viability. She concluded that the results suggest that each of these loci "makes a unique con-

tribution to the growth and multiplication of primordial germ cells," and that it is possible that *W/W*;+/+ and +/+;*Sl/Sl* are defective at different points in a reaction sequence, each indispensable for the production of a further developmental stage (Mintz, 1960).

[4]In the mouse the blood-islands of the yolk sac are the sole source of erythrocytes from the eighth through the eleventh day of gestation. From the twelfth through the sixteenth day of gestation the liver is the only hematopoietic organ and remains an important source of new erythrocytes up to birth. Bone marrow hemopoiesis first appears on the sixteenth day of development (E. Russell and Bernstein, 1966).

[5]Not only are there about half the normal number of megakaryocytes in the marrow of Sl/Sl^d mice but the cells are larger than normal (Ebbe et al., 1973). Moreover, even though these mice have a normal complement of circulating platelets, they may produce a thrombopoietic substance that stimulates the production of larger than normal megakaryocytes. Evidence for this stems from the observation that when normal (+/+) mice are parabiosed to syngeneic Sl/Sl^d animals, megakaryocytopoiesis is stimulated in the +/+ partner and the cells which are formed are larger than normal (Ebbe et al., 1978).

[6]More recently Chui and Loyer (1975b) have shown that, in contrast to the unresponsiveness of Sl/Sl^d mice to erythropoietin *in vivo*, there are cells in the marrow of these mice capable of responding *in vitro* to erythropoietin in a normal fashion. They have also shown that erythropoietin present in Sl/Sl^d serum is biologically active *in vitro*.

[7]Sl/Sl^d mice also have more than 300-times normal specific activities of the enzyme nucleoside deaminase in their circulating erythrocytes (Rothman et al., 1970; Harrison et al., 1975). There is also evidence that lymphocytic neoplasms arise much more frequently in the thymus of these anemics than in their heterozygous or normal littermates (E. Murphy, 1969). On the other hand, there is no evidence that either *Sl* or *W* adversely affects the differentiation of the immune system (Mekori and R.A. Phillips, 1969; but see Chapter 10, note 27).

[8]As in the case of *W*-genotypes which are sterile (see Chapter 10, note 24), the germ cell-deficient ovaries of Sl^d/Sl^d and Sl/Sl^d genotypes develop tumors in old age (M.C. Green, 1966a).

[9]Melanin pigmentation has been found on the ovary, cervix, and related structures of Sl^{con}/Sl^{con} females (Beechey and Searle, 1971).

[10]Recently it has been shown that *Sl/Sl* (or Sl/Sl^d) mouse embryos can be reliably identified in segregating litters at 11 and 12 days gestation by grafting skin from either the dorsal region between the levels of the fore- and hind limbs, or from the dorsal region just anterior to the level of the fore limbs, to the testes of histocompatible adults. The grafts are maintained in this location for 2–3 weeks and the *absence* of pigment in the hairs which form provide an excellent indication of their *Sl/Sl* (or Sl/Sl^d) genotype (Chui et al., 1976).

[11]It should be noted that when 9-day-old +/+neural tube was combined with 13- to 18-day-old embryonic albino (*c/c*) skin, 20 of 23 of the recovered grafts lacked pigment although they possessed a full complement of hairs. Mayer believes these grafts failed to become pigmented because they already were populated by albino (*c/c*) melanoblasts. To support this contention he cites that there was always a large

population of pigment cells in the tissues of the host chick embryo immediately surrounding these grafts (see Mayer, 1973b).

[12]At this stage of development both the dermal mesoderm and the epidermal ectoderm are populated by melanoblasts—at least in pigmented and albino genotypes. This was shown to be the case by Mayer (1973b) who separated C57BL/6 and congenic albino (*c*/*c*) embryonic skin, obtained from the dorsolateral side of the trunk midway between the limb buds, into their epidermal and dermal components and, after recombining the dermis of one with the epidermis of the other, placed them in the chick coelom. Whereas recombined skin from embryos 11 days of age formed pigment only when the mesodermal component stemmed from a genetically black embryo, i.e., black ectoderm–albino mesoderm combinations always failed to become pigmented, skin recombinations made from 12-day-old embryos indicated that the ectodermal component had been populated with melanoblasts and by 13 and 14 days this was even more evident. From these observations Mayer concludes that the migration of melanoblasts into the developing skin occurs through the dermal mesoderm, "with a later secondary migration from the dermis into the overlying epidermal ectoderm." These results are in accord with Rawles' (1940, 1947) observations as well as with the observations of A. Zimmermann and Becker (1959) who examined the positions of pigment cells in fetal Negro skin. They are not in accord with the observations of Weston (1963) who studied the migration of labeled neural crest cells in the chick and concluded that these cells entered the ectoderm almost immediately after leaving the neural tube.

[13]Mayer (1975) also attempted to determine if melanoblasts were present in the skin of 13-day-old *Sl*/*Sl* embryos by assuming that if they were they might be induced to express themselves if their environment was "enriched" with dominant spotting (*W*/*W*) skin cells. Accordingly he dissociated skin from both *Sl*/*Sl* and *W*/*W* embryos, mixed them together in various proportions, and grew them as aggregates for 2 weeks in the chick coelom. Because this procedure failed to reveal the presence of any melanocytes, he concluded that either steel prevents "the migration of melanoblasts into the mutant tissue environment, perhaps at the time of initial migration of melanoblasts from the neural crest," or *Sl*/*Sl* melanoblasts have a normal migratory pattern but are in some way eliminated as viable cells in the skin by 13 days.

[14]Although Mayer (1970, 1973a) cites Silvers (1956) as having reported that steel skin lacks amelanotic melanocytes, no *Sl* genotypes were included in Silvers' investigation.

[15]Nevertheless there are some aspects of Mayer's efforts which raise some interesting questions. For example, while some of his observations are interpreted as indicating that *Sl* adversely affects melanoblast differentiation not only within the skin, but in regions adjacent to it, other observations suggest the influence of *Sl* to be quite localized. Thus, whereas his results following combining +/+ melanoblasts with Sl/Sl^d or Sl^d/Sl^d skin suggest that these skins inhibit the differentiation of melanoblasts which it is assumed must have migrated into the surrounding tissues of the host, his mesodermal and ectodermal skin recombinant results indicate a strictly localized effect of *Sl*. Moreover, if Sl/Sl^d skin has such a strong influence in regions adjacent to it, why were some cases found, albeit very few, in which some +/+ melanocytes occurred in Sl^d/Sl^d hairs?

[16]It is likewise difficult to reconcile with the similar synergistic effect that $Sl/+$ (or $Sl^d/+$) has on the amount of white spotting when combined with other spotting determinants (see Figures 10-4c and 10-8d) which are known to act via the melanoblast and which interact with each other as they do with $Sl/+$ (or $Sl^d/+$).

[17]Although each time a hair is produced the same basic dermal–epidermal connection is retained, and dormant melanocytes persist (Silver et al., 1977), almost an entirely new follicle is formed (see Chase and Silver, 1969). For this reason one question concerning hair growth which remained unanswered was whether a melanocyte could deposit pigment in more than one hair generation, or whether it "shoots its wad," so to speak, after it has delivered pigment to a single hair (see Chapter 3, Section I, G, 2)? To resolve this question we determined how the *secondarily* pigmented Sl/Sl^d hairs which arose just outside the border of pigmented ear skin implants responded to plucking. It follows that if after plucking these hairs (and it should be noted that their numbers varied from a few to a dozen) the regenerated hairs were white (nonpigmented) this would indicate that the melanocytes of the previous hair bulbs had been lost, i.e., they were able to function only in one hair. On the other hand, if a pigmented hair replaced the plucked pigmented one, this would provide strong evidence that at least some $a/a;B/B$ melanocytes (or their mitotic descendents) can pigment more than one hair generation. The latter situation prevailed. Indeed, some secondarily pigmented hairs were plucked (at the end of each cycle) as many as 6 times and still regenerated pigmented (Poole and Silvers, unpublished).

[18]Similar findings have recently been duplicated *in vitro*. Thus when W/W^v bone marrow cells were fed to an adherent layer of previously established Sl/Sl^d marrow cells, cell production was markedly less than when either marrow cells from completely normal donors were employed, or when Sl/Sl^d marrow cells were fed to an adherent layer of previously established W/W^v cells (Dexter and Moore, 1977).

[19]Bernstein (1970) also found that hematologic values were elevated when Sl/Sl^d histocompatible spleens were grafted to W/W^v anemics. However, he notes that whereas these transplants are effective because they provide a source of stem cells which can function normally in the hematopoietic organs of the W/W^v host, W/W^v spleens transplanted to Sl/Sl^d mice are effective because they pick up Sl/Sl^d circulating stem cells and provide them with a suitable environment for their reproduction and normal development. He also notes that in the case in which $+/+$ spleens reside in Sl/Sl^d recipients, migration does not appear to be necessary. Rather in this situation an organ which is entirely normal to begin with continues to function after grafting without the necessity of acquiring any new operational capacity. Mintz and Cronmiller (1978) have produced allophenic mice composed of both $Sl/+$ and $+/+$ cells and the clinical blood picture of these mosaics is indistinguishable from that of normal ($+/+$) controls, even when only a small minority of cells in all tissues are of the $+/+$ type. Indeed, because such a surprisingly small complement of normal cells is able to prevent the expression of the anemia, they propose "that relatively short-range diffusible substances, produced by cells in the microenvironment and required for normal erythropoiesis" are operating. For other hematological investigations on steel mice see Cole et al. (1974), McCarthy (1975), McCarthy et al. (1977), and Adler and Trobaugh (1978).

[20]From 12 days gestation to at least 60 days of age *f/f* mice are also smaller than their normal littermates (E. Russell and McFarland, 1966; E. Russell et al., 1968). Whether this is attributable to the transitory anemia, even though this growth defect persists long after the anemia is over, or whether it reflects another independent primary effect of the *f/f* genotype, is not known. There is also evidence that *f/f* mice are more susceptible to induced leukemia than +/− animals. How this effect is mediated also remains to be determined, but it appears to be a specific gene effect on susceptibility, rather than to linkage of susceptibility genes and *f* (Law, 1952).

[21]Unlike the flexures of the tail and the belly spot, which may often fail to manifest themselves on account of inhibitory modifiers and unknown environmental conditions, the anemia appears to be fully penetrant. Indeed, E. Russell and McFarland (1966) report that it is conceivable that at birth one dose of the flexed-tailed gene (*f*/+) has some effect on erythrocyte number.

[22]The erythrocytes of *f/f* mice have a reduced hemoglobin concentration, but they contain "free" iron in granular form as shown by the Prussian Blue test. This type of erythrocyte has been termed a "siderocyte." These cells are not peculiar to *f/f* mice as they are found also in normal mice, albeit in much smaller numbers with significantly less free iron per cell (Grüneberg, 1941, 1952).

[23]Erythrocyte number and size are normal in adult *f/f* animals, although Grüneberg (1952c) reported 3% siderocytes in their blood in contrast to none in normals over 1 week of age.

[24]There is evidence (Dickie, 1964b) that some *Sp*/+ embryos may die *in utero*, with or without any observable anomaly.

[25]Evidently the longevity of *Sp* homozygotes can be selected for as in a stock selected for minimal white such homozygotes lived to term usually with rudimentary tail, sacral spina bifida, and cranial hernia (Hollander, 1959).

[26]The genotypes of these abnormal embryos were confirmed by transplanting ovaries from them, shortly before they were expected to die, to histocompatible (but appropriately genetically marked) +/+ hosts and mating these hosts to normal (+/+) males. The observation that all the offspring produced from such matings were phenotypically identical with *Sp*/+ mice made it evident that the donor ovaries must have been *Sp/Sp* (W. Russell and Gower, 1950).

[27]It is interesting to note that whereas the two *Sp* mutations (Sp^J and Sp^{3J}) which occurred in the C57BL/6J strain were characterized only by a large belly spot, the one (Sp^{2J}) which occurred in the C3H/HeJ strain was characterized by a head blaze as well. This blaze, however, is occasionally not evident in Sp^{2J}/+ C3H/HeJ mice and is never present in Sp^{2J}/+ offspring derived from C3H/HeJ (Sp^{2J}/+) × C57BL/6J (+/+) matings. This occurrence of a head blaze in strain C3H/HeJ and its absence when outcrossed also has been observed with various steel (*Sl*) alleles (Dickie, 1964b).

[28]The other semidominant mutation did not involve the *Sp*-locus. It occurred in a C57BL/6J male and is called "belly spot" (*Bs*). *Bs/Bs* homozygotes die prior to the twelfth day of gestation (Dickie, 1964b).

[29]Searle (1968a) finds it difficult to think of any other explanation for the variegation of *Va* mice than somatic mutation (see Chapter 10, note 28), and, if he is correct,

this would imply that this gene is highly mutable. He cites as possible evidence Schaible's (1963a) observation that *Va*/+ males had significantly more non-*Va* than *Va* offspring on outcrossing. Searle also notes that "an alternative explanation might be the existence of some physiological threshold during development with respect to some aspect of pigment production, so that only very occasionally does a melanoblast acquire the potentialities necessary for full pigment production in its decendant melanocytes." Schaible (1962) concluded that each wild type area of pigment in *Va*/+ (or in W^a/+ or Mi^{wh}/+) mice resulted from a population of cells descended from a *single* altered melanoblast, a contention which certainly seems most reasonable, but whether this alteration is due to something other than a mutation, such as somatic crossing-over, remains to be determined.

[30]If Grüneberg is correct and the influence which *Va* has on pigmentation is secondary to abnormal hair structure one would anticipate most of the hairs of *Va* homozygotes to be grossly abnormal since they are unpigmented. This remains to be determined. Moreover, if the pigmentary effect of *Va* always is mediated via the hair it is somewhat surprising that the small patches of pigment which occur in homozygous animals are of *unaltered* color as one might expect at least some of these hairs to be moderately abnormal and appear grey.

[31]According to Deol (1954) the anomalies of the labyrinth are not congenital as the first defect is not observed until the animals are 4 days old. The tectorial membrane of the cochlea, the first structure to be visibly affected, is delayed in thinning out, loses all contact with the organ of Corti, and eventually shrivels up. During the second week after birth the spiral ganglion and the organ of Corti also begin to display a progressive degeneration and, during the third week, the stria vascularis degenerates. In *Va*/+ mice the cristae ampullares are severely affected and the maculae only slightly, whereas these sense organs are all severely affected in *Va* homozygotes. The average cell size in the vestibular ganglion also is significantly reduced. In general all of these lesions are more pronounced in *Va* homozygotes than in heterozygotes.

Chapter 12

Microphthalmia and Other Considerations

I. Microphthalmia Locus

Besides dominant spotting (*W*) and steel (*Sl*) there is another series of alleles which frequently produce white spotting when heterozygous, and usually completely eliminate pigment from the coat, and sometimes from the eyes as well, when homozygous. These alleles occur at the so-called "microphthalmia locus" (chromosome 6) and include the following.

A. Microphthalmia (*mi*)

This allele was recognized first in the grandchildren of a male which had received 1500 r of X-irradiation (Hertwig, 1942a,b). Because it is not known whether the son of this male carried the gene, or whether it was introduced by the female to whom he was mated, it is uncertain whether the mutation was induced or occurred spontaneously. Its effects have been described in some detail by Grüneberg (1952) and much of the following is based on his account. *mi*, although not written with a capitol first letter, is a semidominant gene which is usually lethal when homozygous.

1. Effect When Heterozygous

When heterozygous with the wild type allele *mi* has no noticeable affect on the intensity of the coat[1] but does influence the pigmentation of the eye; *mi*/+ newborns have less iris pigment than +/+.[2] These heterozygotes also may display some white spotting (Grüneberg, 1948), a manifestation which seems to depend upon their genetic background. Grüneberg found that when an *mi*/+ male in which spotting was not recognized was outcrossed to two +/+ females from a stock which carried no major spotting genes,[3] that most of their *mi*/+ but none of their +/+ offspring exhibited unpigmented tail tips of varying length. Additional spotting on the head and/or belly and chest occurred in *mi*/+ mice of the F_2 and subsequent generations, while some +/+ animals in these populations also displayed some low grade tail spotting. The amount of spotting on the head of the *mi*/+ mice was quite variable. It always occurred near the midline of the head and varied from a few white hairs on the forehead to fairly extensive white wedges which included a large part of the whisker area. The amount of belly spotting also varied. In some instances it was characterized by long and narrow, nearly symmetrical, streaks along the midline of the belly or chest, while in others it covered up to one-quarter of the ventrum. No dorsal spotting was ever observed. An analysis of the frequency of the spotting in these *mi*/+ mice indicated that at least three genes were involved. Some of these were minor spotting genes (see Chapter 9, Section II, F), which are very common in mice (Grüneberg, 1936b, 1948), whose effects were greatly intensified in *mi*/+ animals.

While *mi*/+ mice vary as to whether they display any external spotting, and as to the degree of this spotting, the internal pigmentation patterns of these heterozygotes seem to be more predictable. Thus Deol (1971) observed that pigment was absent from such a large part of their choroid (usually significantly more than half of the total area) that the pigmented areas appeared as black spots on a clear background. These were, in general, more frequent in the area of the optic nerve or along the base of the iris, but aside from this there was no obvious tendency toward a regular pattern. Deol also reported that the number of melanocytes in the harderian gland of these heterozygotes was greatly reduced and their distribution was very uneven. They were missing completely in some places, but not in any particular area. The melanocytes themselves were mostly smaller and less dendritic than in +/+ mice. Although Deol (1970b) did not find any abnormalities in the inner ear of these animals, about a quarter of the ears examined had no pigment in the lateral crista. The effects of *mi*/+ on pigmentation are summarized in Tables 12-1 and 12-2.

2. Effect When Homozygous

Homozygous *mi*/*mi* mice are completely devoid of pigment both in the fur and in the eyes and in this respect they mimic albino animals. However, unlike the albino, the unpigmented condition of these animals is undoubtedly the consequence of an absence of melanocytes. The eyes of these homozygotes are reduced in size and usually covered with cataracts (Tost, 1958); the eyelids never open and there is a coloboma of the retina (Müller, 1950). These anomalies result from the fact that during development the optic vesicle fails to form a proper cup and the choroid fissure remains per-

Table 12-1. Summary of the Effects of the Genes *mi*, Mi^{wh}, W^{r}, and *s* on the Pigmentation of the Coat, the Choroid, the Harderian Gland, and the Inner Ear[a,b]

Genotypes	Coat	Choroid	Harderian gland	Inner ear
+/*mi*	(Spot on head, belly, tail)	Severely spotted	Severely spotted	(Lightly spotted)
$Mi^{wh}/+$	Spot on belly. Diluted. (Spots on back)	Severely, unevenly diluted	Unpigmented	Unpigmented or diluted or spotted
$W^{v}/+$	Spot on belly. Diluted. (Spot on head)	Moderately spotted	Fairly normal	(Lightly spotted)
W^{v}/W^{v}	Unpigmented	Unpigmented	Unpigmented	Unpigmented or severely spotted
s/s	Widespread spots	Severely spotted	Unpigmented or severely spotted	(Moderately spotted)

[a]From M.S. Deol (1971). Reproduced with permission of the author and Cambridge University Press.
[b]Inconstant features are given in parentheses.

Table 12-2. Summary of the Effects of the Genes *mi*, Mi^{wh}, *s*, and W^{v} on the Pigmentation of the Eye[a]

Genotypes	Choroid	Retina	Iris	
			Outer layer	Inner layer
+/+	Pigmented	Pigmented	Pigmented	Pigmented
+/*mi*	Spotted	Pigmented	Pigmented	Pigmented
$Mi^{wh}/+$	Mostly unpigmented	Pigmented	Mostly pigmented	Pigmented
Mi^{wh}/mi	Unpigmented	Unpigmented	Unpigmented	Mostly pigmented
Mi^{wh}/Mi^{wh}	Unpigmented	Unpigmented	Unpigmented	Partly pigmented
s/s	Spotted	Pigmented	Pigmented	Pigmented
$W^{v}/+$	Spotted	Pigmented	Pigmented	Pigmented
W^{v}/W^{v}	Unpigmented	Pigmented	Unpigmented	Pigmented

[a]From M.S. Deol (1973). Reproduced with permission of the author and Cambridge University Press.

manently open (Müller, 1950, 1951; see also Fischer and Tost, 1959; Konyukhov and Sazhina, 1963, 1966; Konyukhov et al., 1965). The inner ear of these homozygotes also displays abnormalities;[4] it also possesses no pigment (Deol, 1970b).

The most severe consequence of *mi* when homozygous is that, like grey-lethal (*gl*) (see Chapter 5, Section I, B), it results in an impairment of secondary bone resorption. This osteopetrosis, although not as severe as in grey-lethal, nevertheless usually is lethal. Thus in the large majority of *mi* homozygotes the incisors never erupt and as a consequence they die on weaning. Occasionally one or two incisors are cut and if they enable the animal to eat solid food the mouse may live for some time. A microphthalmic male of Hertwig's lived to breed. Another influence of *mi*/*mi* mentioned by Hertwig is that it causes a slight curliness of the vibrissae.

The reshaping of the bones in normal mice and the influence of *mi* (and *gl*) on this process has been studied in some detail by N. Bateman (1954). Grüneberg (1948) also investigated the skeleton and teeth of *mi* homozygotes and found that these animals were "an extraordinarily faithful mimic of *gl*." In both homozygotes secondary bone absorption is disturbed profoundly and the teeth of *mi*/*mi* mice are almost exactly like those of the grey-lethal animal. Nevertheless, the dentition of *mi* homozygotes is less extremely abnormal and the failure of secondary bone absorption is less completely disturbed by *mi* than by *gl*. The differences are, however, slight.

Although the precise etiology of the osteopetrotic syndrome in microphthalmic mice is not known, there are no reasons to believe that it is different from the anomaly in grey-lethal animals. Consequently the reader is referred to Chapter 5, Section C for a more detailed treatment of the condition.[5]

B. White (Mi^{wh})

Mi^{wh} was described originally by Grobman and Charles (1947) after it was found in an F_1 female of a C57BL ♀ × DBA ♂ mating.[6] The male had received a total of 221 r of X-rays and hence it is conceivable that the mutation was induced and not spontaneous (Grüneberg, 1952).

1. Effect When Heterozygous with Wild Type

When heterozygous with the wild type allele Mi^{wh} produces a coat color very similar to dilute (*d*/*d*) though it is slightly lighter (Plate 3-H). The eye color of Mi^{wh}/+ mice is a very dark ruby on a black (*B*/–) background and red in *B*/–;*d*/*d* animals (Grobman and Charles, 1947). The coat color of *a*/*a*;*B*/*B*;*d*/*d*;Mi^{wh}/+ has been described as "pastel-grey" and of *a*/*a*;*b*/*b*;*d*/*d*;Mi^{wh}/+ as a very delicate silver (Grobman and Charles, 1947).

Mi^{wh}/+ heterozygotes usually display spotting on their feet, the tip of the tail, and the belly. Small irregular white spots, sometimes consisting of only a few hairs, often occur also on the back in the lumbar region and very occasionally normally pigmented "spots" are found (Schaible, 1969; see Chapter 10, note 28). In a *Mi^wh^* stock derived from an outcross to C57BL, nearly all *Mi^wh^*/+ mice had fairly large belly spots which were often broad and sometimes formed rudimentary belts (Grüneberg, 1953) (see Plate 3-H). Unlike the situation in *mi*/+ animals, head dots or blazes do not occur. Nevertheless, the total amount of white is much greater in *Mi^wh^*/+ than in *mi*/+. Grüneberg found that on similar genetic backgrounds, *Mi^wh^*/+ and *mi*/+ mice differed greatly in the incidence of ventral spots which occurred in most of the former (37 of 41) but in few of the latter (12 of 56). The shape of the belly spots was about the same in both genotypes but those of *Mi^wh^*/+ were on the average larger. Although in the particular populations studied spotting on the head or face was not present in either heterozygote, tail spotting was present in both with more of the tail tending to be unpigmented in *Mi^wh^*/+.

Hairs from the pigmented regions of *Mi^wh^*/+ mice exhibit what E.S. Russell (1946) has termed "pigmentation lag," i.e., there is a delay in the deposition of medullary pigment in the growing hair. There is also reduced cortical and medullary pigment, especially distally, and the medullary granules vary greatly in size and shape (Wolfe and Coleman, 1964). According to Grobman and Charles *Mi^wh^*/+ reduces or restricts medullary pigment in a more regular manner than does dilute (*d*/*d*). They also reported that *Mi^wh^*/+ animals darken slightly with age, the rate of darkening being somewhat less than in *d*/*d* mice.[7]

The inner ear of *Mi^wh^*/+ mice either has no pigment at all or the amount of pigment is greatly reduced in density and found only in certain areas.[8] The inner ear itself displays severe abnormalities with every ear affected. As in the case of *mi*/*mi* no part of the cochlea duct is normal. Indeed, according to Deol (1970b) *Mi^wh^*/+ animals exhibit the same ear abnormalities as *mi*/*mi* mice (see note 4).[9]

Deol also found that the density of the pigment was reduced throughout the choroid of *Mi^wh^*/+ genotypes.[10] This reduction, although far from uniform, was quite striking. In fact, some regions were pigment-free. In general, there was more pigment in the ventral than in the dorsal half of the eye (Deol, 1971).

The harderian gland of *Mi^wh^*/+ mice is unpigmented[11] and the number of melanocytes in the nictitans is somewhat reduced (Markert and Silvers, 1956). The effects of *Mi^wh^*/+ on the pigmentation of the coat, choroid, harderian gland, and inner ear are summarized in Table 12-1. Its effects on the pigmentation of the eye are presented in Table 12-2.[12]

2. Effect When Homozygous

Like *mi*/*mi* homozygotes, *Mi^wh^*/*Mi^wh^* genotypes are completely white and because their eyes appear pinkish they too have been described as "mock-albinos." However, as in the case of *mi*/*mi*, it must be emphasized that the etiology of this unpigmented condition stems from the absence of melanocytes (Silvers, 1956) and not from a biochemical block in melanin synthesis (see Figure 3-11a).[13] These homozygotes also differ from albinos (*c*/*c*) in that the pigmentary layer of the retina is mostly missing (where present it is unpigmented), their eyes are moderately reduced in size, their fertility is somewhat reduced, they usually are smaller than albinos, and a little pigment sometimes occurs in the inner layer of the iris (Grüneberg, 1953; Deol, 1970b, 1973).[14] Unlike *mi* homozygotes the skeleton of these mice seems to be quite normal (Grüneberg, 1952). The inner ear of Mi^{wh}/Mi^{wh} displays no pigment and exhibits the same abnormalities as $Mi^{wh}/+$ (and *mi*/*mi*) genotypes (Deol, 1970b).[15]

3. Etiology of Spotting

We already have presented the evidence of Mintz (see Chapter 7, Section VII) which indicates that the complete absence of melanocytes from the coat of Mi^{wh}/Mi^{wh} mice, and their partial absence from the coat of $Mi^{wh}/+$ animals, is due to inviable clones of melanoblasts preprogrammed to die as a consequence of Mi^{wh} acting within these cells and *not* via the skin.[16] Further evidence that Mi^{wh} acts autonomously within the melanoblast is derived from the observation that +/+ host melanoblasts which invade transplants of Mi^{wh}/Mi^{wh} neonatal skin function normally and produce intensely pigmented hairs (Silvers and E. Russell, 1955).

4. Effect when Heterozygous with *mi*

Mi^{wh}/mi heterozygotes also lack pigment in the coat but appear to resemble Mi^{wh} homozygotes more than *mi*/*mi* mice. The eyes of these animals have a pigment ring in the iris, sometimes only slight in extent and easily overlooked, but frequently sufficiently heavy to give the eye a red rather than a pink color (Grüneberg, 1953).[17] According to Deol (1973) both the choroid and the retinal epithelium of these heterozygotes are unpigmented but an occasional retinal cell does possess some melanin granules. These, however, are usually misshapen and unevenly distributed.[18] The effects of Mi^{wh}/mi on the pigmentation of the eye are summarized in Table 12-2.

There is no evidence of any skeletal defects in Mi^{wh}/mi mice nor are their teeth affected. They also appear less microphthalmic than Mi^{wh} and, of course, *mi* homozygotes (Grüneberg, 1952, 1953). It thus appears that, in general, Mi^{wh} is completely dominant over *mi* (but see note 25). Nevertheless, taken together, the relationship between these alleles is somewhat paradoxical. Thus, as pointed out by Grüneberg (1953), the homozygous manifestation of *mi* is much more extreme than Mi^{wh}, even though when heterozygous (with +) Mi^{wh} has a much greater effect on the color of the coat and on spotting than does *mi*.[19]

C. Black-Eyed White (*mi^{bw}*)

This allele arose spontaneously in the C3H strain (Kreitner, 1957). It was discovered first in 1954 and was known as *bw* until its allelism with Mi^{wh} was demonstrated by Hollander in 1961. mi^{bw} is completely recessive to wild type. When homozygous it usually eliminates all the pigment in the coat but not the eyes, which remain black. Very occasionally a small pigmented spot(s) ("ticking") may be present on the dorsum or on the head. These may be composed of only a few hairs or comprise a pigmented area about 2 cm in diameter (Kreitner, 1957; Hollander, personal communication). These spots have been described as black with a greyish tone "produced by variegation, and not by incomplete pigmentation of the individual hairs" (Kreitner, 1957). According to Schaible (1963a,b) Mi^{wh}/mi^{bw} heterozygotes are white with some pale yellow spots, of the same shade on a^e,a,+ and A^y backgrounds, which fade in the adult. This compound is similar to that of Mi^{wh}/mi^{sp} (see below) except that the pigmented spots apparently are more dilute and more restricted in size. It should be stressed that since Mi^{wh} homozygotes *never* display any pigment in the coat and mi^{bw}/mi^{bw} mice only rarely display small pigmented spots, it is surprising that Mi^{wh}/mi^{bw} heterozygotes show any spots at all. These heterozygotes have normal eyes except for the red pupil which is characteristic of Mi^{wh} heterozygotes. Very rarely (one of 136 heterozygotes examined) microphthalmia occurs (Schaible, 1963b).

mi^{bw} when homozygous is evidently also a dominance modifier of *P*. Thus whereas $mi^{bw}/mi^{bw};P/P$ mice have black eyes and $mi^{bw}/mi^{bw};p/p$ animals are phenotypically indistinguishable from albinos, $mi^{bw}/mi^{bw};P/p$ genotypes are "ruby-eyed" (Kreitner, 1957). mi^{bw}/mi^{bw} mice have normal viability, size, prolificness, and litter size.

1. Etiology of Spotting

Like $Mi^{wh}/Mi^{wh} \leftrightarrow +/+$ allophenics, $mi^{bw}/mi^{bw} \leftrightarrow +/+$ allophenics are either fully pigmented or display white spotting. Thus Mintz (1971a) found that, in the presence of skin mosaicism, 50% of such allophenics showed both black and white coat areas. These results indicate that, like Mi^{wh}, the white coat of mi^{bw}/mi^{bw} mice is due to the preprogrammed death of melanoblast clones (see Chapter 7, Section VII). Further evidence that this is the case stems from some experiments of Markert (1960). He implanted tissues (which normally contain melanoblasts) from either +/+ or mi^{bw}/mi^{bw} embryos into the anterior chamber of the eye of albino hosts. Whereas such implants always gave rise to many melanocytes which migrated over the inner surface of the host eye when they were derived from +/+ embryos, in no instance did melanocytes occur in or near the mi^{bw}/mi^{bw} implants. It thus appears that for some reason the neural crest of mi^{bw}/mi^{bw} mice does not possess cells capable of forming pigment, even in a favorable milieu.

D. Microphthalmia-Spotted (mi^{sp})

The behavior of this allele was described briefly by Wolfe (1962) and subsequently in much more detail by Wolfe and Coleman (1964); the following is based on their observations. mi^{sp} is one of the most interesting of the *mi*-series because it does not express itself either when heterozygous with wild-type or when homozygous, both $mi^{sp}/+$ and mi^{sp}/mi^{sp} genotypes appearing indistinguishable from +/+. However, when heterozygous with either Mi^{wh} or *mi*, new phenotypes are produced.[20] On an *a*/*a* background, Mi^{wh}/mi^{sp} mice are pale yellow with white spots on both their dorsum and ventrum, and with pigmented eyes. The yellow areas become "sooty" at the first moult. mi/mi^{sp} mice are nearly all-white except for a variable number of faint irregular patches of pigment usually expressed as "flecks" of pigmented hairs on the dorsum.[21] The pigmentation of their eyes is less than that of *mi*/+ and nearly indistinguishable from that of C57BL/6J-*ru*/*ru* (ruby-eye) mice. The mutant was expressed first when a cross between a C57BL/6J ♀ and a C57BL/6J-$Mi^{wh}/+$ ♂ produced five deviants.

1. Tyrosinase Activity

Although $mi^{sp}/+$ and mi^{sp} homozygotes are phenotypically indistinguishable from +/+ mice, tyrosinase activity in skin slices from 5-day-old animals of these genotypes (all on C57BL/6 backgrounds) are different. Thus, as known in Figure 12-1, +/+ displays the most activity, followed in turn by

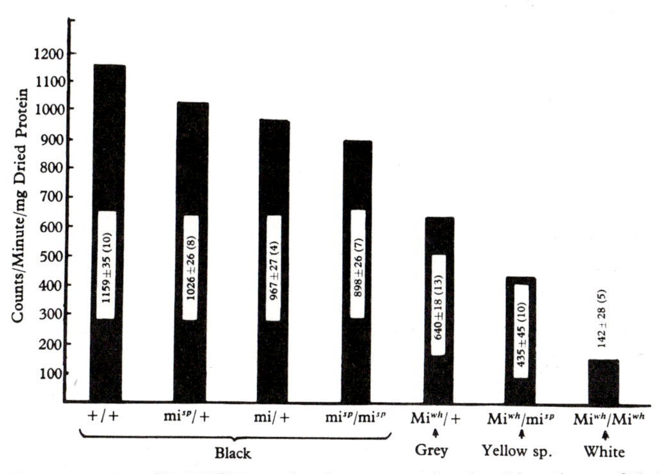

Figure 12-1. [2-^{14}C]Tyrosine incorporation in skin slices of 5-day-old mice differing at the microphthalmia locus. Means, standard errors, and number of determinations are shown for each genotype. All mice were nonagouti (*a*/*a*). From H.G. Wolfe and D.L. Coleman (1964). Reproduced with permission of the authors and Cambridge University Press.

$mi^{sp}/+$, $mi/+$, and mi^{sp}/mi^{sp}. $mi^{sp}/+$ mice differ significantly from mi^{sp}/mi^{sp} and both differ significantly from $+/+$. On the other hand, $mi/+$ does not differ from either $mi^{sp}/+$ or mi^{sp}/mi^{sp}, but does differ from $Mi^{wh}/+$ and, as already noted, $+/+$.[22]

2. Hair Granules

In spite of the fact that mi^{sp}/mi^{sp} and $mi^{sp}/+$ genotypes have a lower tyrosinase level than $+/+$ no differences were observed in the number, kind, or arrangement of cortical granules in these genotypes, or in $mi/+$ mice. Nor were any differences observed in the distance to the first cortical pigment measured from the tip of the hairs. Because medullary granules were compactly grouped and could not be counted without a large error they were not compared. However, the granules in all of these deviants did not appear to differ from $+/+$ (in this case nonagouti, black) which are long oval and intense black (E. Russell, 1946).[23]

Mi^{wh}/Mi^{sp} hairs displayed a pigmentation lag even more pronounced than $Mi^{wh}/+$ and in some hairs no cortical granules occurred at all. Thus most of the pigmentation of the coat of these mice stems from granules located in the proximal portion of the shaft. The medullary granules vary greatly in shape and size (from 0.4 μm to as much as 5 μm in greatest diameter, presumably because of clumping). The granules seem less dense than $+/+$ and have a yellowish appearance when observed by transmitted white light.[24] The sizes and shapes of granules do not vary as much in postjuvenile hairs and there are more of them in both the cortex and the medulla, an observation in accord with the observations of Grobman and Charles for hairs of $Mi^{wh}/+$ mice.

E. White Spot (mi^{ws})

mi^{ws}, a semidominant, occurred in a C57BL/6 strain which had, over a 5-year period, been exposed to chronic γ-irradiation at a dosage of 0.2 mrad/hr (D.S. Miller, 1963). When heterozygous with the wild-type allele it frequently, though not invariably, produces a white spot (often in the form of a diamond) on the belly. The toes and tail tip also are commonly white and adults may become grizzled, but not diluted (Hollander, personal communication). When homozygous mi^{ws} eliminates all pigment from the coat; the eyes are pink (a "mock-albino"). Homozygotes are viable and fertile with little, if any, microphthalmia (Hollander, 1964; personal communication). The mi^{bw}/mi^{ws} heterozygote is a black-eyed white but the Mi^{wh}/mi^{ws} genotype (unlike Mi^{wh} or mi^{ws} homozygotes which are completely white) is obviously spotted, often checkerboard-like, with yellowish-brown (fawn) to grey. These spots gradually molt lighter so that older mice look practically black-eyed white (Hollander, personal communication). Moreover, also unlike either Mi^{wh} or mi^{ws} homozygotes (whose eyes appear pink),

Mi^{wh}/mi^{ws} heterozygotes have dark eyes of normal size.[25] Indeed, as pointed out by Hollander (1968) their phenotype is about what one might expect from the interaction of nonalleles. Nevertheless, no crossing over has been detected. These observations, therefore, give further impetus to the possibility already discussed (see notes 16 and 20) that *mi* is a complex locus, somewhat analogous to the brachyury (*T*) series of alleles.

F. Other Alleles

In addition to the 5 *mi*-alleles noted above there are five others: three which have been only briefly described, red-eyed white (mi^{rw}), microphthalmia-Oak Ridge (Mi^{or}), and microphthalmia-brownish (Mi^{b}), and two, mi^{x} and eyeless *white* (mi^{ew}), which have been assigned provisional symbols.

mi^{rw} is a recessive allele. It occurred in the CBA/J strain (Southard, 1974). "The original deviant was white with a pigmented ring around the neck and small red eyes." Mi^{wh}/mi^{rw} heterozygotes are predominantly white with tan spots and red eyes; mi^{sp}/mi^{rw} genotypes are wild-type but with a white belly spot; and mi/mi^{rw} animals are completely unpigmented either with no eyes or with very small eyes.

The allele Mi^{or} has been described by Stelzner (1964, 1966). It was noticed first in an offspring of a male whose spermatogonia had received γ-irradiation. $Mi^{or}/+$ heterozygotes have a slightly diluted coat and pale, pinkish ears and tail. A white belly streak and/or white head spot also is frequently present. The eyes of these heterozygotes display reduced pigmentation at birth (resembling *b*/*b*) but appear full-colored in adults. Their fertility appears normal. Mi^{or} homozygotes have white fur. The eyes are either absent or they are so reduced in size that they *appear* absent. The eyelids never open and the incisors often fail to erupt or are poorly formed. Like *mi*/*mi*, animals with this latter condition usually die after weaning unless given powdered food. Males are fully fertile, but all six females tested proved sterile. This mutant seems to resemble *mi* in some ways and Mi^{wh} in others. Thus $Mi^{or}/+$ heterozygotes, like $Mi^{wh}/+$ animals, and unlike *mi*/+ mice, have a dilute pigmentation, reduced eye pigment and (usually) white spotting. On the other hand, the effect which Mi^{or} has when homozygous much more closely resembles the situation in *mi*/*mi* than in Mi^{wh}/Mi^{wh} mice.

Mi^{or}/mi heterozygotes resemble Mi^{or}/Mi^{or}; they are white, appear to be eyeless, and, so far, all have been toothless. Males are fertile.

The allele Mi^{b} also occurred at Oak Ridge (Larsen, 1966), this one spontaneously in an offspring of a $(101 \times C3H)F_1$ ♂ × T (multiple recessive) ♀. The coat of $Mi^{b}/+$ heterozygotes is diluted, displaying a brownish cast; their tail and ears also are paler than +/+. Homozygotes are white with reduced eye pigment; they do not appear to be microphthalmic. The teeth are normal and both sexes are fertile. $Mi^{wh}/Mi^{b};A/A$ mice are light cream with white spots and ruby eyes; Mi^{or}/Mi^{b} animals are indistinguishable from Mi^{b}/Mi^{b}.

mi^{x}, a recessive, occurred in the NZB/Mac strain (Munford, 1965). Animals homozygous for this mutation are black-eyed whites. *Mi*^{wh}/*mi*^{x} heterozygotes are also black-eyed whites; reduced eye size and eyelid closure of variable extent also have been observed in these heterozygotes.

mi^{ew}, another recessive, occurred in the C57BL/6Bn stock at Lawrence, Kansas (Miner, 1968). *mi*^{ew}/+ heterozygotes are normally pigmented with black eyes. *mi*^{ew} homozygotes are white with undeveloped eyes (histological studies show an almost complete absence of eye development in adults) and eyelids which never open; their teeth appear normal and their fertility is somewhat reduced. *Mi*^{wh}/*mi*^{ew} heterozygotes are white with slightly reduced dark red eyes and with eyelids which are slightly closed; *mi*^{bw}/*mi*^{ew} animals are black-eyed whites; and *mi*^{sp}/*mi*^{ew} mice are white with black (on a nonagouti background) spots and normal black eyes (Wood and Miner, 1969; H.G. Wolfe, personal communication).[26]

The effects of the various *mi*-locus genotypes considered above are summarized in Table 12-3.

Table 12-3. Descriptions of Various *mi*-Locus Genotypes[a]

Genotype	Description
mi/+	Sometimes spot on head, belly, tail; eye pigment reduced; inner ear normal
mi/*mi*	White; no eye pigment; severe microphthalmia; osteopetrosis; curley whiskers; severe inner ear abnormalities; usually lethal
Mi^{wh}/+	Spot on belly; diluted; sometimes white or intensely pigmented spots on dorsum; eye pigment reduced; severe inner ear abnormalities; skeleton normal
Mi^{wh}/*Mi*^{wh}	White; eye pigment greatly reduced; moderate microphthalmia; fertility reduced; severe inner ear abnormalities; skeleton normal
Mi^{wh}/*mi*	White; eye pigment greatly reduced; less microphthalmic than *Mi*^{wh}/*Mi*^{wh}
mi^{bw}/+	Like wild type
mi^{bw}/*mi*^{bw}	White with occasional pigmented spot(s) ("ticking"); dark eyes; normal viability and fertility
Mi^{wh}/*mi*^{bw}	White with some fawn-colored spots which fade in the adult; dark eyes
mi^{sp}/+	Like wild type
mi^{sp}/*mi*^{sp}	Like wild type
mi/*mi*^{sp}	White but on some backgrounds there are a variable number of faintly pigmented "flecks" on dorsum; eye pigment less than *mi*/+ (like *ru*/*ru*)
Mi^{wh}/*mi*^{sp}	Pale yellow with white spots (yellow becomes sooty); dark eyes
mi^{ws}/+	Like wild type but frequently has belly spot. The toes and tail tip also are commonly white and adults may become grizzled
mi^{ws}/*mi*^{ws}	White with pink eyes; viable and fertile with little, if any, microphthalmia

Table 12-3. *continued*

Genotype	Description
mi^{bw}/mi^{ws}	White with dark eyes
Mi^{wh}/mi^{ws}	Spotted, often checkerboard-like, with fawn to grey-colored spots which gradually molt lighter; dark eyes
$mi^{rw}/+$	Like wild type
mi^{rw}/mi^{rw}	White with pigmented ring around neck; red eyes; moderate microphthalmia
Mi^{wh}/mi^{rw}	Predominantly white with tan spots; red eyes
mi^{sp}/mi^{rw}	Like wild type but with a white belly spot
mi/mi^{rw}	White; severe microphthalmia
$Mi^{or}/+$	Diluted; pinkish ears and tail; white belly streak and/or white head spot frequently present; eye pigment reduced at birth but appears full-colored in adult; fertility normal
Mi^{or}/Mi^{or}	White; severe microphthalmia; sometimes incisors fail to erupt or are poorly formed; males fertile but females may all be sterile
Mi^{or}/mi	White; severe microphthalmia; toothless; males fertile
$Mi^{b}/+$	Diluted (showing brownish cast); tail and ears paler than normal
Mi^{b}/Mi^{b}	White; eye pigment reduced; not microphthalmic; teeth normal; fertile
Mi^{wh}/Mi^{b}	Light cream with white spots; ruby eyes
Mi^{or}/Mi^{b}	Same as Mi^{b}/Mi^{b}
$mi^{x}/+$	Like wild type
mi^{x}/mi^{x}	Black-eyed white
Mi^{wh}/mi^{x}	Black-eyed white; reduced eye size and eyelid closure of variable extent
$mi^{ew}/+$	Like wild type
mi^{ew}/mi^{ew}	White; severe microphthalmia; eyelids never open; teeth normal; fertility slightly reduced
Mi^{wh}/mi^{ew}	White; dark red eyes, slightly reduced; eye lids slightly closed
mi^{bw}/mi^{ew}	Black-eyed white
mi^{sp}/mi^{ew}	White with pigmented spots; normal black eyes

[a]Only effects noted in literature are included.

II. Other Determinants Associated with White Spotting

A. Fleck (*Fk*)

Fleck is an autosomal dominant which is lethal when homozygous. The mutation, which has not as yet been mapped, was first observed by Sheridan (1968) and what follows is based on his account. Fleck arose in the CBA strain. It is characterized by a white belly spot, white tail tip and often white back paws. Although animals expressing this completely penetrant character are both viable and fertile, when mutant males are mated to mutant females the rate of intrauterine death at the time of implantation is about 25% greater than when mutant females are mated to normal males, an observation which suggests that the *Fk*/*Fk* genotype is lethal.

In spite of the fact that there is no indication that *Fk* when heterozygous affects the well being or life expectancy of its carriers, it does influence their breeding performance. Thus *Fk*/+ males had a significantly lower mating prowess than their normal brothers when mated to normal CBA females. The size of the first and second (but not the third) litters of mutant females mated to normal CBA males also was smaller than their normal sisters. Moreover, the *Fk*/+ mothers were more likely to lose their litters completely. Whether this is due to cannibalism, or to the inability of the parent to care for the litter, has not yet been determined but it is consistent with the observation that smaller litters are more apt to be completely lost (Lüning, cited by Sheridan). The reduction in litter size from fleck females could be due to any one of several different factors or some combination of them: for example, a smaller number of eggs ovulated, a reduced rate of implantation, or an increased rate of intrauterine death.[27]

Despite the higher prenatal mortality in the first two litters of *Fk*/+ ♀ × +/+ ♂ matings, there is no evidence that this mortality is directed against the heterozygous embryos. Indeed, the fact that the data from weaned litters showed a good agreement with the expected 1:1 ratio argues against such selection.

Finally, although *Fk* in some respects mimics the effect of splotch (*Sp*) no evidence was found in either dissection studies or in progeny tests that these mutants are allelic. *Fk* has been shown also not to be allelic with viable dominant spotting (W^v) or steel (Sl^d).

B. Belly Spot and Tail (B^{st})

This dominant mutation (linkage unknown) is of recent origin (Southard and Eicher, 1977) and its gene symbol is therefore provisional. It occurred in strain C57BL/KsJ and causes a white belly spot, a short, kinky tail, and a polydactylous condition affecting one to three feet. Both sexes are fertile and viable.

III. Influence of Some Coat-Color Determinants on White Spotting

It has long been known that lethal yellow (A^y) may significantly reduce the amount of white spotting. This was first noted by Little (1917a,b) who observed that $A^y/a;W/+;s/+$ and $A^y/a;W/+;s/s$ mice displayed much less white in their coats than the corresponding nonagouti (*a*/*a*) genotypes. It was subsequently shown (Dunn et al., 1937) that this was due entirely to the effect of A^y on dominant spotting (*W*/+) as A^y does not appear to have any influence on the genes of the piebald group (*s* and the "*k*" genes); nor is there any evidence that either white-bellied agouti (A^w) or black-and-tan (a^t) have any influence on these genes (Carroll, 1934; Hughes, 1934; both quoted from Dunn et al., 1937).

In contrast to the lack of influence which A^y has on piebald (*s*/*s*) spotting (see also Lamoreux, 1973), recessive yellow, piebald (*e*/*e*;*s*/*s*) mice consistently showed less spotting than their black or brown piebald sibs (Hauschka et al., 1968), and both $A^y/a;s/s^l$ and $e/e;s/s^l$ yellow animals had more pigment than $a/a;E/E;s/s^l$ mice (Lamoreux, 1973). Indeed, Lamoreux found that lethal spotting (*ls*/*ls*), belted (*bt*/*bt*), splotch (*Sp*/+), Jay's dominant spotting (W^j/+), flexed-tailed (*f*/*f*), and white (Mi^{wh}/+) all produced significantly less white spotting when combined with lethal yellow (A^y/–) than when associated with nonagouti (*a*/*a*).[28] She also found that *ls*/*ls* and Mi^{wh}/+ (but *not* *Sp*/+) produced less spotting when combined with recessive yellow (*e*/*e*), and Hauschka *et al.* (1968) observed that *e*/*e* also reduced the amount of white spotting produced by belted (*bt*/*bt*) (see Table 12-4).

Since both lethal yellow and recessive yellow appear, with few exceptions, to reduced the amount of white spotting it might seem logical to conclude that A^y/– and *e*/*e* for some reason are better able than *a*/*a* and *E*/– to support melanoblast differentiation. However, if Mintz's hypothesis is correct these genes could have what is essentially the opposite effect, i.e., the smaller spots in yellow mice could be due to the fact that A^y/– and *e*/*e* frequently cause the melanoblasts of inviable clones to die a little sooner and, by so doing, provide more time for the melanoblasts of viable clones to take their place. If this is the case it is likely that A^y produces this adverse affect on melanoblast viability via the environment while *e*/*e* probably acts within the melanoblast itself. Thus, the inverse relationship which exists between the amount of white spotting (produced by W^j/+) and the amount of yellow in the coat of viable yellow (A^{vy}/–) mice (Lamoreux, 1973) could be due to the fact that the melanoblasts of inviable clones do not survive as long in "yellow" as in "agouti" phenoclones.

IV. Mutation Rates

Although some genes of the mouse seem more unstable than others, to estimate their relative mutation rates requires the surveillance of a large breeding colony over an extended period. It is scarcely surprising, therefore, that one of the most comprehensive studies on the frequency of occurrence of spontaneous mutations was carried out at the Jackson Laboratory between

Table 12-4. Influence of Recessive Yellow (*e*) on Degree of Expression of Belted (*bt*) in 10-Week-Old Mice of genotypes *E/E;a/a;bt/bt* and *e/e;a/a;bt/bt*[a]

Number phenotype, and sex of mice measured	Mean Weight (g)	Mean belt width (range of widths)		Percentage complete 360° belts	Percentage of belts interrupted by pigment	Percentage partial belts		
		Widest part (cm)	Narrowest part (cm)			1/4 belts	1/2 belts	3/4 belts
63 black belted ♂	27.6	2.12 ± 0.19 (1.27 to 3.07)	0.52 ± 0.14 (0.25 to 1.02)	69.8 ± 5.8	14.3	0	6.3	9.5
91 yellow belted ♂	28.2	1.60 ± 0.11 (0.95 to 2.37)	0.40 ± 0.08 (0.20 to 0.88)	33.0 ± 4.9	15.4	18.7	9.9	23.0
51 black belted ♀	21.2	2.11 ± 0.18 (1.27 to 3.14)	0.52 ± 0.19 (0.29 to 1.33)	68.7 ± 6.5	13.7	0	7.8	9.8
80 yellow belted ♀	21.1	1.51 ± 0.13 (0.77 to 2.08)	0.39 ± 0.06 (0.25 to 0.64)	18.8 ± 4.5	25.0	16.2	15.0	25.0

[a]From T.S. Hauschka et al. (1968). Reproduced with permission of the American Genetic Association.

1963 and 1969 (Schlager and Dickie, 1966, 1967, 1971). Of special interest to us is the fact that it included a number of coat color determinants, paying particular attention to the spontaneous mutation rates, both forward and reverse, at the agouti (*a*), brown (*b*), albino (*c*), dilute (*d*) and leaden (*ln*) loci. The number of mutations that were observed at these loci, along with their calculated mutation rates and 95% confidence limits (Stevens, 1942), are presented in Table 12-5. Attention is drawn to the following:

1. The overall spontaneous mutation rates were 11×10^{-6} per locus per gamete for mutations from wild type (forward mutations) and 2.5×10^{-6} for mutations from recessive alleles (reverse mutations). The 95% confidence limits of these two estimates do not overlap.
2. *A* and *C* had relatively high mutation rates while the rates of *a* and *d* were higher than *b*, *c*, and *ln*. Indeed, *b* and *c* appear to be very stable since no mutations occurred in over 3 million opportunities.[29]
3. The only actual example of a forward and reverse mutation at the same locus occurred at the *d*-locus where *D* mutated to *d* (at a rate of 1.2×10^{-6}) and *d* mutated to *D* (at a rate of 3.9×10^{-6}).[30]
4. Of the five specific loci studied the highest rate of spontaneous mutation from wild type was recorded for the agouti locus, an observation which contrasts with the low mutation rate recorded for this locus under acute spermatogonial X-irradiation. Thus W. Russell and L. Russell (1959) observed only 2 mutations out of a total of 174 under dosages ranging from 300 to 1000 rad, and Lyon and Morris (1966) found no *a*-locus mutations in over 16,000 offspring sired by males whose hind quarters had been exposed to 600 rad. Indeed, when one compares the *spontaneous* mutation rates of the *a*, *b*, *c*, and *d*-loci, as recorded by Schlager and Dickie, with those of the Russells' irradiation study, there appears to be an inverse relationship between the two rates in rank order: $a > c > d > b$ for spontaneous mutations and $b > d > c > a$ under irradiation. The basis for this relationship, if in fact it is a real one, remains to be elucidated.
5. The results noted in Table 12-5 are in general accord with previous estimates of spontaneous mutation rates in mice. Thus the 95% confidence limits ($7.3-16.6 \times 10^{-6}$) of Schlager and Dickie's results encompass the 7.5×10^{-6} per locus per gamete spontaneous forward mutation rate cited by W. Russell (1963) for seven loci, including the *a*, *b*, *c*, *d*, *p*, and *s* loci, and the 10×10^{-6} forward mutation rate reported by Carter et al. (1958) for these same loci.[31]

Schlager and Dickie also observed some spontaneous dominant coat-color mutations and have estimated the mutation rates at the loci involved (Table 12-6). Again it is obvious that there are differences.

Of considerable interest also is Schlager and Dickie's (1967) observation

Table 12-5: Mutations and Mutation Rates at Five Specific Coat-Color Loci[a]

Locus	Number of gametes tested	Number of mutations	Mutation rate × 10^6	95% confidence limits × 10^6	Mutations
Mutations from wild-type (forward)					
A	67,395	3	44.5	9.2–130.1	$+ \to a$, $+ \to A^{iy}$, $+ \to A^{y}$
B	919,699	3	3.3	0.7– 9.5	(3) $+ \to b$
C	150,391	5	33.2	10.8– 77.6	(4) $+ \to c$, $+ \to c^{p}$
D	839,447	10	11.9	5.2– 21.9	(7) $+ \to d^{l}$, (2) $+ \to d^{s}$, $+ \to d$
Ln	243,444	4	16.4	4.5– 42.1	$+ \to ln$
Total	2,220,376	25	11.2	7.3– 16.6	
Dominant mutations (reverse)					
a	8,167,854	34	4.2	2.9– 5.8	(23) $a \to a^{t}$, (10) $a \to A^{w}$, $a \to A^{sy}$
b	3,092,806	0	0	0– 1.2	
c	3,423,724	0	0	0– 1.1	
d	2,286,472	9	3.9	1.8– 7.5	(9) $d \to +$
ln	266,122	0	0	0– 13.9	
Total	17,236,978	43	2.5	1.8– 3.4	

[a]Modified from Schlager and Dickie (1971).

Table 12-6: Spontaneous Mutation Rate at Some Coat-Color Loci at Which Dominant Mutations Occurred[a]

Name	Symbol	Number of mutants found	Mutation rate × 10^6
Dominant spotting	W	23	2.20
Splotch	*Sp*	12	1.15
Steel	*Sl*	5	0.48
Tortoise	To^b	6	0.77
Brindled	Mo^{brb}	1	0.13

[a]Data from Schlager and Dickie (1967). Based on 5,226,531 mice.
[b]Although the X-linked traits *brindled* and *tortoise* were considered as distinct loci by Schlager and Dickie, they undoubtedly are alleles (see Chapter 8, Section II) and if considered as such yield a mutation rate at the *Mo* locus of 0.89 × 10^{-6} (assuming a 50:50 sex ratio).

that the incidence of spontaneous mutations appears to vary greatly from strain to strain (Table 12-7). For example, of 18 *W* locus mutations, 10 occurred in the C57BL/6J strain and 3 appeared in the C3H/HeJ strain. The mutation rates of the *W*-locus in these two strains are therefore considerably higher than in other nonalbino strains. The C57BL/6J strain also displays a high incidence of reverse mutation at the *a*-locus, from *a* to a^t and to A^w (see Chapter 2, note 14). Indeed, the high frequency of mutations in the C57BL/6J strain raises the possibility that one or more mutator genes, i.e., genes thought to produce a substance that induces transition mutants, are involved. Such genes are believed to occur in maize (McClintock, 1951), *Drosophila* (Demerec, 1937; Ives, 1950), and *Salmonella* (Kirchner, 1960).[32]

Notes

[1]Deol (1970b) notes that there is a slight dilution of the coat in young heterozygotes, but that this disappears later on. Moreover, it should be noted that Wolfe and Coleman (1964) found that tyrosine incorporation was less in 5-day-old *mi*/+ than +/+ skin (see Figure 12-1).

[2]Although it is not possible to classify *mi*/+ and +/+ mice as easily later on, Grüneberg (1948, 1952) states it can be done provided the eyes are examined with a strong light under magnification. However, according to him this classification "is complicated by the fact that contrary to statements in the literature, the color genes *a*, *b* and *d* also influence eye color very markedly." "In a normal (*A*/*A*;*B*/*B*) mouse, the choroidal vessels are hidden from view by the choroidal pigment; the pupil is jet-black and the sclera next to the cornea is opaque and slate-grey." On the other hand, in cinnamons (*A*/*A*;*b*/*b*) and in dilute (*d*/*d*) animals the choroidal pigmentation near

Table 12-7: Frequency of Mutations in Different Mouse Strains[a]

Strain	Coat-color genotype	Number mice examined	Number mutations[b]
A/HeJ	*a/a;b/b;c/c*	145,942	0
A/J	*a/a;b/b;c/c*	387,153	4
AKR/J	*a/a;c/c*	637,141	13
BALB/cJ	*b/b;c/c*	274,469	7
CBA/J		245,379	5
CE/J	$A^w/A^w;c^e/c^e$	4,657	0
C3H/HeJ		683,156	29
C3HeB/FeJ		117,566	4
C57BL/6J	*a/a*	1,555,246	61
C57BL/10J	*a/a*	47,821	2
C57BR/cdJ	*a/a;b/b*	26,662	1
C57L/J	*a/a;b/b;ln/ln*	133,061	5
C58/J	*a/a*	34,574	2
DBA/1J	*a/a;b/b;d/d*	128,729	1
DBA/2J	*a/a;b/b;d/d*	1,014,507	24
LP/J	$A^w/A^w;s/s$	3,025	0
MA/J	*c/c*	3,367	0
CBA/CA		875	0
RF/J	*c/c*	40,521	3
B10D2/Sn	*a/a*	40,138	4
SJL/J	*c/c;p/p*	68,018	0
SM/J	A^w/a or *a/a*	1,687	0
ST/bJ	*a/a;b/b;c/c*	9,115	0
SWR/J	*c/c*	32,476	1
129/J	$A^w/A^w;c^{ch}/c;p/p$	67,847	0
HRS	*b/b;c/c;d/d*	8,625	1
C57BL/KsFc	*a/a*	7,441	1
SEC	$a/a;b/b;c^{ch}/c^{ch}$	4,492	3
C3HeB/Fe × BL/Ks	*a/+*	1,124	0
AKD2F$_1$	*a/a;b/+;c/+;d/+*	70,139	8
B6AF$_1$	*a/a;b/+;c/+*	80,252	1
B6D2F$_1$	*a/a;b/+;d/+*	701,913	15
CAF$_1$	*a/+;b/b;c/c*	122,775	1
C3D2F$_1$	*a/+;b/+;d/+*	67,395	0
LAF$_1$	*a/a;b/b;c/+;ln/+*	243,444	10
Total		7,010,732	206

[a]Based on Schlager and Dickie (1971).

[b]Actually a total of 249 mutations was recovered in this study but only 206, involving both pigmentary and nonpigmentary traits, were characterized.

the fundus is somewhat diluted so that even adult animals show a dull red light reflex in their pupils. Grüneberg also states that *a/a* increases the amount of choroidal pigment; hence a chocolate mouse (*a/a;b/b*) has much more pigment than a cinnamon (*A/A;b/b*) animal. The influence which *mi/+* has on the eye is similar to, but much stronger than, the diluting effect of *b/b*. "Thus a *mi/+;A/A;B/B* mouse shows a red pupillary reflex, and often the pigmentation is reduced in the periphery of the fundus so that the choroidal vessels are visible through a semi-transparent sclera near the corneal margin." In *mi/+;A/A;b/b* animals the pigmentation of the iris is also diluted so "that a ruby-colored reflex is generally visible in ordinary daylight with the naked eye." In Grüneberg's experience *mi/+* can be distinguished from +/+ in the presence of *A*, *a*, *B*, *b*, *D*, and *d* on the basis of choroidal and iris pigmentation "except for rare errors in *A/A;B/B* or *a/a;B/B* mice" (Grüneberg, 1952).

[3]Because the spotting effect of *mi/+* was discovered accidentally, Grüneberg's analysis of this condition was, in part, a retrospective one. Consequently he was not absolutely certain whether any of the three animals he originally mated showed minor spotting. It should be emphasized, however, that the *mi/+* male employed in this mating came from Hertwig's original strain, a strain in which tail spotting or spotting elsewhere does not seem to have occurred at all (Grüneberg, 1948).

[4]Deol (1970b) found that all 20 ears he examined were affected. The abnormalities always were severe in the cochlea, but the saccule sometimes exhibited a fairly normal appearance, particularly in very young animals. In no instance was any section of the cochlea duct found to be normal, although the abnormalities were far from uniform throughout its length. The stria vascularis was abnormal in its entirety but small groups of hair cells occurred in the organ of Corti at scattered points. Degeneration of the spiral ganglion was not particularly severe, but this could have been attributed to the fact that no old animals were examined. Deol also observed that "severe dedifferentiation and cellular migrations occurred in the cochlear duct and the saccule in the majority of ears."

[5]Briefly "the microphthalmic mouse has a mild hypocalcaemia, hypophosphatemia, an increased serum alkaline phosphatase and greatly increased bone and serum citrate" (H. Murphy, 1973). Although some animals are found with normal levels of these parameters, the severe osteopetrotic lesion is always present histologically. According to Murphy, 28% of the mice have a life expectancy in excess of 3 months which is much greater than the life expectancy of grey-lethal animals. As in the case of grey-lethal both thyroid and parathyroid endocrinopathies seem to be involved; at least many of the morphological, histological, and biochemical observations on these mutants can be explained by an increased secretion of calcitonin eliciting a compensatory increase in parathyroid hormone (H. Murphy, 1973). The osteopetrotic condition in microphthalmic mice can also be improved by parabiosis with normal mice (Walker, 1972, 1973; Barnes et al., 1975) or by injecting cells from the spleen or marrow of normal animals into irradiated mutants (Walker, 1975a,c). Loutit and Sansom (1976) have been able to overcome the osteopetrotic defect of *mi/mi* mice by injecting them intraperitoneally at birth, or intravenously at weaning or maturity, with cell suspensions containing hematopoietic stem cells from normal syngeneic (or H-2 compatible allogeneic) donors. Thus following this treatment resolution of much of the osteopetrosis but none of the other effects of *mi/mi* occurred within a few months in the majority of cases. Moreover, irradiation of the recipients was *not* necessary to accomplish this "cure" (Loutit, 1977). These investigators believe that "osteoclasis of scaffold-type woven bone is

impaired in *mi*/*mi*" and "that osteoclastic cells are derived through circulating monocytes from hematopoietic stem cells." More recently Nisbet et al. (1978) have shown that if microphthalmic mice are either inoculated intravenously with bone marrow cells from allogeneic (but H-2 compatible) donors (bearing a T-6 marker), or are placed in temporary parabiosis with such animals, some improvement of their condition occurs (blindness and failure of eruption of teeth are unaffected) despite the fact that *no* donor cells can subsequently be found. On the basis of these findings they believe that during their limited period of residence the normal cells may cause considerable resorption of bone by providing osteoclastic procursors, or, they may trigger "the recipient's osteoclasts into effective function, comparable to the T–B cell–cell interactions of immunology." In contrast to the situation when normal cells are given to *mi*/*mi* mice, osteopetrosis can be produced by inoculating spleen cells from mutants into lethally irradiated normal recipients (Walker, 1975b). Raisz and his colleagues (1972) have determined the response of *mi*/*mi* (and normal) long bones in organ culture to some known stimulators of osteoclastic bone resorption. Their results indicate that the congenital osteopetrotic condition stems from "a generalized defect in the function and hormonal response of osteoclasts and suggest that this cell line is separate from the osteoblast cell line which shows no impairment of hormonal response."

[6]The gene originally was given the symbol *Wh* but was changed to Mi^{wh} when its allelism with *mi* was demonstrated (Grüneberg, 1953).

[7]When Mi^{wh}/+ is combined with splotch (*Sp*/+) on a nonagouti background it produces a decided yellowing effect on the ventrum, almost like a^t, as well as some increase in white area. Although the combination Mi^{wh}/+;W^a/+ produces typically a black-eyed white phenotype on some genetic backgrounds some pigment occurs, primarily on the rump and around the ears (but demonstrably less than in Mi^{wh}/+;*W*/+) (Hollander, 1959).

[8]Deol (1970b) found that pigment was present in about two-thirds of Mi^{wh}/+ ears but even in these cases it was severely reduced and mostly found in the utricle.

[9]The fact that Mi^{wh}/+ mice, which possess a good deal of pigment in the coat, display much severer inner ear abnormalities than W^v/W^v animals, which are completely white, provides an excellent example of an *exception* to the rule that, in general, the abnormalities of the inner ear are related to coat color (Deol, 1970b). On the other hand, as already noted (see Chapter 10, note 19), the correlation between abnormalities of the inner ear and pigmentation of the inner ear is quite good (Deol, 1970b).

[10]Markert and Silvers (1956) did not observe any choroidal pigment in either *a*/*a*;Mi^{wh}/+or a^t/a^t;Mi^{wh}/+ mice (see Table 10-1). However, only a few animals were examined. Indeed, Deol (1973) too found that in some Mi^{wh}/+ mice the entire choroid was unpigmented.

[11]Deol (1971) found no melanocytes in the harderian gland of Mi^{wh}/+ mice, "although in sections a few granules could occasionally be seen which might have been lightly melanized melanosomes."

[12]The situation in the eye of Mi^{wh}/+ provides further evidence that the tissue environment can influence the amount of pigment produced (see Chapter 10, Section I, E). Thus while the pigment cells in the choroid of these genotypes form little or no

pigment, their kindred cells in the outer layer of the iris in the same eye produce considerable amounts, enough to give it a normal appearance in most instances. As a consequence of these observations Deol (1973) concludes that there is evidently something in the tissue environment of the outer layer, presumably some melanogenesis-promoting factor or factors, which brings about a change in the behavior of the pigment cells contained therein.

[13]Nevertheless, according to Wolfe and Coleman (1964) there is some tyrosinase activity in 5-day-old Mi^{wh}/Mi^{wh} skin (see Figure 12-1) since the activity in this skin is greater than in albino (c/c) and extreme dilution (c^e/c^e) skin, skin in which tyrosinase is known to be absent or at very low concentration (Coleman, 1962).

[14]Grüneberg (1953) points out that when the iris is pigmented, "careful inspection usually reveals the presence of a very slender pigment ring even through the closed eyelids at birth." He also draws attention to the fact that "the situation is like that in a normal 12-day embryo which has a pigment ring in the iris, but no melanin anywhere else."

[15]As pointed out by Deol (1970b) the fact that Mi^{wh}/Mi^{wh} mice are like W^v/W^v genotypes in terms of "external spotting" but much more like $Mi^{wh}/+$ as regards "internal spotting" again indicates that there is often little or no relationship between the two.

[16]Because $Mi^{wh}/+$ mice very occasionally display intensely pigmented "spots" (see Chapter 10, note 28), and because Hertwig (1942a,b) noted that the tips of the whiskers of *mi/mi* homozygotes tend to be bent, one must consider the possibility that the *mi*-locus is a complex one involving, in addition to inviable melanoblast clones, two phenotypic colors of viable clones and two hair follicle phenoclones (see Mintz, 1971a). This conclusion is supported further by the interaction of certain *mi*-alleles, namely, Mi^{wh} with mi^{bw}, mi^{sp}, and mi^{ws}.

[17]Deol (1973) reports that "in the iris the outer layer was unpigmented, but the inner layer always had some pigment, although it was unevenly distributed and much below normal in density." He also draws attention to the fact that the cells of this inner layer are derived from the retina which is almost completely unpigmented.

[18]Deol (1973) emphasizes the fact that because the identification of the amelanotic *retinal* pigment cell is beyond doubt, and very few of these cells in Mi^{wh}/mi heterozygotes form any pigment, that the widespread assumption that melanoblasts which fail to differentiate do not survive does not appear to be well founded, or at least is not invariably correct. While Deol's conclusion cannot be refuted this particular argument is not very convincing. Thus it is difficult to accept his premise that merely because the retinal cells of Mi^{wh}/mi fail to produce melanin they must be considered undifferentiated.

[19]Grüneberg points out that a similar situation is found in the dominant spotting (W) series (Chapter 10, Section I). The anemia of W/W animals is much more severe than that of W^v/W^v mice; yet $W/+$ heterozygotes are not anemic while $W^v/+$ mice are slightly so. Furthermore, W^v "dilutes" the fur when heterozygous with the normal allele and W does not. The basis for these different effects is not known.

[20]As pointed out by Wolfe and Coleman (see also Hollander, 1968), the fact that mi^{sp} is expressed only when heterozygous with other alleles (but not +) makes it appear, at least superficially, like the t-T (tailless-Brachyury) relationship. The T/t

complex is composed of a class of semidominant *T* mutations which, when heterozygous, produce a short tail and when homozygous are lethal. *T* mutations interact with recessive alleles at this locus to produce a tailless (*T*/*t*) phenotype; this interaction makes it possible to detect recessive *t* alleles that would otherwise go unnoticed (Artzt and D. Bennett, 1975; see D. Bennett, 1975).

[21]Wolfe and Coleman draw attention to the fact that *mi* was in the process of being backcrossed onto the C57BL/6J background when employed in these crosses, and that the faint dorsal pigmentation of mi/mi^{sp} animals present in early crosses was not observed in animals of this same genotype subsequent to generation 4, all hair being devoid of pigment at this stage.

[22]Although it appears in Figure 12-1 that $Mi^{wh}/+$ may differ significantly from Mi^{wh}/mi^{sp}, this difference could be due to the fact that in some cases the pieces of Mi^{wh}/mi^{sp} dorsal skin assayed overlapped some faintly visible white spots which may have reduced the counts. Indeed, counts from Mi^{wh}/mi^{sp} were about the same as those from $Mi^{wh}/+$ in three samples in which white spotted areas were carefully avoided.

[23]Since there are apparently no differences in the attributes of the pigment granules in $mi^{sp}/+$, mi^{sp}/mi^{sp}, and +/+ genotypes, one must account for the differences in the activity of tyrosinase in the skins of these animals. To account for this Wolfe and Coleman suggest that "there is a threshold level for tyrosine incorporation, above which full pigmentation can occur." A similar suggestion was made by Coleman (1962) to account for the activity of the alleles at the albino (*c*) locus. The data suggest that this threshold falls between 640 and 898 cpm (see Figure 12-1) compared to 600 cpm for the albino-locus series of alleles (see Table 3-5).

[24]It should be stressed that although the pigment granules in Mi^{wh}/mi^{sp} hairs are yellow-brown when viewed by transmitted white light, they are nevertheless noticeably different from the small, round yellow granules produced by agouti locus alleles and by recessive yellow (*e*/*e*).

[25]Another example where *mi*-locus alleles when heterozygous produce an effect different than when either allele is homozygous occurs in Mi^{wh}/mi animals. Thus these heterozygotes are *less* microphthalmic than either parental homozygote (Grüneberg, 1952).

[26]An *mi* mutant also appeared spontaneously in the CBA/CaCrc inbred colony. Animals heterozygous for this mutation have less iris pigment than normal at birth. They also have nonpigmented tail tips and may display white spotting on the belly, but not the head. Homozygotes are devoid of pigment in the hair and eyes. The eyes are small and the eyelids do not open. The teeth of homozygotes fail to erupt and they are osteopetrotic. There is also a significant deviation from the expected segregation ratio when heterozygotes are mated, a deviation which appears to be due to the death of microphthalmic female embryos before or soon after implantation. Homozygotes of both sexes are sterile. In matings of heterozygotes with *mi*/+ mice, microphthalmic young occur which are indistinguishable from animals homozygous for the new mutation. It therefore seems likely that the new deviant represents a remutation to *mi*. Accordingly, it has been designated mi^{Crc} (Hetherington, 1976).

[27]In this regard it is of interest to note that in the tests with the *Fk*/+ males, both a slightly reduced implantation frequency and a somewhat higher rate of intrauterine death were observed. These, however, were *not* statistically significant.

[28]One of the most significant influences on spotting occurred when A^y was substituted for *a* in the presence of $W^j/+$ as such a substitution reduced the amount of white in the coat from approximately 26 to about 4% (Lamoreux and E. Russell, 1971). In the case of flexed-tailed, 95% of *a/a;f/f* mice displayed white spotting while only 23% of $A^y/a;f/f$ animals had spots (Lamoreux, 1973). Lamoreux also noted that neither lethal yellow nor recessive yellow had any influence on the amount of white spotting produced by patch (*Ph*/+) and that $a/a;p^{un}/p^{un};bt/bt$ mice displayed less spotting than *a/a*;+/+;*bt/bt* animals.

[29]Nevertheless, one mutation from $c \rightarrow +$ did occur in a research stock not included in the analysis. This mutation is important because it provides strong evidence that albinism is not the consequence of a deletion.

[30]Because the mutation of wild type to *d* at the dilute locus has a higher reversion than forward rate, Schlager and Dickie (1971) suggest that $+ \rightarrow d \rightarrow +$ changes are mediated by a base-pair change, whereas the $+ \rightarrow a$ or $+ \rightarrow c$ change may involve longer segments of the DNA chain.

[31]On the other hand, Lyon and Morris (1966) found no mutations in a different set of specific loci in over 9000 offspring. They suggest that the mutation rate of the specific loci usually used may be higher than average since the irradiation-induced rate was significantly lower in this new set of specific loci.

[32]Fahrig (1975, 1977, 1978) has described an *in vivo* method, based on the pioneer experiments of L. Russell and Major (1957), for detecting genetic alterations in the somatic cells of mice. This test, which is known as "the mammalian spot test", involves treating mouse embryos, heterozygous for different recessive coat color genes, in utero, at 7-10 days gestation, with mutagens. These agents either are injected into the peritoneal cavity of the mother or are given to her orally. If this treatment results in an alteration, or loss, in a melanoblast of the wild-type allele of one of the genes under study a color spot is produced in the adult coat. Employing this test it has been found that (1) the frequencies of color spots in mutagen-treated animals depend upon the mutagens employed (L. Russell and Major, 1957; L. Russell, 1977; Davidson and Dawson, 1976, 1977; Fahrig, 1975, 1977); (2) the position of a color spot seems to depend to a certain extent on its color; and (3) the more white-grey spots that are induced by a mutagen, the more spots are located on the ventrum. Fahrig (1978) also notes that the spontaneous frequency of color spots is very low and appears to be similar for different hybrids. For example, among 891 ($a/a;b/+;c^{ch}p/+\ +;d/+;s/+$) F_1 animals whose adult coat were examined, only 6 displayed spots (one a light brown head spot; one a white-grey spot on its underside; and four midventral white spots).

References

Adler, S.S., and Trobaugh, F.E., Jr.: Pluripotent (CFU-S) and granulocyte-commited (CFU-C) stem cells in intact and ^{89}Sr marrow-ablated *Sl/Sl^d* mice. *Cell Tissue Kinet.* **11:** 555–566, 1978.

Altman, K.I., and Russell, E.S.: Heme synthesis in normal and genetically anemic mice. *J. Cell. Comp. Physiol.* **64:** 293–301, 1964.

Altus, M.S., Bernstein, S.E., Russell, E.S., Carsten, A.L., and Upton, A.C.: Defect extrinsic to stem cells in spleens of steel anemic mice. *Proc. Soc. Exp. Biol. Med.* **138:** 985–988, 1971.

Andrews, E.J., White, W.J., and Bullock, L.P.: Spontaneous aortic aneurysms in *blotchy* mice. *Amer. J. Pathol.* **78:** 199–208, 1975.

Artzt, K., and Bennett, D.: Analogies beyween embryonic (*T/t*) antigens and adult major histocompatibility (H-2) antigens. *Nature* (*London*) *256:* 545—547, 1975.

Attfield, M.: Inherited macrocytic anemias in the house mouse. III. Red blood cell diameters. *J. Genet* **50:** 250–263, 1951.

Auerbach, R.: Analysis of the developmental effects of a lethal mutation in the house mouse. *J. Exp. Zool.* **127:** 305–329, 1954.

Bagnara, J.T., Matsumoto, J., Ferris, W., Frost, S.K., Turner, W.A., Jr., Tchen, T.T., and Taylor, J.D.: Common origin of pigment cells. *Science* **203:** 410–415, 1979.

Ballantyne, J., Bock, F.G., Strong, L.C., and Quevedo, W.C., Jr.: Another allele at the *W* locus of the mouse. *J. Hered.* **52:** 200–202, 1961.

Bangham, J.W.: Private communication. *Mouse News Letter* **38:** 31, 1968.

Barnes, D.W.H., Loutit, J.F., and Sansom, J.M.: Histocompatible cells for the resolution of osteopetrosis in microphthalmic mice. *Proc. Roy. Soc.* (*London*) **B 188:** 501–505, 1975.

Barnicot, N.A.: Studies on the factors involved in bone absorption. I. The effect of subcutaneous transplantation of bones of the grey-lethal house mouse into normal hosts and of normal bones into grey-lethal hosts. *Amer. J. Anat.* **68:** 497–531, 1941.

Barnicot, N.A.: Some data on the effect of parathormone on the grey-lethal mouse. *J. Anat.* **79:** 83–91, 1945.

Barnicot, N.A.: The local action of the parathyroid and other tissues on bone in intracerebral grafts. *J. Anat.* **82:** 233–248, 1948.

Barrows, E.F.: Modification of the dominance of agouti to nonagouti in the mouse. *J. Genet.* **29:** 9–15, 1934.

Barrows, E.F.: Selection for tail-spotting in the house mouse. *J. Exp. Zool.* **80:** 107–111, 1939.

Batchelor, A.L., Phillips, R.J.S., and Searle, A.G.: A comparison of the mutagenic effectiveness of chronic neutron-and γ-irradiation of mouse spermatogonia. *Mutat. Res.* **3:** 218–229, 1966.

Bateman, A.E., and Cole, R.J.: Colony forming cells in the livers of prenatal flexed (*f/f*) anaemic mice. *Cell Tissue Kinet.* **5:** 165–173, 1972.

Bateman, A.E., Cole, R.J., Regan, T., and Tarbutt, R.G.: The role of erythropoietin in prenatal erythropoiesis of congenitally anaemic flexed-tailed (*f/f*) mice. *Brit. J. Haematol.* **22:** 415–427, 1972.

Bateman, A.J.: A probable case of mitotic crossing-over in the mouse. *Genet. Res. (Camb.)* **9:** 375, 1967.

Bateman, N.: Bone growth: A study of the grey-lethal and microphthalmic mutants of the mouse. *J. Anat.* **88:** 212–262, 1954.

Bateman, N.: Private communication. *Mouse News Letter* **16:** 7, 1957.

Bateman, N.: Sombre, a viable dominant mutant in the house mouse. *J. Hered.* **52:** 186–189, 1961.

Beechey, C.V., and Searle, A.G.: Private communication. *Mouse News Letter* **44:** 27, 1971.

Beechey, C.V., and Searle, A.G.: Private communication. *Mouse News Letter* **46:** 27, 1972.

Beechey, C.V., and Searle, A.G.: Private communication. *Mouse News Letter* **55:** 15, 1976.

Beechey, C.V., and Searle, A.G.: Private communication. *Mouse News Letter* **59:** 19, 1978.

Benestad, H.B., Bøyum, A., and Warhuus, K.: Haematopoietic defects of W/W^v mice studied with the spleen colony, agar colony, and diffusion chamber techniques. *Scand. J. Haematol.* **15:** 219–227, 1975.

Bennett, D.: Developmental analysis of a mutant with pleiotropic effects in the mouse. *J. Morphol.* **98:** 199–234, 1956.

Bennett, D.: The T-locus of the mouse. *Cell* **6:** 441–454, 1975.

Bennett, J.M., Blume, R.S., and Wolff, S.M.: Characterization and significance of abnormal leukocyte granules in the beige mouse. A possible homologue for Chediak–Higashi Aleutian trait. *J. Lab. Clin. Med.* **73:** 235–243, 1969.

Bennett, M., Cudkowicz, G., Foster, R.A., and Metcalf, D.: Hemopoietic progenitor cells of *W* anemic mice studied *in vivo* and *in vitro*. *J. Cell. Physiol.* **71:** 211–226, 1968a.

Bennett, M., Steeves, R.A., Cudkowicz, G., Mirand, E.A., and Russell, L.B.: Mutant *Sl* alleles of mice affect susceptibility to Friend spleen focus-forming virus. *Science* **162:** 564–565, 1968b.

Bernstein, S.E.: Private communication. *Mouse News Letter* **23:** 33–34, 1960.

Bernstein, S.E.: Acute radiosensitivity in mice of differing *W* genotype. *Science* **137:** 428–429, 1962.

Bernstein, S.E.: Modification of radiosensitivity of genetically anemic mice by implantation of blood-forming tissue. *Radiat. Res.* **20:** 695–702, 1963.

Bernstein, S.E.: Tissue transplantation as an analytic and therapeutic tool in hereditary anemias. *Amer. J. Surg.* **119:** 448–451, 1970.

Bernstein, S.E., and Russell, E.S.: Implantation of normal blood-forming tissue in genetically anemic mice, without x-irradiation of host. *Proc. Soc. Exp. Biol. Med.* **101:** 769–773, 1959.

Bernstein, S.E., Russell, E.S., and Keighley, G.: Two hereditary mouse anemias (Sl/Sl^d and W/W^v) deficient in response to erythropoietin. *Ann. N.Y. Acad. Sci.* **149:** 475–485, 1968.

Bhat, N.R.: A dominant mutant mosaic house mouse. *Heredity* **3:** 243–248, 1949.

Bielschowsky, M., and Schofield, G.C.: Studies on the inheritance and neurohistology of megacolon in mice. *Proc. Univ. Otago Med. Sch.* **38:** 14–15, 1960.

Bielschowsky, M., and Schofield, G.C.: Studies on megacolon in piebald mice. *Aust. J. Exp. Biol. Med. Sci.* **40:** 395–404, 1962.

Billingham, R.E., and Silvers, W.K.: The melanocytes of mammals. *Quart. Rev. Biol.* **35:** 1–40, 1960.

Bolois, M.S., Zahlan, A.B., and Maling, J.E.: Electron spin-resonance studies on melanin. *Biophys. J.* **4:** 471–490, 1964.

Bloom, J.L., and Falconer, D.S.: "Grizzled", a mutant in linkage group X of the mouse. *Genet. Res. (Camb.)* **7:** 159–167, 1966.

Blume, R.S., and Wolff, S.M.: The Chediak–Higashi syndrome. Studies in four patients and a review of the literature. *Medicine* **51:** 247–280, 1972.

Blume, R.S., Bennett, J.M., Yankee, R.A., and Wolff, S.M.: Defective granulocyte regulation in the Chediak–Higashi syndrome. *New Eng. J. Med.* **279:** 1009–1015, 1968.

Blume, R.S., Padgett, G.A., Wolff, S.M., and Bennett, J.M.: Giant neutrophil granules in the Chediak–Higashi syndrome of man, mink, cattle and mice. *Canad. J. Comp. Med.* **33:** 271–274, 1969.

Boggs, S.S., Wilson, S.M., and Smith, W.W.: Effects of endotoxin on hematopoiesis in irradiated and nonirradiated W/W^v mice. *Radiat. Res.* **56:** 481–493, 1973.

Borghese, E.: Gonads of *W* mice cultured *in vitro*. *Anat. Rec.* **124:** 481–482, 1956 (Abstract).

Borghese, E.: The present state of research on *WW* mice. *Acta Anat.* **36:** 185–220, 1959.

Boxer, L.A., Watanabe, A.M., Rister, M., Besch, H.R., Jr., Allen, J., and Baehner, R.L.: Correction of leukocyte function in Chediak-Higashi syndrome by ascorbate. *New Eng. J. Med.* **295:** 1041–1045, 1976.

Brandt, E.J., and Swank, R.T.: The Chediak–Higashi (beige) mutation in two mouse strains. *Amer. J. Pathol.* **82:** 573–588, 1976.

Brandt, E.S., Elliot, R.W., and Swank, R.T.: Defective lysosomal enzyme secretion in kidneys of Chediak–Higashi (beige) mice. *J. Cell Biol.* **67:** 774–788, 1975.

Brauch, L.R., and Russell, W.L.: Colorimetric measurement of the effects of the major coat color genes in the mouse on the quantity of yellow pigment in extracts. *Genetics* **31:** 212, 1946 (Abstract).

Bulfield, G.: Private communication. *Mouse News Letter* **50:** 35, 1974.

Burnett, J.B.: The tyrosinases of mouse melanoma. Isolation and molecular properties. *J. Biol. Chem.* **246:** 3079–3091, 1971.

Burnett, J.B., and Seiler, H.: Separation and properties of solubilized tyrosinase from mouse melanoma. *Fed. Proc.* **25:** 294, 1966 (Abstract).

Burnett, J.B., Seiler, H., and Brown, I.V.: Separation and characterization of multiple forms of tyrosinase from mouse melanoma. *Cancer Res.* **27:** 880–889, 1967.

Burnett, J.B., Holstein, T.J., and Quevedo, W.C., Jr.: Electrophoretic variations of tyrosinase in follicular melanocytes during the hair growth cycle in mice. *J. Exp. Zool.* **171:** 369–376, 1969.

Burrows, E.F.: Selection for tail-spotting in the house mouse. *J. Exp. Zool.* **80:** 107–111, 1939.

Butler, J., and Lyon, M.F.: Private Communication. *Mouse News Letter* **40:** 25, 1969.

Butler, L.: The effect of the coat colour dilution gene on body size in the mouse. *Heredity* **8:** 275–278, 1954.

Cacheiro, N.L.A., and Russell, L.B.: Evidence that linkage group IV as well as linkage group X of the mouse are in chromosome 10. *Genet. Res.* (*Camb.*) **25:** 193–195, 1975.

Calarco, P.G., and Pedersen, R.A.: Ultrastructural observations of lethal yellow (A^y/A^y) mouse embryos. *J. Embryol. Exp. Morphol.* **35:** 73–80, 1976.

Carnes, W.H.: Copper and connective tissue metabolism. *Int. Rev. Connect. Tissue Res.* **4:** 197–232, 1968.

Carter, T.C.: A mosaic mouse with an anomalous segregation ratio. *J. Genet.* **51:** 1–6, 1952.

Carter, T.C.: Private communication. *Mouse News Letter* **21:** 40, 1959.

Carter, T.C., Lyon, M.F., and Phillips, R.J.S.: Genetic hazard of ionizing radiations. *Nature* (*London*) **182:** 409, 1958.

Castle, W.E.: *Mammalian Genetics.* Harvard Univ. Press, Cambridge, 1940.

Castle, W.E.: Influence of certain color mutations on body size in mice, rats, and rabbits. *Genetics* **26:** 177–191, 1941.

Castle, W.E.: Size genes of mice. *Proc. Nat. Acad. Sci. USA* **28:** 69–72, 1942.

Castle, W.E., and Allen, G.M.: The heredity of albinism. *Proc. Amer. Acad. Arts Sci.* **38:** 603–622, 1903.

Castle, W.E., and Little C.C.: On a modified Mendelian ratio among yellow mice. *Science* **32:** 868–870, 1910.

Castle, W.E., Gates, W.H., Reed, S.C., and Law, L.W.: Studies of a size cross in mice, II. *Genetics* **21:** 310–323, 1936.

Cattanach, B.M.: A chemically-induced variegation-type position effect in the mouse. *Z. Vererbungslehre* **92:** 165–182, 1961.

Cattanach, B.M.: Position effect variegation in the mouse. *Genet. Res.* (*Camb.*) **23:** 291–306, 1974.

Cattanach, B.M.: Private Communication. *Mouse News Letter* **59:** 18, 1978.

Cattanach, B.M., and Williams, C.E.: Private communication. *Mouse News Letter* **47:** 34–35, 1972.

Cattanach, B.M., Pollard, C.E., and Perez, J.N.: Controlling elements in the mouse X-chromosome. I. Interaction with the X-linked genes. *Genet. Res.* (*Camb.*) **14:** 223–235, 1969.

Cattanach, B.M., Wolfe, H.G., and Lyon, M.F.: A comparative study of the coats of chimeric mice and those of heterozygotes for X-linked genes. *Genet. Res.* (*Camb.*) **19:** 213–228, 1972.

Charles, D.R.: Studies on spotting patterns. IV. Pattern variation and its developmental significance. *Genetics* **23:** 523–547, 1938.

Chase, H.B.: Studies on the tricolor pattern of the guinea pig. II. The distribution of black and yellow as affected by white spotting and by imperfect dominance in the tortoise shell series of alleles. *Genetics* **24:** 622–643, 1939.

Chase, H.B.: Greying of hair. I. Effects produced by single doses of x-rays on mice. *J. Morphol.* **84:** 57–80, 1949.

Chase, H.B.: Number of entities inactivated by x-rays in greying of hair. *Science* **113:** 714–716, 1951.

Chase, H.B.: Growth of the hair. *Physiol. Rev.* **34:** 113–126, 1954.

Chase, H.B.: The behavior of pigment cells and epithelial cells in the hair follicle. In *The Biology of Hair Growth*, W. Montagna and R.A. Ellis (eds.), pp. 229–237, Academic Press, New York, 1958.

Chase, H.B.: Private communication. *Mouse News Letter* **21:** 21, 1959.

Chase, H.B.: Private communication. *Mouse News Letter* **33:** 17, 1965.

Chase, H.B., and Mann, S.J.: Phenogenetic aspects of some hair and pigment mutants. *J. Cell. Comp. Physiol.* **56:** 103–112, 1960.

Chase, H.B., and Rauch, H.: Greying of hair. II. Response of individual hairs in mice to variations in x-radiation. *J. Morphol.* **87:** 381–392, 1950.

Chase, H.B., and Silver, A.F.: The biology of hair growth. In *The Biological Basis of Medicine*, Vol. 6, Part I. *Hair and Skin*, E.E. Bittar (ed.), pp. 3–19, Academic Press, London, 1969.

Chase, H.B., Rauch, H., and Smith, V.W.: Critical stages of hair development and pigmentation in the mouse. *Physiol. Zool.* **24:** 1–8, 1951.

Chervenick, P.A., and Boggs, D.A.: Decreased neutrophils and megakaryocytes in anemic mice of genotype W/W^v. *J. Cell Physiol.* **73:** 25–30, 1969.

Chi, E.Y., and Lagunoff, D.: Abnormal mast cell granules in the beige (Chediak–Higashi syndrome) mouse. *J. Histochem. Cytochem.* **23:** 117–122, 1975.

Chi, E.Y., Prueitt, J.L., and Lagunoff, D.: Abnormal lameller bodies in type II pneumocytes and increased lung surface active material in the beige mouse. *J. Histochem. Cytochem.* **23:** 863–866, 1975.

Chi, E.Y., Lagunoff, D., and Koehler, J.K.: Abnormally large lamellar bodies in type II pneumocytes in Chediak–Higashi syndrome in beige mice. *Lab. Invest.* **34:** 166–173, 1976.

Chi, E.Y., Ignácio, E., and Lagunoff, D.: Mast cell granule formation in the beige mouse. *J. Histochem. Cytochem.* **26:** 131–137, 1978.

Chian, L.T.Y., and Wilgram, G.F.: Tyrosinase inhibition: Its role in suntanning and in albinism. *Science* **155:** 198–200, 1967.

Chiquoine, A.D.: The identification, origin and migration of the primordial germ cells in the mouse embryo. *Anat. Rec.* **118:** 135–146, 1954.

Chui, D.H.K., and Loyer, B.V.: Foetal erythropoiesis in steel mutant mice. II. Haemopoietic stem cells in foetal livers during development. *Brit. J. Haematol.* **29:** 553–565, 1975a.

Chui, D.H.K., and Loyer, B.V.: Erythropoiesis in steel mutant mice: Effects of erythropoietin *in vitro*. *Blood* **45:** 427–433, 1975b.

Chui, D.H.K., and Russell, E.S.: Fetal erythropoiesis in steel mutant mice. I. A morphological study of erythroid cell development in fetal liver. *Develop. Biol.* **40:** 256–269, 1974.

Chui, D.H.K., Loyer, B.V., and Russell, E.S.: Steel (*Sl*) mutation in mice. Identification of mutant embryos early in development. *Develop. Biol.* **49:** 300–303, 1976.

Chui, D.H.K., Sweeney, G.D., Patterson, M., and Russell, E.S.: Hemoglobin synthesis in siderocytes of flexed-tailed mutant (*f/f*) fetal mice. *Blood* **50:** 165–177, 1977.

Cizadlo, G.R., and Granholm, N.H.: *In vivo* development of the lethal yellow (A^y/A^y) mouse embryo at 105 hours post coitum. *Genetica* **48:** 89–93, 1978a.

Cizadlo, G.R., and Granholm, N.H.: Ultrastructural analysis of preimplantation lethal yellow (A^y/A^y) mouse embryos. *J. Embryol. Exp. Morphol.* **45:** 13–24, 1978b.

Clark, F.H.: The inheritance and linkage relations of a new recessive spotting in the house mouse. *Genetics* **19:** 365–393, 1934.

Clarke, W.E.: Notes on the mice of St. Kilda. *Scot. Natural.* (*Edinburgh*) pp. 124–128, 1914.

Cleffmann, G.: Untersuchen über die Fellzeichnung des Wildkaninchens Ein Betrag zur Wirkungsweise des Agutifaktors. *Z. indukt. Abstamm.-u. VererbLehre* **85:** 137–162, 1953.

Cleffmann, G.: Über den Einfluss des Milieus *in situ* and *in vitro* auf die Manifestierung der Agutizeichung. In *Verhandl. Deutsch. Zool. Gesellsch.*, pp. 263–268, Geest and Portig K.-G., Leipzig, 1960.

Cleffmann, G.: Agouti pigment cells *in situ* and *in vitro*. *Ann. N.Y. Acad. Sci.* **100:**749–760, 1963.

Cleffmann, G.: Function-specific changes in the metabolism of agouti pigment cells. *Exp. Cell Res.* **35:** 590–600, 1964.

Cloudman, A.M., and Bunker, L.E.: The varitint-waddler mouse. A dominant mutation in *Mus musculus*. J. Hered. **36:** 259–263, 1945.

Cole, R.J., Tarbutt, R.G., Cheek, E.M., and White, S.L.: Expression of congenital defects in the haemopoietic micro-environment: Pre-natal erythropoiesis in anaemic "steel" (Sl^j/Sl^j) mice. *Cell Tissue Kinet.* **7:** 463–477, 1974.

Coleman, D.L.: Phenylalanine hydroxylase activity in dilute and non-dilute strains of mice. *Arch. Biochem. Biophys.* **91:** 300–306, 1960.

Coleman, D.L.: Effect of genic substitution on the incorporation of tyrosine into the melanin of mouse skin. *Arch. Biochem. Biophys.* **96:** 562–568, 1962.

Coleman, D.L.: Tactics in pigment cell research. (Discussion). In *Methodology in Mammalian Genetics*, W.J. Burdette (ed.), pp. 342, Holden-Day, San Francisco, 1963.

Coleman, D.L., Russell, E.S., and Levin, E.Y.: Enzymatic studies of the hemopoietic defect in flexed mice. *Genetics* **61:** 631–642, 1969.

Cotzias, G.C., Papavasiliov, P.S., and Miller, S.T.: Manganese in melanin. *Nature (London)* **201:** 1228–1229, 1964.

Coulombre, J.L., and Russell, E.S.: Analysis of the pleiotropism at the *W*-locus in the mouse: The effects of *W* and W^v substitution upon postnatal development of germ cells. *J. Exp. Zool.* **126:** 277–296, 1954.

Creel, D.J.: Visual system anomaly associated with albinism in the cat. *Nature (London)* **231:** 465–466, 1971.

Cuénot, L.: La loi de Mendel et l'héréditié de la pigmentation chez la souris. *Arch. Zool. exp. gén., 3e sér.*, **10:** xxvii–xxx, 1902.

Cuénot, L.: Les races pures et leurs combinaisons chez les souris (4e note). *Arch. Zool. exp. gén., 4e sér.*, **3:** cxxiii–cxxxii, 1905.

Dancis, J., Jansen, V., Brown, G.F., Gorstein, F., and Balis, M.E.: Treatment of hypoplastic anemia in mice with placental transplants. *Blood* **50:** 663–670, 1977.

Danks, D.M., Campbell, P.E., Stevens, B.J., Mayne, V., and Cartwright, E.: Menkes's kinky hair syndrome: An inherited defect in copper absorption with wide-spread effects. *Pediatrics* **50:** 188–201, 1972a.

Danks, D.M., Campbell, P.E., Walker-Smith, J., Stevens, B.J., Gillespie, J.M., Blomfield, J., and Turner, B.: Menkes' kinky-hair syndrome. *Lancet* **ii:** 1100–1103, 1972b.

Davidson, G.E., and Dawson, G.W.P.: Chemically induced presumed somatic mutations in the mouse. *Mutat. Res.* **38:** 151–154, 1976.

Davidson, G.E., and Dawson, G.W.P.: The induction of somatic mutations in mouse embryos by benzo(a)pyrene. *Arch. Toxicol.* **38:** 99–103, 1977.

Davis, W.C., and Douglas, S.D.: Defective granule formation and function in the

Chediak-Higashi syndrome in man and animals. *Semin. Hematol.* **9:** 431–450, 1972.

Davisson, M.T.: Private communication. *Mouse News Letter* **57:** 22, 1977.

deAberle, S.: A study of hereditary anemia in mice. *Amer. J. Anat.* **40:** 219–247, 1927.

Demerec, M.: Frequency of spontaneous mutations in certain stocks of *Drosophila melanogaster*. *Genetics* **22:** 469–478, 1937.

Deol, M.S.: The anomalies of the labyrinth of the mutants varitint-waddler, shaker-2 and jerker in the mouse. *J. Genet.* **52:** 562–588, 1954.

Deol, M.S.: Inheritance of coat color in laboratory rodents. In *Animals for Research*, W. Lane-Petter (ed.), pp. 177–198, Academic Press, London, 1963.

Deol, M.S.: The neural crest and the acoustic ganglion. *J. Embryol. Exp. Morphol.* **17:** 533–541, 1967.

Deol, M.S.: Inherited diseases of the inner ear in man in the light of studies on the mouse. *J. Med. Genet.* **5:** 137–158, 1968.

Deol, M.S.: The determination and distribution of coat colour variation in the house mouse. *Symp. Zool. Soc. London* **26:** 239–250, 1970a.

Deol, M.S.: The relationship between abnormalities of pigmentation and of the inner ear. *Proc. Roy. Soc. (London)* **A 175:** 201–217, 1970b.

Deol, M.S.: The origin of the acoustic ganglion and effects of the gene dominant spotting (W^v) in the mouse. *J. Embryol. Exp. Morphol.* **23:** 773–784, 1970c.

Deol, M.S.: Spotting genes and internal pigmentation patterns in the mouse. *J. Embryol. Exp. Morphol.* **26:** 123–133, 1971.

Deol, M.S.: The role of the tissue environment in the expression of spotting genes in the mouse. *J. Embryol. Exp. Morphol.* **30:** 483–489, 1973.

Deol, M.S., and Whitten, W.K.: Time of X chromosome inactivation in retinal melanocytes of the mouse. *Nature New Biol.* **238:** 159–160, 1972.

Deringer, M.K.: Influence of the lethal yellow (A^y) gene on the development of reticular neoplasms. *J. Nat. Cancer Inst.* **45:** 1205–1210, 1970.

Detlefsen, J.A.: A new mutation in the house mouse. *Amer. Nat.* **55:** 469–473, 1921.

Dexter, T.M., and Moore, M.A.S.: *In vitro* duplication and "cure" of haemopoietic defects in genetically anaemic mice. *Nature (London)* **269:** 412–414, 1977.

Dickerson, G.E., and Gowen, J.W.: Food utilization in genetic obesity of mice. *Genetics* **31:** 214–215, 1946 (*Abstract*).

Dickerson, G.E., and Gowen, J.W.: Hereditary obesity and efficient food utilization in mice. *Science* **105:** 496–498, 1947.

Dickie, M.M.: The tortoise shell house mouse. *J. Hered.* **45:** 158–159, 1954.

Dickie, M.M.: A new viable yellow mutation in the house mouse. *J. Hered.* **53:** 84–86, 1962a.

Dickie, M.M.: Private communication. *Mouse News Letter* **27:** 37, 1962b.

Dickie, M.M.: Private communication. *Mouse News Letter* **30:** 30, 1964a.

Dickie, M.M.: New splotch alleles in the mouse. *J. Hered.* **55:** 97–101, 1964b.

Dickie, M.M.: Private communication. *Mouse News Letter* **32:** 43, 1965a.

Dickie, M.M.: Private communication. *Mouse News Letter* **32:** 44, 1965b.

Dickie, M.M.: Private communication. *Mouse News Letter* **34:** 30, 1966a.

Dickie, M.M.: Private communication. *Mouse News Letter* **35:** 31, 1966b.

Dickie, M.M.: Private communication. *Mouse News Letter* **36:** 39, 1967a.

Dickie, M.M.: Private communication. *Mouse News Letter* **37:** 33, 1967b.

Dickie, M.M.: Mutations at the agouti locus in the mouse. *J. Hered.* **60:** 20–25, 1969a.

Dickie, M.M.: Private communication. *Mouse News Letter* **41:** 31, 1969b.

Doolittle, C.H., and Rauch, H.: Epinephrin and norepinephrin levels in dilute-lethal mice. *Biochem. Biophys. Res. Commun.* **18:** 43–47, 1965.

Doykos, J.D., Cohen, M.M., and Shklar, G.: Physical, histological and roentgenographic characteristics of the grey lethal mouse. *Amer. J. Anat.* **121:** 29–40, 1967.

Dry, F.W.: The coat of the mouse. *J. Genet.* **16:** 287–340, 1926.

Dry, F.W.: The agouti coloration of the mouse (*Mus musculus*) and the rat (*Mus norvegicus*). *J. Genet.* **20:** 131–144, 1928.

Dunn, L.C.: The genetic behavior of mice of the color varieties "black-and-tan" and "red." *Amer. Nat.* **50:** 664–675, 1916.

Dunn, L.C.: The sable varieties of mice. *Amer. Nat.* **54:** 247–261, 1920a.

Dunn, L.C.: Types of white spotting in mice. *Amer. Nat.* **54:** 465–495, 1920b.

Dunn, L.C.: A fifth allelomorph in the agouti series of the house mouse. *Proc. Nat. Acad. Sci. USA* **14:** 816–819, 1928.

Dunn, L.C.: Analysis of a case of mosaicism in the house mouse. *J. Genet.* **29:** 317–326, 1934.

Dunn, L.C.: Studies of multiple allelomorphic series in the house mouse. I. Description of agouti and albino series of allelomorphs. *J. Genet.* **33:** 443–453, 1936.

Dunn, L.C.: Studies on spotting patterns. II. Genetic analysis of variegated spotting in the house mouse. *Genetics* **22:** 43–64, 1937.

Dunn, L.C.: Studies of spotting patterns. V. Further analysis of minor spotting genes in the house mouse. *Genetics* **27:** 258–267, 1942.

Dunn, L.C.: A new eye color mutant in the mouse with asymmetrical expression. *Proc. Nat. Acad. Sci. USA* **31:** 343–346, 1945.

Dunn, L.C., and Charles, D.R.: Studies on spotting patterns. I. Analysis of quantitative variations in the pied spotting of the house mouse. *Genetics* **22:** 14–42, 1937.

Dunn, L.C., and Einsele, W.: Studies of multiple allelomorphic series in the house mouse. IV. Quantitative comparisons of melanins from members of the albino series. *J. Genet.* **36:** 145–152, 1938.

Dunn, L.C., and Mohr, J.: A association of hereditary eye defects with white spotting. *Proc. Nat. Acad. Sci. USA* **38:** 872–875, 1952.

Dunn, L.C., and Thigpen, L.W.: The silver mouse, a recessive color variation. *J. Hered.* **21:** 495–498, 1930.

Dunn, L.C., MacDowell, E.C., and Lebedeff, G.A.: Studies on spotting patterns. III. Interaction between genes affecting white spotting and those affecting color in the mouse. *Genetics* **22:** 307–318, 1937.

Durham, F.M.: A preliminary account of the inheritance of coat colour in mice. *Rep. Evol. Com. Roy. Soc. Rep.* **4:** 41–53, 1908.

Durham, F.M.: Further experiments on the inheritance of coat colour in mice. *J. Genet.* **1:** 159–178, 1911.

Eaton, G.J.: Stimulation of trophoblastic giant cell differentiation in the homozygous yellow mouse embryo. *Genetica* **39:** 371–378, 1968.

Eaton, G.J., and Green, M.M.: Implantation and lethality of the yellow mouse. *Genetica* **33:** 106–112, 1962.

Eaton, G.J., and Green, M.M.: Giant cell differentiation and lethality of homozygous yellow mouse embryos. *Genetica* **34:** 155–161, 1963.

Ebbe, S., and Phalen, E.: Regulation of megakaryocytes in W/W^v mice. *J. Cell Physiol.* **96:** 73–79, 1978.

Ebbe, S., Phalen, E., and Stohlman, F., Jr.: Abnormalities of megakaryocytes in W/W^v mice. *Blood* **42:** 857–864, 1973a.

Ebbe, S., Phalen, E., and Stohlman, F., Jr.: Abnormalities of megakaryocytes in Sl/Sl^d mice. *Blood* **42:** 865–871, 1973b.

Ebbe, S., Phalen, E., and Ryan, M.K.: The production of megakaryocytic macrocytosis by systemic factors in Sl/Sl^d mice. *Proc. Soc. Exp. Biol. Med.* **155:** 243–246, 1977.

Ebbe, S., Phalen, E., and Howard, D.: Parabiotic demonstration of a humoral factor affecting megakaryocyte size in Sl/Sl^d mice. *Proc. Soc. Exp. Biol. Med.* **158:** 637–642, 1978.

Eicher, E.M.: The position of *ru-2* and *qv* with respect to the *flecked* translocation in the mouse. *Genetics* **64:** 495–510, 1970a.

Eicher, E.M.: X-autosome translocations in the mouse: Total inactivation versus partial inactivation of the X chromosome. *Adv. Genetics* **15:** 175–259, 1970b.

Eicher, E.M.: Private communication. *Mouse News Letter* **47:** 36, 1972.

Eicher, E.M., and Fox, S.: Private communication. *Mouse News Letter* **56:** 42, 1977.

Erickson, R.P., Gluecksohn-Waelsch, S., and Cori, C.F.: Glucose-6-phosphatase deficiency caused by radiation induced alleles at the albino locus in the mouse. *Proc. Nat. Acad. Sci. USA* **59:** 437–444, 1968.

Erickson, R.P., Eicher, E.M., and Gluecksohn-Waelsch, S.: Demonstration in mouse of x-ray induced deletions for a known enzyme structural locus. *Nature (London)* **248:** 416–418, 1974a.

Erickson, R.P., Siekevitz, P., Jacobs, K., and Gluecksohn-Waelsch, S.: Chemical and immunological studies of liver microsomes from mouse mutants with ultrastructurally abnormal hepatic endoplasmic reticulum. *Biochem. Genet.* **12:** 81–95, 1974b.

Erway, L., Hurley, L.S., and Fraser, A.: Neurological defect: Manganese in phenocopy and prevention of a genetic abnormality of inner ear. *Science* **152:** 1766–1768, 1966.

Erway, L., Hurley, L.S., and Fraser, A.: Congenital ataxia and otolith defects due to manganese deficiency in mice. *J. Nutrit.* **100:** 643–654, 1970.

Erway, L.C., Fraser, A.S., and Hurley, L.S.: Prevention of congenital otolith defect in pallid mutant mice by maganese supplementation. *Genetics* **67:** 97–108, 1971.

Essner, E., and Oliver, C.: A hereditary alteration in kidneys of mice with Chediak–Higashi syndrome. *Amer. J. Pathol.* **73:** 217–232, 1973.

Essner, E., and Oliver, C.: Lysosome formation in hepatocytes of mice with Chediak–Higashi syndrome. *Lab. Invest.* **30:** 596–607, 1974.

Evans, E.P., and Phillips, R.J.S.: Private communication. *Mouse News Letter* **58:** 44–45, 1978.

Fahrig, R.: A mammalian spot test: Induction of genetic alterations in pigment cells of mouse embryos with x-rays and chemical mutagens. *Mol. Gen. Genet.* **138:** 309–314, 1975.

Fahrig, R.: The mamalian spot test (Fellfleckentest) with mice. *Arch. Toxicol.* **38:** 87–98, 1977.

Fahrig, R.: The mammalian spot test: A sensitive *in vivo* method for the detection of genetic alterations of somatic cells of mice. In *Chemical Mutagens*, Vol. 5, A. Hollaender and F.J. de Serres (eds.), pp. 151–176, Plenum, New York, 1978.

Falconer, D.S.: Total sex linkage in the house mouse. *Z. indukt. Abstamm.-u. VererbLehre* **85:** 210–219, 1953.

Falconer, D.S.: Private communication. *Mouse News Letter* **15:** 23, 1956a.

Falconer, D.S.: Private communication. *Mouse News Letter* **15:** 24–25, 1956b.

Falconer, D.S., and Isaacson, J.H.: Private communication. *Mouse News Letter* **27:** 30, 1962.

Falconer, D.S., and Isaacson, J.H.: Selection for expression of a sex linked gene (brindled) in mice. *Heredity* **24:** 180, 1969 (Abstract).

Falconer, D.S., and Isaacson, J.H.: Sex-linked variegation modified by selection in brindled mice. *Genet. Res.* (*Camb.*) **20:** 291–316, 1972.

Fattorusso, E., Minale, L., De Stefano, S., Cimino, G., and Nicolaus, R.A.: Struttura e biogenesi della feomelanine. IX. Feomelanine biosintetiche. *Gazzetta Chim. Ital.* **99:** 969–992, 1969.

Feldman, H.W.: A fourth allelomorph in the albino series in mice. *Amer. Nat.* **56:** 573–574, 1922.

Feldman, H.W.: A fifth allelomorph in the albino series of the house mouse. *J. Mammal.* **16:** 207–210, 1935.

Ferguson, J.M., and Wallace, M.E.: Private communication. *Mouse News Letter* **57:** 11, 1977.

Fielder, J.H.: The taupe mouse. *J. Hered.* **43:** 75–76, 1952.

Fischer, H. and Tost, M.: Missbildungen im fasciculus opticus beim hereditären mikropthalmus der Hausmaus. *Biol. Zentralbl.* **78:** 759–776, 1959.

Fisher, R.A., and Landauer, W.: Sex differences of crossing-over in close linkage. *Amer. Nat.* **87:** 116, 1953.

Fisher, R.A., and Mather, K.: A linkage test with mice. *Ann. Eugen.* **7:** 265–280, 1936.

Fitzpatrick, T.B., and Breathnach, A.S.: Das epidermale Melanin-Einheit-System. *Dermatol. Wochenschrift* **147:** 481–489, 1963.

Fitzpatrick, T.B., and Kukita, A.: Tyrosinase activity in vertebrate melanocytes. In *Pigment Cell Biology*, M. Gordon (ed.), pp. 489–524, Academic Press, New York, 1959.

Fitzpatrick, T.B., and Lerner, A.B.: Biochemical basis of human melanin pigmentation. *Arch. Dermatol. Syphilol.* **69:** 133–149, 1954.

Fitzpatrick, T.B., Brunet, P., and Kukita, A.: The nature of hair pigment. In *The Biology of Hair Growth*, W. Montagna and R.A. Ellis (eds.), pp. 255–303, Academic Press, New York, 1958.

Fitzpatrick, T.B., Miyamoto, M., and Ishikawa, K.: The evolution of concepts of melanin biology. In *The Pigmentary System, Advances in Biology of Skin, Vol. 8*, W. Montagna and F. Hu (eds.), pp. 1–30, Pergamon Press, Oxford, 1967.

Flaherty, L., Cantor, L., Zimmerman, D., and Bennett, D.: Cell surface antigens on erythroid cells. A comparison of normal and anemic (*W*/*W*) mice. *Develop. Biol.* **59:** 237–240, 1977.

Flesch, P.: Studies of the red pigmentary system. *Arch. Dermatol.* **101:** 475–481, 1970.

Foster, M.: Enzymatic studies of pigment-forming abilities in mouse skin. *J. Exp. Zool.* **117:** 211–246, 1951.

Foster, M.: Physiological studies of melanogenesis. In *Pigment Cell Biology*, M. Gordon (ed.), pp. 301–314, Academic Press, New York, 1959.

Foster, M.: Tactics in pigment-cell research. (Discussion). In *Methodology in Mammalian Genetics*, W.J. Burdette (ed.), pp. 336–339, Holden-Day, San Francisco, 1963.

Foster, M.: Mammalian pigment genetics. *Adv. Genet.* **13:** 311–339, 1965.

Foster, M.: Genetic aspects of mammalian melanogenesis. In *The Pigmentary System, Advances in Biology of Skin, Vol. 8*, W. Montagna and F. Hu (eds.), pp. 467–477, Pergamon Press, Oxford, 1967.

Foster, M., and Thomson, L.: The effects of substitution at the leaden (*ln*) locus on melanogenic attributes of the mouse. *Proc. X Int. Congr. Genet.* **2:** 84, 1958 (Abstract).

Foster, M., Barto, E., and Thomson, L.: Genetic control of mammalian melanogenesis. In *Pigmentation: Its Genesis and Biologic Control*, V. Riley (ed.), pp. 387–400, Appleton-Century-Crofts, New York, 1972.

Foulks, J.G.: An analysis of the source of melanophores in regenerating feathers. *Physiol. Zool.* **16:** 351–380, 1943.

Fowler, J.H., Till, J.E., McCulloch, E.A., and Siminovitch, L.: The cellular basis for the defect in haemopoiesis in flexed-tailed mice. II. The specificity of the defect for erythropoiesis. *Brit. J. Haematol.* **13:** 256–264, 1967.

Fox, S., and Eicher, E.M.: Private communication. *Mouse News Letter* **58:** 47, 1978.

Frankel, F.R., Tucker, R.W., Bruce, J., and Stenberg, R.: Fibroblasts and macrophages of mice with the Chediak-Higashi-like syndrome have microtubules and actin cables. *J. Cell Biol.* **79:** 401–408, 1978.

Fraser, A.S., Sobey, S., and Spicer, C.C.: Mottled, a sex-modified lethal in the house mouse. *J. Genet.* **51:** 217–221, 1953.

Fried, W., Rishpon-Meyerstein, N., and Gurney, C.W.: The effect of testosterone on erythropoiesis of W/W^v mice. *J. Lab. Clin. Med.* **70:** 813–819, 1967.

Fried, W., Chamberlain, W., Knospe, W.H., Husseini, S., and Trobaugh, F.E., Jr.: Studies on the defective hematopoietic microenvironment of Sl/Sl^d mice. *Brit. J. Haematol.* **24:** 643–650, 1973.

Fry, R.J.M., Slaughter, B., Grahn, D., Wasserman, F., Manelis, F., Hamilton, K.F., and Staffeldt, E.: Inherited connective tissue defect in tortoise mice. *US AEC Argonne Nat. Lab.* **7409:** 144–117, 1967.

Fuller, J.L.: Effects of the albino gene upon behaviour of mice. *Anim. Behav.* **15:** 467–470, 1967.

Galbraith, D.B.: The agouti pigment pattern of the mouse: A quantitative and experimental study. *J. Exp. Zool.* **155:** 71–90, 1964.

Galbraith, D.B.: Cell mass, hair type and expression of the agouti gene. *Nature (London)* **222:** 288–290, 1969.

Galbraith, D.B.: Expression of genes at the agouti locus and mitotic activity of the hair bulb of the mouse. *Genetics* **67:** 559–568, 1971.

Galbraith, D.B., and Arceci, R.J.: Melanocyte populations of yellow and black hair bulbs in the mouse. *J. Hered.* **65:** 381–382, 1974.

Galbraith, D.B., and Patrignani, A.M.: Sulfhydryl compounds in melanocytes of yellow (A^y/a), nonagouti (a/a), and agouti (A/A) mice. *Genetics* **84:** 587–591, 1976.

Galbraith, D.B., and Wolff, G.L.: Aberrant regulation of the agouti pigment pattern in the viable yellow mouse. *J. Hered.* **65:** 137–140, 1974

Gallin, J.I., Bujak, J.S., Patten, E., and Wolff, S.M.: Granulocyte function in the Chediak-Higashi syndrome of mice. *Blood* **43:** 201–206, 1974.

Gardner, R.L., and Lyon, M.F.: X chromosome inactivation studied by injection of a single cell into the mouse blastocyst. *Nature* (*London*) **231:** 385–386, 1971.

Garland, R.C., Satrustegui, J., Gluecksohn-Waelsch, S., and Cori, C.F.: Deficiency in plasma protein synthesis caused by x-ray-induced lethal albino alleles in mouse. *Proc. Nat. Acad. Sci. USA* **73:** 3376–3380, 1976.

Gasser, D.L., and Fischgrund, T.: Genetic control of the immune response in mice. IV. Relationship between graft vs host reactivity and possession of the high tumor genotypes A^ya and $A^{vy}a$. *J. Immunol.* **110:** 305–308, 1973.

Gates, W.H.: The Japanese waltzing mouse; its origin, heredity and relation to the genetic characters of other varieties of mice. *Pub. Carnegie Inst. Wash. No.* **337:** 83–138, 1926.

Geissler, E.N., and Russell, E.S.: Private communication. *Mouse News Letter* **59:** 26, 1978.

Gerson, D.E., and Szabó, G.: Effect of single gene substitution on the melanocyte system of the C57BL mouse: Quantitative and qualitative histology. *Nature* (*London*) **218:** 381–382, 1968.

Geschwind, I.I.: Change in hair color in mice induced by injection of α MSH. *Endocrinology* **79:** 1165–1167, 1966.

Geschwind, I.I., and Huseby, R.A.: Melanocyte-stimulating activity in a transplantable mouse pituitary tumor. *Endocrinology* **79:** 97–105, 1966.

Geschwind, I.I., and Huseby, R.A.: Hormonal modification of coat color in the laboratory mouse. In *Pigmentation: Its Gensis and Biologic Control*, V. Riley (ed.), pp. 207–214, Appleton-Century-Crofts, New York, 1972.

Geschwind, I.I., Huseby, R.A., and Nishioka, R.: The effect of melanocyte-stimulating hormone on coat color in the mouse. *Recent Prog. Hormone Res.* **28:** 91–130, 1972.

Gluecksohn-Waelsch, S.: Private communication. *Mouse News Letter* **30:** 14, 1964a.

Gluecksohn-Waelsch, S.: Private communication. *Mouse News Letter* **31:** 10, 1964b.

Gluecksohn-Waelsch, S.: Private communication. *Mouse News Letter* **32:** 15, 1965.

Gluecksohn-Waelsch, S., Schiffman, M.B., Thorndike, J., and Cori, C.F.: Complementation studies of lethal alleles in the mouse causing deficiencies of glucose-6-phosphatase, tyrosine aminotransferase, and serine dehydratase. *Proc. Nat. Acad. Sci. USA* **71:** 825–829, 1974.

Gluecksohn-Waelsch, S., Schiffman, M.B., and Moscona, M.H.,: Glutamine synthetase in newborn mice homozygous for lethal albino alleles. *Develop. Biol.* **45:** 369–371, 1975.

Goka, T.J., Stevenson, R.E., Hefferan, P.M., and Howell, R.R.: Menkes disease: A biochemical abnormality in cultured human fibroblasts. *Proc. Nat. Acad. Sci. USA* **73:** 604–606, 1976.

Goodale, H.D.: Evidence that size of head-spot (headdot, Keeler) in the mouse is not controlled by modifiers distributed among many chromosomes. *Genetics* **22:** 193, 1937a (Abstract).

Goodale, H.D.: Can artificial selection produce unlimited change? *Amer. Nat.* **71:** 433–459, 1937b.

Goodale, H.D.: Further progress with artificial selection. *Amer. Nat.* **76:** 515–519, 1942.

Goodale, H.D.: New extreme variants in a selection experiment. *Genetics* **28:** 75, 1943 (Abstract).

Goodale, H.D.: Transformation of black mice into white mice. *Genetics* **33:** 106, 1948 (Abstract).

Goodman, R.M.: Effect of W^v locus in mouse on differential excretion of isomers of several amino acids. *Proc. Soc. Exp. Biol. Med.* **88:** 283–287, 1955.

Goodman, R.M.: The effect of the W^v allele in the mouse on the differential excretion of the optical isomers of several amino acids. *J. Exp. Zool.* **132:** 189–217, 1956.

Goodman, R.M.: *In vitro* amino acid metabolism of tissues from a mouse mutant showing differential patterns of amino acid excretion. *Fed. Proc.* **17:** 57, 1958 (Abstract).

Gordon, J.: Failure of XX cells containing the sex reversed gene to produce gametes in allophenic mice. *J. Exp. Zool.* **198:** 367–373, 1976.

Gordon, J: Modification of pigmentation patterns in allophenic mice by the *W* gene. *Differentiation* **9:** 19–28, 1977.

Grahn, D., and Craggs, R.: Mammalian genetics: An unusual occurrence of XXY tortoise males and presumed XY/XY mosaic. *Argonne National Laboratory Biological and Medical Division Annual Report 7409:* 111–113, 1967.

Grahn, D., Fry, R.J.M., and Hamilton, K.F.: Genetic and pathologic analysis of the sex-linked alleleic series, mottled, in the mouse. *Genetics* **61:** s22–s23, 1969a (Abstract).

Grahn, D., Allen, K.H., Fry, R.J.M., and Hulesch, J.: Genetics of the "mottled" alleles on the X-chromosome of the mouse. *Argonne National Laboratory Biological and Medical Research Division Annual Report 7635:* 154–156, 1969b.

Grahn D., Fry, R.J.M., and Allen, K.: Private communication. *Mouse News Letter* **44:** 16, 1971.

Grahn, D., Fry, R.J.M., and Hulesch, J.L.: Private communication. *Mouse News Letter* **47:** 20, 1972.

Granholm, N.H. and Johnson, P.M.: Enhanced identification of lethal yellow (A^y/A^y) mouse embryos by means of delayed development of four-cell stages. *J. Exp. Zool.* **205:** 327–333, 1978.

Green, M.C.: Himalayan, a new allele of albino in the mouse. *J. Hered.* **52:** 73–75, 1961.

Green, M.C.: Mutant genes and linkages. In *Biology of the Laboratory Mouse*, 2nd ed., E.L. Green (ed.), pp. 87–150, McGraw-Hill, New York, 1966a.

Green, M.C.: Private communication. *Mouse News Letter* **34:** 31, 1966b.

Green, M.C.: Private communication. *Mouse News Letter* **44:** 30, 1971.

Green, M.C.: Private communication. *Mouse News Letter* **47:** 36, 1972.

Green, M.C.: Private communication. *Mouse News Letter* **49:** 32, 1973.

Green, M.C., and Sweet, H.O.: Private communication. *Mouse News Letter* **49:** 32, 1973.

Gregory, C.J., McCulloch, E.A., and Till, J.E.: The cellular basis for the defect in haemopoiesis in flexed-tailed mice. III. Restriction of the defect to erythropoietic progenitors capable of transient colony formation *in vivo*. *Brit. J. Haematol.* **30:** 401–410, 1975.

Grobman, A.B., and Charles, D.R.: Mutant white mice. *J. Hered.* **38:** 381–384, 1947.

Gropp, A., Tettenborn, U., and von Lehmann, E.: Chromosomenuntersuchungen bei der Tobakmaus (*M. poschiavinus*) und bei Tobakmaus-Hybriden. *Experientia* **25:** 875–876, 1969.

Gropp, A., Tettenborn, U., and von Lehmann, E.: Chromosomenvariation vom Roberton'schen Typus bei der Tabakmaus, *M. poschiavinus*, und ihren Hybriden mit der Laboratoriumsmaus. *Cytogentics* **9:** 9–23, 1970.

Grüneberg, H.: A new sub-lethal colour mutation in the house mouse. *Proc. Roy. Soc. (London)* **B 118:** 321–342, 1935.

Grüneberg, H.: Grey-lethal, a new mutation in the house mouse. *J. Hered.* **27:** 105–109, 1936a.

Grüneberg, H.: The inheritance of tail tip pigmentation in the house mouse. *J. Genet* **33:** 343–345, 1936b.

Grüneberg, H.: Some new data on the grey-lethal mouse. *J. Genet* **36:** 153–170, 1938.

Grüneberg, H.: Inherited macrocytic anemias in the house mouse. *Genet* **24:** 777–810, 1939.

Grüneberg, H.: Siderocytes: A new kind of erythrocytes. *Nature (London)* **148:** 114–115, 1941.

Grüneberg, H.: Inherited macrocytic anaemias in the house mouse. II. Dominance relationships. *J. Genet* **43:** 285–293, 1942a.

Grüneberg, H.: The anemia of flexed-tailed mice (*Mus musculus* L.). I. Static and dynamic hematology. *J. Genet* **43:** 45–68, 1942b.

Grüneberg, H.: The anemia of flexed-tailed mice (*Mus musculus* L.). II. Siderocytes *J. Genet* **44:** 246–271, 1942c.

Grüneberg, H.: Some observations on the microphthalmia gene in the mouse. *J. Genet* **49:** 1–13, 1948.

Grüneberg, H.: *The Genetics of the Mouse*, 2nd ed., Nijhoff, The Hague, 1952.

Grüneberg, H.: The relations of microphthalmia and white in the mouse. *J. Genet* **51:** 359–362, 1953.

Grüneberg, H.: *The Pathology of Development; a Study of Inherited Skeletal Disorders in Animals*, Wiley, New York, 1963.

Grüneberg, H.: The case for somatic crossing over in the mouse. *Genet. Res. (Camb.)* **7:** 58–75, 1966a.

Grüneberg, H.: The molars of the tabby mouse, and a test of the 'single-active X-chromosome' hypothesis. *J. Embryol. Exp. Morphol.* **15:** 223–244, 1966b.

Grüneberg, H.: More about the tabby mouse and about the Lyon hypothesis. *J. Embryol, Exp. Morphol.* **16:** 569–590, 1966c.

Grüneberg, H.: Gene action in the mammalian X-chromosome. *Genet. Res. (Camb.)* **9:** 343–357, 1967a.

Grüneberg, H.: Sex-linked genes in man and the Lyon hypothesis. *Ann. Human Genet.* **30:** 239–237, 1967b.

Grüneberg, H.: Threshold phenomena versus cell heredity in the manifestation of sex-linked genes in mammals. *J. Embryol. Exp. Morphol.* **22:** 145–179, 1969.

Grüneberg, H.: The tabby syndrome in the mouse. *Proc. Roy. Soc.* (*London*) **B. 179:** 139–156, 1971a.

Grüneberg, H.: The glandular aspects of the tabby syndrome in the mouse. *J. Embryol. Exp. Morphol.* **25:** 1–19, 1971b.

Grüneberg, H., and Truslove, G.M.: Two closely linked genes in the mouse. *Genet. Res.* (*Camb.*) **1:** 69–90, 1960.

Grüneberg, H., Cattanach, B.M., McLaren, A., Wolfe, H.G., and Bowman, P.: The molars of tabby chimeras in the mouse. *Proc. Roy. Soc.* (*London*) **B 182:** 183–192, 1972.

Guénet, J.L., and Mercier-Balaz, M.: Private communication. *Mouse News Letter* **53:** 57–58, 1975.

Guénet, J.L., Marchal, G., Milon, G., Tambourin, P., and Wendling, F.: Fertile dominant spotting (W^f): A new allele at the *W* locus. *J. Hered.*, *in press.*

Guillery, R.W., Amorn, C.S., and Eighmy, B.B.: Mutants with abnormal visual pathways: An explanation of anomalous geniculate laminae. *Science* **174:** 831–832, 1971.

Guillery, R.W., Scott, G.L., Cattanach, B.M., and Deol, M.S.: Genetic mechanisms determining the central visual pathways of mice. *Science* **179:** 1014–1016, 1973.

Guy-Grand, D., Griscelli, C., and Vassalli, P.: The mouse gut T lymphocyte a novel type of T cell. Nature, origin, and traffic in mice in normal and graft-versus-host conditions. *J. Exp. Med.* **148:** 1661–1677, 1978.

Hadley, M.E., and Quevedo, W.C., Jr.: Vertebrate epidermal melanin unit. *Nature* (*London*) **209:** 1334–1335, 1966.

Hagedoorn, A.L.: The genetic factors in the development of the house mouse which influence coat color. *Z. indukt. Abstamm.-u. VererbLehre* **6:** 97–136, 1912.

Håkansson, E.M., and Lundin, L.G.: Effect of a coat color locus on kidney lysosomal glycosidases in the house mouse. *Biochem. Genet.* **15:** 75–85, 1977.

Haldane, J.B.S., Sprunt, A.D., and Haldane, N.M.: Reduplication in mice. *J. Genet.* **5:** 133–135, 1915.

Hance, R.T.: Detection of heterozygotes with x-rays. *J. Hered.* **19:** 481–485, 1928.

Harrison, D.E.: Normal function of transplanted mouse erythrocyte precursors for 21 months beyond donor life spans. *Nature New Biol.* **237:** 220–222, 1972a.

Harrison, D.E.: Lifesparing ability (in lethally irradiated mice) of W/W^v mouse marrow with no macroscopic colonies. *Radiat. Res.* **52:** 553–563, 1972b.

Harrison, D.E.: Normal function of transplanted marrow cell lines from aged mice. *J. Gerontol.* **30:** 279–285, 1975a.

Harrison, D.E.: Defective erythropoietic responses of aged mice not improved by young marrow. *J. Gerontol.* **30:** 286–288, 1975b.

Harrison, D.E., and Cherry, M.: Survival of marrow allografts in W/W^v anemic mice: Effect of disparity at the *Ea-2* locus. *Immunogenet.* **2:** 219–229, 1975.

Harrison, D.E., and Russell, E.S.: The response of W/W^v and Sl/Sl^d anaemic mice to haemopoietic stimuli. *Brit. J. Haematol.* **22:** 155–168, 1972.

Harrison, D.E., Malathi, V.G., and Silber, R.: Elevated erythrocyte nucleoside deaminase levels in genetically anemic W/W^v and Sl/Sl^d mice. *Blood Cells* **1:** 605–614, 1975.

Hauschka, T.S., Jacobs, B.B., and Holdridge, B.A.: Recessive yellow and its interaction with belted in the mouse. *J. Hered.* **59:** 339–341, 1968.

Hearing, V.J.: Tyrosinase activity in subcellular fractions of black and albino mice. *Nature New Biol.* **245:** 81–83, 1973.

Hearing, V.J., Phillips, P., and Lutzner, M.A.: The fine structure of melanogenesis in coat color mutants of the mouse. *J. Ultrastruct. Res.* **43:** 88–106, 1973.

Heath, J.: Private communication. *Mouse News Letter* **58:** 66, 1978.

Hertwig, P.: Neue Mutationen und koppelungsgruppen bei der Hausmaus. *Z. indukt. Abstamm.-u. VererbLehre* **80:** 220–246, 1942a.

Hertwig, P.: Sechs neue Mutationen bei der Hausmaus in ihrer Bedeutung für allgemeine Vererbungsfragen. *Z. menschl. Vererbungs-u. KonstL.* **26:** 1–21, 1942b.

Heston, W.E.: Relationship between the lethal yellow (A^y) gene of the mouse and susceptibility to induced pulmonary tumors. *J. Nat. Cancer Inst.* **3:** 303–308, 1942.

Heston, W.E., and Deringer, M.K.: Relationship between the lethal yellow (A^y) gene of the mouse and susceptibility to spontaneous pulmonary tumors. *J. Nat. Cancer Inst.* **7:** 463–465, 1947.

Heston, W.E., and Vlahakis, G.: Influence of the A^y gene on mammary-gland tumors, hepatomas, and normal growth in mice. *J. Nat. Cancer Inst.* **26:** 969–983, 1961a.

Heston, W.E., and Vlahakis, G.: Elimination of the effect of the A^y gene on pulmonary tumors in mice by alteration of its effect on normal growth. *J. Nat. Cancer Inst.* **27:** 1189–1196, 1961b.

Heston, W.E., and Vlahakis, G.: C3H-A^{vy}—a high hepatoma and high mammary tumor strain of mice. *J. Nat. Cancer Inst.* **40:** 1161–1166, 1968.

Hetherington, C.: Private communication. *Mouse News Letter* **54:** 34, 1976.

Hirobe, T., and Takeuchi, T.: Induction of melanogenesis in the epidermal melanoblasts of newborn mouse skin by MSH. *J. Embryol. Exp. Morphol.* **37:** 79–90, 1977a.

Hirobe, T., and Takeuchi, T.: Induction of melanogenesis *in vitro* in the epidermal melanoblasts of newborn mouse skin by MSH. *In Vitro* **13:** 311–315, 1977b.

Hirobe, T., and Takeuchi, T.: Changes of organelles associated with the differentiation of epidermal melanocytes in the mouse. *J. Embryol. Exp. Morphol.* **43:** 107–121, 1978.

Hirsch, M.S.: Studies on the response of osteopetrotic bone explants to parathyroid explants *in vitro*. *Bull. Johns Hopkins Hosp.* **110:** 257–263, 1962.

Hoecker, G.: Mutaciones dominantes reversivas en mosaico en la cepa pura de ratones C_{58} negra. *Biológica, Santiago* **12:** 25–37, 1950.

Holland, J.M.: Serotonin deficiency and prolonged bleeding in beige mice. *Proc. Soc. Exp. Biol. Med.* **151:** 32–39, 1976.

Hollander, W.F.: Private communication. *Mouse News Letter* **15:** 29, 1956.

Hollander, W.F.: Private communication. *Mouse News Letter* **20:** 34, 1959.

Hollander, W.F.: Private communication. *Mouse News Letter* **25:** 9, 1961.

Hollander, W.F.: Private communication. *Mouse News Letter* **30:** 29, 1964.

Hollander, W.F.: Complementary alleles at the *mi*-locus in the mouse. *Genetics* **60:** 189, 1968 (Abstract).

Hollander, W.F., and Gowen, J.W.: An extreme non-agouti mutant in the mouse. *J. Hered.* **47:** 221–224, 1956.

Hollander, W.F., Bryan, J.H.D., and Gowen, J.W.: Pleiotropic effects of a mutant at the P locus from x-irradiated mice. *Genetics* **45:** 413–418, 1960a.

Hollander. W.F., Bryan, J.H.D., and Gowen, J.W.: A male sterile pink-eyed mutant type in the mouse. *Fertil. Steril.* **11:** 316–324, 1960b.

Hollinshead, M.B., and Schneider, L.C.: Identification of neonatal grey lethal mice. *Anat. Rec.* **176:** 273–278, 1973.

Hollinshead, M.B., Schneider, L.C., and Smith, M.E.: Prenatal development of the grey lethal mouse. I. Teeth and jaws. *Anat. Rec.* **182:** 305–320, 1975.

Holstein, T.J., Burnett, J.B., and Quevedo, W.C., Jr.: Genetic regulation of multiple forms of tyrosinase in mice: Action of *a* and *b* loci. *Proc. Soc. Exp. Biol. Med.* **126:** 415–418, 1967.

Holstein, T.J., Quevedo, W.C., Jr., and Burnett, J.B.: Multiple forms of tyrosinase in rodents and lagomorphs with special reference to their genetic control in mice. *J. Exp. Zool.* **177:** 173–184, 1971.

Hrubant, E.H.: Urinary amino acid differences in C57BL/6 and C3Heb/Fe inbred mice, and their F_1 hybrid. *Canad. J. Genet. Cytol.* **7:** 530–535, 1965.

Hsu, C.Y., and van Dyke, J.H.: An analysis of growth rates in neural epithelium of normal and spina bifidous (myeloschisis) mouse embryos. *Anat. Rec.* **100:** 745, 1948 (Abstract).

Huff, S.D., and Fuller, J.L.: Audiogenic seizures, the dilute locus, and phenylalanine hydroxylase in DBA/1 mice. *Science* **144:** 304–305, 1964.

Huff, S.D., and Huff, R.L.: Dilute locus and audiogenic seizures in mice. *Science* **136:** 318–319, 1962.

Hummel, K.P., Coleman, D.L., and Lane, P.W.: The influence of genetic background on expression of mutations at the diabetes locus in the mouse. I. C57BL/KsJ and C57BL/6J strains. *Biochem. Genet.* **7:** 1–13, 1972.

Hunsicker, P.R.: Private communication. *Mouse News Letter* **38:** 31, 1968.

Hunsicker, P.R.: Private communication. *Mouse News Letter* **40:** 41, 1969.

Hunt, D.M.: Primary defect in copper transport underlies mottled mutants in the mouse. *Nature (London)* **249:** 852–854, 1974a.

Hunt, D.M.: Private communication. *Mouse News Letter* **50:** 36, 1974b.

Hunt, D.M., and Johnson, D.R.: Abnormal spermiogenesis in two pink-eyed sterile mutants in the mouse. *J. Embryol. Exp. Morphol.* **26:** 111–121, 1971.

Hunt, D.M., and Johnson, D.R.: Aromatic amino acid metabolism in brindled (Mo^{br}) and viable-brindled (Mo^{vbr}), two alleles at the mottled locus in the mouse. *Biochem. Genet.* **6:** 31–40, 1972a.

Hunt, D.M., and Johnson, D.R.: An inherited deficiency in noradrenalin biosynthesis in the brindled mouse. *J. Neurochem.* **19:** 2811–2819, 1972b.

Hunt, D.M., and Skinner, D.F.: Private communication. *Mouse News Letter* **54:** 42, 1976.

Hunt, H.R.: The flexed tailed mouse. *Proc. VI Int. Congr. Genet.* **2:** 91–93, 1932 (Abstract).

Hunt, H.R., Mixter, R., and Permar, D.: Flexed tail in the mouse, *Mus Musculus*. *Genetics* **18:** 335–366, 1933.

Hurley, L.S., and Everson, G.J.: Influence of timing of short-term supplementation during gestation on congenital abnormalities of manganese-deficient rats. *J. Nutrit.* **79:** 23–27, 1963.

Hurley, L.S., Everson, G.J., and Geiger, J.F.: Manganese deficiency in rats. Congenital nature of ataxia. *J. Nutrit.* **66:** 309–320, 1958.

Hurley, L.S., Theriault, L.L., and Dreosti, I.E.: Liver mitochondria from manganese-deficient and pallid mice: Function and ultrastructure. *Science* **170:** 1316–1318, 1970.

Ibsen, H.L., and Steigleder, E.: Evidence for the death in utero of the homozygous yellow mouse. *Amer. Nat.* **51:** 740–752, 1917.

Ikejima, T., and Takeuchi, T.: Fluorescence spectrophotometric analysis of melanins in the house mouse. *Biochem. Genet.* **16:** 673–679, 1978.

Iljin, N.A., and Iljin, V.N.: Temperature effects on the color of the siamese cat. *J. Hered.* **21:** 309–318, 1930.

Isherwood, J.E., Strong, L.C., and Quevedo, W.C., Jr.: A new mutation of *a* to a^t in the mouse. *J. Hered.* **51:** 121, 135, 1960.

Ives, P.T.: The importance of mutation rate genes in evolution. *Evolution* **4:** 236–252, 1950.

Jimbow, K., Quevedo, W.C., Jr., Fitzpatrick, T.B., and Szabó, G.: Some aspects of melanin biology: 1950–1975. *J. Invest. Dermatol.* **67:** 72–89, 1976.

Johnson, D.R., and Hunt, D.M.: Private communication. *Mouse News Letter* **47:** 52, 1972.

Kales, A.N., Fried, W., and Gurney, C.W.: Mechanism of the hereditary anemia of Sl^m mutant mice *Blood* **28:** 387–397, 1966.

Kaliss, N.: The morphogenesis of pigment in the hair follicles of the house mouse. *J. Morphol.* **70:** 209–219, 1942.

Kamenoff, R.J.: Effects of the flexed-tailed gene on the development of the house mouse. *J. Morphol.* **58:** 117–155, 1935.

Kamenoff, R.J.: A cytological study of the embryonic livers (16-18 days) of normal and flexed-tailed (anemic) mice. *Genetics* **27:** 150, 1942 (Abstract).

Kaplan, S.S., Boggs, S.S., Nardi, M.A., Basford, R.E., and Holland, J.M.: Leukocyte-platelet interactions in a murine model of Chediak-Higashi syndrome. *Blood* **52:** 719–725, 1978.

Keeler, C.E.: A probable new mutation to white-belly in the house mouse. *Mus musculus*, *Proc. Nat. Acad. Sci. USA* **17:** 700–703, 1931.

Keeler, C.E.: Akhissar spotting in the mouse. *Proc. Nat. Acad. Sci. USA* **19:** 477–481, 1933.

Keeler, C.E.: Headdot: An incompletely recessive white spotting character of the house mouse. *Proc. Nat. Acad. Sci. USA* **21:** 379–383, 1935.

Keeler, C.E., and Goodale, H.D.: A second occurrence of headdot spotting in the house mouse. *J. Mammal.* **17:** 263–265, 1936.

Keighley, G., Russell, E.S., and Lowy, P.H.: Response of normal and genetically anaemic mice to erythropoietic stimuli. *Brit. J. Haematol.* **8:** 429–441, 1962.

Keighley, G.H., Lowy, P., Russell, E.S., and Thompson, M.W.: Analysis of homeostatic mechanisms in normal and genetically anaemic mice. *Brit. J. Haematol.* **12:** 461–477, 1966.

Kelly, E.M.: Private communication. *Mouse News Letter* **16:** 36, 1957.

Kelly, E.M.: Private communication. *Mouse News Letter* **38:** 31, 1968.

Kelly, E.M.: Private communication. *Mouse News Letter* **43:** 59, 1970.

Kelly, E.M.: Private communication. *Mouse News Letter* **50:** 52, 1974.

Kelly, E.M.: Private communication. *Mouse News Letter* **52:** 46, 1975.

Kelton, D.E., and Rauch, H.: Myelination and myelin degeneration in the central nervous system of dilute-lethal mice. *Exp. Neurol.* **6:** 252–262, 1962.

Kindred, B.: Some observations on the skin and hair of tabby mice. *J. Hered.* **58:** 197–199, 1967.

King, R.A., and Rush, W.A.: Alcohol sensitivity in the albino mouse. In *Pigment Cell*, V. Riley (ed.), Vol. 3, pp. 211–219, Karger, Basal, 1976.

Kirby, G.C.: Greying with age: A coat-color variant in wild Australian populations of mice. *J. Hered.* **65:** 126–128, 1974.

Kirchner, C.E.J.: The effects of the mutator gene on molecular changes and mutations in *Salmonella Typhimurium*. *J. Mol. Biol.* **2:** 331–338, 1960.

Kirkham, W.B.: The fate of homozygous yellow mice. *J. Exp. Zool.* **28:** 125–135, 1919.

Kitamura, Y., Shimada, M., Hatanaka, K., and Miyano, Y.: Development of mast cells from grafted bone marrow in irradiated mice. *Nature* (*London*) **268:** 442–443, 1977.

Kitamura, Y., Go, S., and Hatanaka, K.: Decrease of mast cells in W/W^v mice and their increase by bone marrow transplantation. *Blood* **52:** 447–452, 1978.

Klein, A., and Sitarz, K.: The influence of the mosaic (*Ms*) mutation on the amino acid composition of blood plasma and the osmotic resistance of erythrocytes. *Acta Biol. Cracov. Ser. Zool.* **14:** 129–136, 1971.

Klein, A., and Styrna, J.: Ninhydrin positive substances present in the urine of mice carrying the lethal mosaic (*Ms*) mutation. *Acta Biol. Cracov. Ser. Zool.* **14:** 121–127, 1971.

Klein, J.: *Biology of the Mouse Histocompatibility-2 Complex*. Springer-Verlag, New York, 1975.

Knisely, A.S., Gasser, D.L., and Silvers, W.K.: Expression in organ culture of agouti locus genes of the mouse. *Genetics* **79:** 471–475, 1975.

Knospe, W.H., Hinrichs, B., Fried, W., Robinson, W., and Trobaugh, F.E., Jr.: Normal colony stimulating factor (CSF) production by bone marrow stromal cells and abnormal granulopoiesis with decreased CFUc in Sl/Sl^d mice. *Exp. Hematol.* **4:** 125–130, 1976.

Konyukhov, B.V., and Sazhina, M.V.: Eye development in mutant mice strain microphthalmia, *J. Gen. Biol.* **24:** 285–295, 1963 (in Russian).

Konyukhov, B.V., and Sazhina, M.V.: Interaction of the genes of ocular retardation and microphthalmia in mice. *Folia Biol.* (*Praha*) **12:** 116–123, 1966.

Konyukhov, B.V., Osipov, V.V., Vachruscheva, M.P.: Damage of derivatives of neural crest in mice of mutant lines Microphthalmia and White. *Arch. Anat. Histol. Embryol.* **8:** 100–107, 1965 (in Russian).

Kramer, J.W., Davis, W.C., and Prieur, D.J.: The Chediak–Higashi syndrome of cats. *Lab. Invest.* **36:** 554–562, 1977.

Kreitner, P.C.: Linkage studies in a new black-eyed white mutation. *J. Hered.* **48:** 300–304, 1957.

Krzanowska, H.: Private communication. *Mouse News Letter* **35:** 35, 1966.

Krzanowska, H.: Private communication. *Mouse News Letter* **38:** 25, 1968.

Krzanowska, H., and Wabik, B.: Selection for expression of sex linked gene *Ms* (mosaic) in heterozygous mice. *Genet. Pol.* **12:** 537–544, 1971.

Krzanowska, H., and Wabik, B.: Further studies on the expression of sex-linked gene *Ms* (mosaic) in heterozygous mice. *Genet. Pol.* **14:** 193–198, 1973.

Lamoreux, M.L.: *A Study of Gene Interactions Using Coat Color Mutants in the Mouse and Selected Mammals*. Ph.D. thesis, University of Maine, Orono, 1973.

Lamoreux, M.L., and Mayer, T.C.: Site of gene action in the development of hair pigment in recessive yellow (*e/e*) mice. *Develop. Biol.* **46:** 160–166, 1975.

Lamoreux, M.L., and Russell, E.S.: Effects of agouti-locus alleles (A^y/a vs a/a) on amount of white spotting when combined with two different spotting genotypes (s/s^l and $W^{j2}/+$). *Genetics* **68:** s36, 1971 (Abstract).

Lane, P.W.: Private communication. *Mouse News Letter* **22:** 35, 1960a.

Lane, P.W.: Private communication. *Mouse News Letter* **23,** 36, 1960b.

Lane, P.W.: Private communication. *Mouse News Letter* **26:** 35, 1962.

Lane, P.W.: Private communication. *Mouse News Letter* **32:** 47, 1965.

Lane, P.W.: Association of megacolon with two recessive spotting genes in the mouse. *J. Hered.* **57:** 29–31, 1966.

Lane, P.W.: Private communication. *Mouse News Letter* **44:** 30, 1971.

Lane, P.W.: Two new mutations in linkage group XVI of the house mouse. Flaky tail and varitint-waddler-J. *J. Hered.* **63:** 135–140, 1972.

Lane, P.W., and Deol, M.S.: Mocha, a new coat color and behavior mutation on chromosome 10 of the mouse. *J. Hered.* **65:** 362–364, 1974.

Lane, P.W., and Green, M.C.: Mahogany, a recessive color mutation in linkage group V of the mouse. *J. Hered.* **51:** 228–230, 1960.

Lane, P.W., and Green, E.L.: Pale ear and light ear in the house mouse. Mimic mutations in linkage groups XII and XVII. *J. Hered.* **58:** 17–20, 1967.

Lane, P.W., and Murphy, E.D.: Susceptibility to spontaneous pneumonitis in an inbred strain of beige and satin mice. *Genetics* **72:** 451–460, 1972.

Lane, P.W., and Womack, J.E.: Private communication. *Mouse News Letter* **57:** 18, 1977.

Larsen, M.M.: Private communication. *Mouse News Letter* **34:** 41, 1966.

LaVail, M.M., and Sidman, R.L.: C57BL/6J mice with inherited retinal degeneration. *Arch. Ophthalmol.* **91:** 394–400, 1974.

Law, L.W.: The flexed-tail-anemia gene (*f*) and induced leukemia in mice. *J. Nat. Cancer Inst.* **12:** 1119–1126, 1952.

Lerner, A.B.: Melanin pigmentation. *Amer. J. Med.* **19:** 902–924, 1955.

Lerner, A.B., and Case, J.D.: Pigment cell regulatory factors. *J. Invest. Dermatol.* **32:** 211–221, 1959.

Lewis, E.B.: Pseudoallelism and gene evolution. *Cold Spring Harbor Symp. Quant. Biol.* **16:** 159–174, 1961.

Lewis, J.P., O'Grady, L.F., Bernstein, S.E., Russell, E.S., and Trobaugh, F.E. Jr.: Growth and differentiation of transplanted W/W^v marrow. *Blood* **30:** 601–616, 1967.

Lewis, S.E., Turchin, H.A., and Gluecksohn-Waelsch, S.: The developmental analysis of an embryonic lethal (c^{6H}) in the mouse. *J. Embryol. Exp. Morphol.* **36:** 363–371, 1976.

Lilly, F.: Private communication. *Mouse News Letter* **34:** 14, 1966.

Little, C.C.: Experimental studies of the inheritance of color in mice. *Publ. Carneg. Instn., No.* **179:** 13–102, 1913.
Little, C.C.: "Dominant" and "recessive" spotting in mice. *Amer. Nat.* **48:** 74–82, 1914.
Little, CC.: The inheritance of black-eyed white spotting in mice. *Amer. Nat.* **49:** 727–740, 1915.
Little, CC.: The occurrence of three recognized color mutations in mice. *Amer. Nat.* **50:** 335–349, 1916.
Little, C.C.: The relation of yellow coat color to black-eyed white spotting of mice, in heredity. *Anat. Rec.* **11:** 501, 1917a (Abstract).
Little, C.C.: The relation of yellow coat color and black-eyed white spotting in mice. *Genetics* **2:** 433–444, 1917b.
Little, C.C.: Evidence of multiple factors in mice and rats. *Amer. Nat.* **51:** 457–480, 1917c.
Little, CC.: The genetics of spotting. *Carnegie Inst. Wash. Yearbook* **23:** 42, 1924.
Little, CC.: The inheritance of blaze spotting in mice. *Anat. Rec.* **34:** 171, 1926 (Abstract).
Little, C.C.: Coat color genes in rodents and carnivores. *Quart. Rev. Biol.* **33:** 103–137, 1958.
Little, C.C., and Cloudman, A.M.: The occurrence of a dominant spotting mutation in the house mouse. *Proc. Nat. Acad. Sci. USA* **23:** 535–537, 1937.
Little, C.C., and Hummel, K.P.: A reverse mutation to a "remote" allele in the house mouse. *Proc. Nat. Acad. Sci. USA* **33:** 42–43, 1947.
Littleford, R.A.: Occurrence of piebald spotting in a wild house mouse. *Amer. Nat.* **80:** 283–288, 1946.
Loosli, R.: Tanoid—a new agouti mutant in the mouse. *J. Hered.* **54:** 26–29, 1963.
Loutit, J.F.: Bone marrow grafts in mature osteopetrotic mice. Space not a requirement. *Transplantation* **24:** 299–301, 1977.
Loutit, J.F., and Sansom, J.M.: Osteopetrosis of microphthalmic mice—a defect of the hematopoietic stem cell? *Calcif. Tissue. Res.* **20:** 251–259, 1976.
Lund, R.D.: Uncrossed visual pathways of hooded and albino rats. *Science* **149:** 1506–1507, 1965.
Lutzner, M.: Ultrastructure of giant melanin granules in the biege mouse during ontogeny. *J. Invest. Dermatol.* **54:** 91, 1970 (Abstract).
Lutzner, M.A., and Lowrie, C.T.: Ultrastructure of the development of the normal black and giant beige melanin granules in the mouse. In *Pigmentation: Its Genesis and Biologic Control*, V. Riley (ed.), pp. 89–105, Appleton-Century-Crofts, New York, 1972.
Lutzner, M.A., Lowrie, C.T., and Jordan, H.W.: Giant granules in leukocytes of the beige mouse. *J. Hered.* **58:** 299–300, 1967.
Lyon, M.F.: Hereditary absence of otoliths in the house mouse. *J. Physiol.* **114:** 410–418, 1951.
Lyon, M.F.: Absence of otoliths in the mouse: An effect of the pallid mutant. *J. Genet.* **51:** 638–650, 1953.
Lyon, M.F.: Stage of action of the litter-size effect on absence of otoliths in mice. *Z. indukt. Abstamm.-u. VererbLehre* **86:** 289–292, 1954.
Lyon, M.F.: The development of otoliths in the mouse. *J. Embryol. Exp. Morphol.* **3:** 213–229, 1955a.

Lyon, M.F.: The developmental origin of hereditary absence of otoliths in mice. *J. Embryol. Exp. Morphol.* **3:** 230–241, 1955b.

Lyon, M.F.: A further mutation of the mottled type. *J. Hered.* **51:** 116–120, 1960.

Lyon, M.F.: Gene action in the X-chromosome of the mouse (*Mus musculus* L.). *Nature* (*London*) **190:** 372–373, 1961.

Lyon, M.F.: Attempts to test the inactive-X theory of dosage compensation in mammals. *Genet. Res.* (*Camb.*) **4:** 93–103, 1963.

Lyon, M.F.: Private communication. *Mouse News Letter* **34:** 28, 1966.

Lyon, M.F.: Chromosomal and subchromosomal inactivation. *Ann. Rev. Genet.* **2:** 31–52, 1968.

Lyon, M.F.: Genetic activity of sex chromosomes in somatic cells of mammals. *Phil. Trans. Roy. Soc.* (*London*) **B 259:** 41–53, 1970.

Lyon, M.F.: Possible mechanisms of X chromosome inactivation. *Nature New Biol.* **232:** 229–232, 1971.

Lyon, M.F.: X-chromosome inactivation and developmental patterns in mammals. *Biol. Rev.* **47:** 1–35, 1972a.

Lyon, M.F.: Private communication. *Mouse News Letter* **47:** 34, 1972b.

Lyon, M.F.: Mechanisms and evolutionary origins of variable X-chromosome activity in mammals. *Proc. Roy. Soc.* (*London*) **B 187:** 243–268, 1974.

Lyon, M.F.: Private communication. *Mouse News Letter* **53:** 29, 1975.

Lyon, M.F., and Glenister, P.: Private communication. *Mouse News Letter* **45:** 24, 1971.

Lyon, M.F., and Glenister, P.H.: Private communication. *Mouse News Letter* **48:** 31, 1973.

Lyon, M.F., and Glenister, P.H.: Private communication. *Mouse News Letter* **59:** 18, 1978.

Lyon, M.F., and Meredith, R.: Private communication. *Mouse News Letter* **32:** 38, 1965a.

Lyon, M.F., and Meredith, R.: Private communication. *Mouse News Letter* **33:** 29, 1965b.

Lyon, M.F., and Meredith, R.: *Muted*, a new mutant affecting coat color and otoliths of the mouse, and its position in linkage group XIV. *Genet. Res.* (*Camb.*) **14:** 163–166, 1969.

Lyon, M.F., and Morris, T.: Mutation rates at a new set of specific loci in the mouse. *Genet. Res.* (*Camb.*) **7:** 12–17, 1966.

Lyons, R.T. and Pitot, H.C.: Protein degradation in normal and beige (Chediak–Higashi) mice. *J. Clin. Invest.* **61:** 260–268, 1978.

MacDowell, E.C.: "Light"—a new mouse color. *J. Hered.* **41:** 35–36, 1950.

Maloney, M.A., Dorie, M.J., Lamela, R.A., Rogers, Z.R., and Patt, H.M.: Hematopoietic stem cell regulatory volumes as revealed in studies of the bg^J/bg^J:W/W^v chimera. *J. Exp. Med.* **147:** 1189–1197, 1978.

Margolis, F.L. and Russell, E.S.: Delta-amino-levulinate dehydratase activity in mice with hereditary anemia. *Science* **144:** 844–846, 1965.

Markert, C.L.: Substrate utilization in cell differentiation. *Ann. N.Y. Acad. Sci.* **60:** 1003–1014, 1955.

Markert, C.L.: Biochemical embryology and genetics. In *Symposium on Normal and Abnormal Differentiation and Development.* N. Kaliss (ed.) *Nat. Cancer Inst. Monogr.* **2:** 3–17, 1960.

Markert, C.L. and Petters, R.M.: Manufactured hexaparental mice show that adults are derived from three embryonic cells. *Science* **202:** 56–58, 1978.

Markert, C.L., and Silvers, W.K.: The effects of genotype and cell environment on melanoblast differentiation in the house mouse. *Genetics* **41:** 429–450, 1956.

Markert, C.L., and Silvers, W.K.: Effects of genotype and cellular environment on melanocyte morphology. In *Pigment Cell Biology*, M. Gordon (ed.), pp. 241–248, Academic Press, New York, 1959.

Marks, S.C. Jr.: Studies of the mechanism of spleen cell cure for osteopetrosis in *ia* rats: Appearance of osteoclasts with ruffled borders. *Amer. J. Anat.* **151:** 119–130, 1978a.

Marks, S.C. Jr.: Studies of the cellular cure for osteopetrosis by transplanted cells: Specificity of the cell type in *ia* rats. *Amer. J. Anat.* **151:** 131–137, 1978b.

Marks, S.C., Jr., and Lane, P.W.: Osteopetrosis, a new recessive skeletal mutation on chromosome 12 of the mouse. *J. Hered.* **67:** 11–18, 1976.

Marks, S.C., Jr., and Walker, D.G.: The role of the parafollicular cell of the thyroid gland in the pathogenesis of congenital osteopetrosis in mice. *Amer. J. Anat.* **126:** 299–314, 1969.

Marks, S.C., Jr., and Walker, D.G.: Mammalian osteopetrosis: A model for studying cellular and humoral factors in bone resorption. In *The Biochemistry and Physiology of Bone*, Chap. 6, G. Bourne (ed.), pp. 227–301, Academic Press, New York, 1976.

Mather, K., and North, S.B.: Umbrous: A case of dominance modification in mice. *J. Genet.* **40:** 229–241, 1940.

Mauer, I., and Sideman, M.B.: Phenylalanine metabolism in a dilute-lethal strain of mice. *J. Hered.* **58:** 14–16, 1967.

Mayer, T.C.: The development of piebald spotting in mice. *Develop. Biol.* **11:** 319–334, 1965.

Mayer, T.C.: Pigment cell migration in piebald mice. *Develop. Biol.* **15:** 521–535, 1967a.

Mayer, T.C.: Temporal skin factors influencing the development of melanoblasts in piebald mice. *J. Exp. Zool.* **166:** 397–403, 1967b.

Mayer, T.C.: A comparison of pigment cell development in albino, steel, and dominant spotting mutant mouse embryos. *Develop. Biol.* **23:** 297–309, 1970.

Mayer, T.C.: Site of gene action in steel mice. Analysis of the pigment defect by mesoderm–ectoderm recombinations. *J. Exp. Zool.* **184:** 345–352, 1973a.

Mayer, T.C.: The migratory pathway of neural crest cells into the skin of mouse embryos. *Develop. Biol.* **34:** 39–46, 1973b.

Mayer, T.C.: Tissue environmental influences on the development of melanoblasts in steel mice. In *Extracellular Matrix Influences on Gene Expression*, H.C. Slavkin and R.C. Grevlich (eds.), pp. 555–560, Academic Press, New York, 1975.

Mayer, T.C., and Green, M.C.: An experimental analysis of the pigment defect caused by mutations at the *W* and *Sl* loci in mice. *Develop. Biol.* **18:** 62–75, 1968.

Mayer, T.C., and Fishbane, J.L.: Mesoderm–ectoderm interaction in the production of the agouti pigmentation pattern in mice. *Genetics* **71:** 297–303, 1972.

Mayer, T.C., and Maltby, E.L.: An experimental investigation of pattern development in lethal spotting and belted mouse embryos. *Develop. Biol.* **9:** 269–286, 1964.

Mayer, T.C., and Oddis, L.: Pigment cell differentiation in embryonic mouse skin and isolated epidermis: An *in vitro* study. *J. Exp. Zool.* **202:** 415–424, 1977.

Mayer, T.C., and Reams, W.M., Jr.: An experimental analysis and description of the melanocytes in the leg musculature of the PET strain of mice. *Anat. Rec.* **142:** 431–442, 1962.

McCarthy, K.F.: *In vivo* colony formation by hematopoietic cells from mice of genotype Sl/Sl^d. *Cell Tissue Kinet.* **8:** 397–398, 1975.

McCarthy, K.F., Ledney, G.D., and Mitchell, R.: A deficiency of hematopoietic stem cells in steel mice. *Cell Tissue Kinet.* **10:** 121–126, 1977.

McClintock, B.: Chromosome organization and genic expression. *Cold Spring Harbor Symp. Quant. Biol.* **16:** 13–47, 1951.

McCoshen, J.A., and McCallion, D.J.: A study of primoridal germ cells during their migratory phase in steel mutant mice. *Experientia* **31:** 589–590, 1975.

McCulloch, E.A., Siminovitch, L., and Till, J.E.: Spleen-colony formation in anemic mice of genotype W/W^v. *Science* **144:** 844–846, 1964.

McCulloch, E.A., Siminovitch, L., Till, J.E., Russell, E.S., and Bernstein, S.E.: The cellular basis of the genetically determined hemopoietic defect in anemic mice of genotype Sl/Sl^d. *Blood* **26:** 399–410, 1965.

McCuskey, R.S., and Meineke, H.A.: Studies of the hemopoietic micro-environment. III. Differences in the splenic microvascular system and stroma between Sl/Sl^d and W/W^v anemic mice. *Amer. J. Anat.* **137:** 187–198, 1973.

McGrath, E.P., and Quevedo, W.C., Jr.: Genetic regulation of melanocyte function during hair growth in the mouse: Cellular events leading to "pigment clumping" within developing hairs. In *Biology of the Skin and Hair Growth*, A.G. Lyne and B.F. Short (eds.), pp. 727–745, Angus and Robertson, Sydney, 1956.

McLaren, A., and Bowman, P.: Mouse chimeras derived from fusion of embryos differing by nine genetic factors. *Nature (London)* **224:** 238–240, 1969.

McLaren, A., Gauld, I.K., and Bowman, P.: Comparison between mice chimaeric and heterozygous for the X-linked gene *tabby*. *Nature New Biol.* **241:** 180–183, 1973.

Meisler, M.H.: Synthesis and secretion of kidney β-galactosidase in mutant *le/le* mice. *J. Biol. Chem.* **253:** 3129–3134, 1978.

Mekori, T., and Phillips, R.A.: The immune response in mice of genotypes W/W^v and Sl/Sl^d. *Proc. Soc. Exp. Biol. Med.* **132:** 115–119, 1969.

Melvold, R.W.: Spontaneous somatic reversion in mice. Effects of parental genotype on stability at the *p*-locus. *Mutation Res.* **12:** 171–174, 1971.

Melvold, R.W.: Private communication. *Mouse News Letter* **46:** 32, 1972.

Melvold, R.W.: The effects of mutant *p*-alleles on the reproductive system in mice. *Genet. Res. (Camb.)* **23:** 319–325, 1974.

Menkes, J.H., Alter, M., Steigleder, G.K., Weakley, D.R., and Sung, J.H.: A sex-linked, recessive disorder with retardation of growth, peculiar hair, and focal cerebral and cellular degeneration. *Pediatrics* **29:** 764–779, 1962.

Milhaud, G., Labat, M., Parant, M., Damais, C., and Chedid, L.: Immunological defect and its correction in the osteopetrotic mutant rat. *Proc. Nat. Acad. Sci. USA* **74:** 339–342, 1977.

Miller, D.A., Dev, V.G., Tantravahi, R., Miller, O.J., Schiffman, M.B., Yates, R.A., and Gluecksohn-Waelsch, S.: Cytological detection of the C^{25H} deletion involving the albino (*c*) locus on chromosome 7 in the mouse. *Genetics* **78:** 905–910, 1974.

Miller, D.S.: Coat color and behavior mutations in inbred mice under chronic low-level γ-irradiation. *Radiat. Res.* **19:** 184–185, 1963 (Abstract).

Miller, D.S., and Potas, M.Z.: Cordovan, a new allele of black and brown color in the mouse. *J. Hered.* **46:** 293–296, 1955.

Miner, G.: Private communication. *Mouse News Letter* **38:** 25, 1968.

Mintz, B.: Embryological development of primordial germ-cells in the mouse: Influence of a new mutation, W^j. *J. Embryol. Exp. Morphol.* **5:** 396–403, 1957a.

Mintz, B.: Interaction between two allelic series modifying primordial germ cell development in the mouse embryo. *Anat. Rec.* **128:** 591, 1957b (Abstract).

Mintz, B.: Embryological phases of mammalian gametogenesis. *J. Cell Comp. Physiol.* **56** (suppl. 1): 31–48, 1960.

Mintz, B.: Gene control of mammalian pigmentary differentiation, I. Clonal origin of melanocytes. *Proc. Nat. Acad. Sci. USA* **58:** 344–351, 1967.

Mintz, B.: Developmental mechanisms found in allophenic mice with sex chromosomal and pigmentary mosaicism. In *Birth Defects: Orig. Art. Ser.* 5, *First Conf. Clin. Delineation Birth Defects*, D. Bergsma and V. McKusick (eds.), pp. 11–22, National Found., New York, 1969a.

Mintz, B.: Gene control of the mouse pigmentary system. *Genetics* **61:** s41, 1969b (Abstract).

Mintz, B.: Gene expression in allophenic mice. In *Control Mech. Expression Cellular Phenotypes*, *Symp. Int. Soc. Cell Biol*, H. Padylkula (ed.), pp. 15–42, Academic Press, New York, 1970.

Mintz, B.: Clonal basis of mammalian differentiation. In: *Control Mechanisms of Growth and Differentiation. 25th Symp. Soc. Exp. Biol.*, D.D. Davies and M. Balls (eds.), pp. 345–369, University Press, Cambridge, 1971a.

Mintz, B.: Allophenic mice of multi-embryo origin. In *Methods in Mammalian Embryology*, J. Daniel, Jr. (ed.), pp. 186–214, Freeman, San Francisco, 1971b.

Mintz. B.: Genetic mosaicism *in vivo:* Development and disease in allophenic mice. *Fed. Proc.* **30:** 935–943, 1971c.

Mintz, B.: Gene control of mammalian differentiation. *Ann. Rev. Genet.* **8:** 411–470, 1974.

Mintz, B., and Cronmiller, C.: Normal blood cells of anemic genotype in teratocarcinoma-derived mosaic mice. *Proc. Nat. Acad. Sci. USA* **75:** 6247–6251, 1978.

Mintz, B., and Palm, J.: Gene control of hematopoiesis. I. Erythrocyte mosaicism and permanent immunological tolerance in allophenic mice. *J. Exp. Med.* **129:** 1013–1027, 1969.

Mintz, B., and Russell, E.S.: Gene-induced embryological modifications of primordial germ cells in the mouse. *J. Exp. Zool.* **134:** 207–237, 1957.

Mintz, B., and Silvers, W.K.: Histocompatibility antigens on melanoblasts and hair follicle cells. *Transplantation* **9:** 497–505, 1970.

Misuraca, G., Nicolaus, R.A., Prota, G., and Ghiara, G.: A cytochemical study of phaeomelanin formation in feather papillae of New Hampshire chick embryos. *Experientia* **25:** 920–922, 1969.

Mixter, R., and Hunt, H.R.: Anemia in the flexed tailed mouse, *Mus musculus*. *Genetics* **18:** 367–387, 1933.

Mohr, E.: Akromelanismus bei *Mus musculus*. *Zool Anz.* **126:** 45–46, 1939.

Moyer, F.H.: Electron microscope studies on the origin, development, and genetic control of melanin granules in the mouse eye. In *The Structure of the Eye*, C.K. Smelser (ed.), pp. 469–486, Academic Press, New York, 1961.

Moyer, F.H.: Genetic effects on melanosome fine structure and ontogeny in normal and malignant cells. *Ann. N.Y. Acad. Sci.* **100:** 584–606, 1963.

Moyer, F.H.: Genetic variations in the fine structure and ontogeny of mouse melanin granules. *Amer. Zool.* **6:** 43–66, 1966.

Mullen, R.J., and Whitten, W.K.: Relationship of genotype and degree of chimerism in coat color to sex ratios and gametogenesis in chimeric mice. *J. Exp. Zool.* **178:** 165–176, 1971.

Müller, G: Eine entwicklungsgeschichtliche Untersuchung über das erbliche Kolobom mit Mikrophthalmus bei der Hausmaus. *Z. Mikr.-anta. Forsch.* **56:** 520–558, 1950.

Müller, G: Die embryonale Entwicklung eines sich rezessiv vererbenden Merkmals. (Kolobom bei der Hausmaus). *Wiss. Z. Martin-Luther-Univ.* **1:** 27–43, 1951.

Munford, R.E.: Private communication. *Mouse News Letter* **33:** 52, 1965.

Murphy, E.D.: Steel alleles and leukemia. *In 40th Annual Report, The Jackson Laboratory*, 1968–1969, pp. 29–30, 1969.

Murphy, E.D., and Russell, E.S.: Ovarian tumorigenesis following deletion of germ cells in hybrid mice. *Acta Unio Int. Contra Cancrum* **19:** 779–782, 1963.

Murphy, E.D., Harrison, D.E., and Roths, J.B.: Giant granules of beige mice: A quantitative marker for granulocytes in bone marrow transplantation. *Transplantation* **15:** 526–530, 1973.

Murphy, H.M.: Calcium and phosphorus metabolism in the grey-lethal mouse. *Genet. Res. (Camb.)* **11:** 7–14, 1968.

Murphy, H.M.: A review of inherited osteopetrosis in the mouse. *Clin. Orthopaed. Related Res.* **65:** 97–109, 1969.

Murphy, H.M.: Calcitonin-like activity in the circulation of osteopetrotic grey-lethal mice. *J. Endocrinol.* **53:** 139–150, 1972.

Murphy, H.M.: The osteopetrotic syndrome in the microphthalmic mutant mouse. *Calc. Tissue. Res.* **13:** 19–26, 1973.

Murray, J.M.: A preliminary note on the occurrence of a color mutation in the house mouse (*Mus musculus*). *Science* **73:** 482, 1931.

Murray, J.M.: "Leaden", a recent color mutation in the house mouse. *Amer. Nat.* **67:** 278–283, 1933.

Murray, J.M., and Green, C.V.: Inheritance of ventral spotting in mice. *Genetics* **18:** 481–486, 1933.

Murray, J.M., and Snell, G.D.: Belted, a new sixth chromosome mutation in the mouse. *J. Hered.* **36:** 266–268, 1945.

Mystkowska, E.T., and Tarkowski, A.K.: Observations on CBA-*p*/CBA-T6T6 mouse chimeras. *J. Embryol. Exp. Morphol.* **20:** 33–52, 1968.

Nachmias, V.T.: Tryptophan oxidation by yellow mouse skin. *Proc. Soc. Exp. Biol. Med.* **101:** 247–250, 1959.

Nash, D.J.: Private communication. *Mouse News Letter* **29:** 84, 1963.

Newsome, A.E.: A population study of house-mice temporarily inhabiting a South Australian wheat field. *J. Anim. Ecol.* **38:** 341–359, 1969a.

Newsome, A.E.: A population study of house-mice permanently inhabiting a reed-bed in South Australia. *J. Anim. Ecol.* **38:** 361–377, 1969b.

Nicolaus, R.A., and Piattelli, M.: Progress in the chemistry of natural black pigments. *Rendiconti Accad. Sci. Fisiche Matemat.* (*Naples*) **32:** 1–17, 1965.

Nichols, S.E., and Reams, W.M., Jr.: The occurrence and morphogenesis of melanocytes in the connective tissues of the PET/MCV mouse strain. *J. Embryol. Exp. Morphol.* **8:** 24–32, 1960.

Niece, R.L., McFarland, E.C., and Russell, E.S.: Erythroid homeostasis in normal and genetically anemic mice: Reaction to induced polycythemia. *Science* **142:** 1468–1469, 1963.

Nisbet, N.W., Menage, J., and Loutit, J.F.: Host-donor cellular interactions in the treatment of experimental osteopetrosis. *Nature* (*London*) **271:** 464–466, 1978.

Novak, E.K., and Swank, R.T.: Lysosomal dysfunctions associated with mutations at mouse pigment genes. *Genetics*, in press.

Ohno, S., and Cattanach, B.M.: Cytological study of an X-autosome translocation in *Mus musculus*. *Cytogenetics* **1:** 129–140, 1962.

Okun, M.R.: Mast cells and melanocytes. *Int. J. Dermatol.* **15:** 711–722, 1976.

Oliver, C., and Essner, E.: Distribution of anomalous lysosomes in the beige mouse: A homologue of Chediak–Higashi syndrome. *J. Histochem. Cytochem.* **21:** 218–228, 1973.

Oliver, C., and Essner, E.: Formation of anomalous lysosomes in monocytes, neutrophils, and eosinophils from bone marrow of mice with Chediak–Higashi syndrome. *Lab. Invest.* **32:** 17–27, 1975.

Oliver, C., Essner, E., Zimring, A., and Haimes, H.: Age-related accumulation of ceroid-like pigment in mice with Chediak–Higashi syndrome. *Amer. J. Pathol.* **84:** 225–238, 1976.

Oliver, J.M.: Impaired microtubule function correctable by cyclic GMP and cholinergic agonists in the Chediak–Higashi syndrome. *Amer. J. Pathol.* **85:** 395–418, 1976.

Oliver, J.M., Krawiec, J.A., and Berlin, R.D.: Carbamylcholine prevents giant granule formation in cultured fibroblasts from beige (Chediak–Higashi) mice. *J. Cell Biol.* **69:** 205–210, 1976.

Onslow, H.: A contribution to our knowledge of the chemistry of coat-colour in animals and of dominant and recessive whiteness. *Proc. Roy. Soc.* (*London*) **B 89:** 36–58, 1915.

Owens, J.: Private communication. *Mouse News Letter* **47:** 61, 1972.

Padgett, G.A.: The Chediak–Higashi syndrome. *Adv. Vet. Sci.* **12:** 239–284, 1968.

Padgett, G.A., Leader, R.W., Gorham, J.R., and O'Mary, C.C.: The familial occurrence of the Chediak–Higashi syndrome in mink and cattle. *Genetics* **49:** 505–512, 1964.

Padgett, G.A., Holland, J.M., Prieur, D.J., Davis, W.C., and Gorham, J.R.: The Chediak–Higashi syndrome. A review of the disease in man, mink, cattle and mice. In *Animal Models For Biomedical Research*, pp. 1–12, Nat. Acad. Sci. Printing and Publishing Office, Washington, D.C., 1970.

Papaioannou, V., and Mardon, H.: Private communication. *Mouse News Letter* **58:** 66, 1978.

Parsons, P.A.: A balanced four-point linkage experiment for linkage group XIII of the house mouse. *Heredity* **12:** 77–95, 1958.

Patt, H.M., and Maloney, M.A.: The bg^J/bg^J:W/W^v bone marrow chimera. *Blood Cells* **4:** 27–35, 1978.

Pedersen, R.A.: Development of lethal yellow (A^y/A^y) mouse embryos *in vitro*. *J. Exp. Zool.* **188:** 307–320, 1974.

Pedersen, R.A., and Spindle, A.I.: Genetic effects of mammalian development during and after implantation. In *Embryogenesis in Mammals. Ciba Foundation Symposium 40,* I. Elliott and M. O'Conner (eds.), pp. 133–154, Elsevier-North Holland, Amsterdam, 1976.

Phillips, R.J.S.: Private communication. *Mouse News Letter* **19:** 22, 1958.

Phillips, R.J.S.: Private communication. *Mouse News Letter* **21:** 39, 1959.

Phillips, R.J.S.: Private communication. *Mouse News Letter* **22:** 30, 1960.

Phillips, R.J.S.: A comparison of mutation induced by acute x and chronic gamma irradiation in mice. *Brit. J. Radiol.* **34:** 261–264, 1961a.

Phillips, R.J.S.: "Dappled", a new allele at the mottled locus in the house mouse. *Genet. Res.* (*Camb.*) **2:** 290–295, 1961b.

Phillips, R.J.S.: Private communication. *Mouse News Letter* **27:** 34, 1962.

Phillips, R.J.S.: Private communication. *Mouse News Letter* **29:** 38, 1963.

Phillips, R.J.S.: Private communication. *Mouse News Letter* **32:** 39, 1965.

Phillips, R.J.S.: A *cis-trans* position effect at the *A* locus of the house mouse. *Genetics* **54:** 485–495, 1966a.

Phillips, R.J.S.: Private communication. *Mouse News Letter* **34:** 27, 1966b.

Phillips, R.J.S.: Private communication. *Mouse News Letter* **39:** 25, 1968.

Phillips, R.J.S.: Private communication. *Mouse News Letter* **42:** 26, 1970a.

Phillips, R.J.S.: Private communication. *Mouse News Letter* **42:** 27, 1970b.

Phillips, R.J.S.: Private communication. *Mouse News Letter* **45:** 25, 1971.

Phillips, R.J.S.: Private communication. *Mouse News Letter* **48:** 30, 1973.

Phillips, R.J.S.: Private communication. *Mouse News Letter* **50:** 42, 1974.

Phillips, R.J.S.: Private communication. *Mouse News Letter* **53:** 29, 1975.

Phillips, R.J.S.: Private communication. *Mouse News Letter* **55:** 14, 1976.

Phillips, R.J.S.: Private communication. *Mouse News Letter* **56:** 38, 1977.

Phillips, R.J.S., and Hawker, S.G.: Private communication. *Mouse News Letter* **49:** 29, 1973.

Phillips, R.J.S., Hawker, S.G., and Moseley, H.J.: Private communication. *Mouse News Letter* **48:** 30–31, 1973.

Pierro, L.J.: Effects of the *light* mutation of mouse coat color on eye pigmentation. *J. Exp. Zool.* **153:** 81–87, 1963a.

Pierro, L.J.: Pigment granule formation in slate, a coat color mutant in the mouse. *Anat. Rec.* **146:** 365–371, 1963b.

Pierro, L.J.: Private communication. *Mouse News Letter* **32:** 82, 1965.

Pierro, L.J., and Chase, H.B.: Slate—a new coat color mutant in the mouse. *J. Hered.* **54:** 46–50, 1963.

Pierro, L.J., and Chase, H.B.: Temporary hair loss associated with the slate mutation of coat colour in the mouse. *Nature* (*London*) **205:** 579–580, 1965.

Pierro, L.J., and Spiggle, J.: Gene interaction and hair pigmentation in the mouse. *J. Invest. Dermatol.* **54:** 95, 1970 (Abstract).

Pincus, G.: A spontaneous mutation in the house mouse. *Proc. Nat. Acad. Sci. USA* **15:** 85–88, 1929.

Pincus, G.: A modifier of piebald spotting in mice. *Amer. Nat.* **65:** 283–286, 1931.

Pizarro, O.: Private communication. *Mouse News Letter* **17:** 95, 1957.

Pomerantz, S.H., and Chuang, L.: Effects of β-MSH, cortisol and ACTH on tyrosinase in the skin of newborn hamsters and mice. *Endocrinology* **87:** 302–310, 1970.

Pomerantz, S.H., and Li, J. P-C.: Tyrosinase in the skin of albino hamsters and mice. *Nature* (*London*) **252:** 241–243, 1974.

Poole, T.W.: Dermal–epidermal interactions and the site of action of the yellow (A^y) and nonagouti (a) coat color genes in the mouse. *Develop. Biol.* **36:** 208–211, 1974.

Poole, T.W.: Dermal–epidermal interactions and the action of alleles at the agouti locus in the mouse. *Develop. Biol.* **42:** 203–210, 1975.

Poole, T.W., and Silvers, W.K.: The development of regional pigmentation patterns in black and tan (a^t) mice. *J. Exp. Zool.* **197:** 115–120, 1976a.

Poole, T.W., and Silvers, W.K.: An experimental analysis of the recessive yellow coat color mutant in the mouse. *Develop. Biol.* **48:** 377–381, 1976b.

Potten, C.S.: Radiation depigmentation of mouse hair: A study of follicular melanocyte populations. *Cell Tissue Kinet.* **1:** 239–254, 1968.

Prieur, D.J., and Collier, L.L.: Animal model of human disease. Chèdiak–Higashi syndrome. *Amer. J. Pathol.* **90:** 533–536, 1978.

Prieur, D.J., Davis, W.C., and Padgett, G.A.: Defective function of renal lysosomes in mice with the Chèdiak–Higashi syndrome. *Amer. J. Pathol.* **67:** 227–235, 1972.

Prota, G.: Structure and biogenesis of phaeomelanins. *J. Invest. Dermatol.* **54:** 95, 1970 (Abstract).

Prota, G.: Structure and biogenesis of phaeomelanins. In *Pigmentation: Its Genesis and Biologic Control*, V. Riley (ed.), pp. 615–630, Appleton-Century-Crofts, New York, 1972.

Prota, G., and Nicolaus, R.A.: On the biogenesis of phaeomelanins. In *The Pigmentary System, Advances in Biology of Skin, Vol. 8,* W. Montagna and F. Hu (eds.), pp. 323–328, Pergamon Press, Oxford, 1967.

Prota, G., and Thomson, R.H.: Melanin pigmentation in mammals. *Endeavor* **35:** 32–38, 1976.

Pullig, T.: A new recessive spotting gene. *J. Hered.* **40:** 229–230, 1949.

Quevedo, W.C., Jr.: Effect of biotin deficiency on follicular melanocytes of mice. *Proc. Soc. Exp. Biol. Med.* **93:** 260–263, 1956.

Quevedo, W.C., Jr.: Loss of clear cells in the hair follicles of x-irradiated albino mice. *Anat. Rec.* **127:** 725–734, 1957.

Quevedo, W.C., Jr.: Genetic regulation of melanocyte responses to U.V. In *Recent Progress in Photobiology*, E.J. Bowen (ed.), pp. 383–386, Blackwell, Oxford, 1965.

Quevedo, W.C., Jr.: The control of color in mammals. *Amer. Zool.* **9:** 531–540, 1969a.

Quevedo, W.C., Jr.: Genetics of mammalian pigmentation. In *The Biologic Effects of Ultraviolet Radiation*, F. Urbach (ed.), pp. 315–324, Pergamon Press, Oxford, 1969b.

Quevedo, W.C., Jr.: Genetic regulation of pigmentation in mammals. In *Biology of Normal and Abnormal Melanocytes*, T. Kawamura, T.B. Fitzpatrick, and M. Seiji (eds.), pp. 99–115, Univ. of Tokyo Press, 1971.

Quevedo, W.C., Jr.: Epidermal melanin units: Melanocyte–keratinocyte interactions. *Amer. Zool.* **12:** 35–41, 1972.

Quevedo, W.C., Jr., and Chase, H.B.: An analysis of the light mutation of coat color in mice. *J. Morphol.* **102:** 329–346, 1958.

Quevedo, W.C., Jr., and McTague, C.F.: Genetic influences on the response of mouse melanocytes to ultraviolet light: The melanocyte system of hair-covered skin. *J. Exp. Zool.* **152:** 159–168, 1963.

Quevedo, W.C., Jr., and Smith, J.A.: Studies on radiation-induced tanning of skin. *Ann. N.Y. Acad. Sci.* **100:** 364–388, 1963.

Quevedo, W.C., Jr., and Smith, J.: Electron microscope observations on the postnatal "loss" of interfollicular epidermal melanocytes in mice. *J. Cell Biol.* **39:** 103a, 1968 (Abstract).

Quevedo, W.C., Jr., Youle, M.C., Rovee, D.T., and Bienieki, T.C.: The developmental fate of melanocytes in murine skin. In *Structure and Control of the Melanocyte*, G. Della Porta (ed.), pp. 228–241, Springer-Verlag, Berlin, 1966.

Quevedo, W.C., Jr., Bienieki, T.C., McMorris, F.A., and Hepinstall, M.J.: Environmental and genetic influences on radiation-induced tanning of murine skin. In *The Pigmentary System, Advances in Biology of Skin, Vol. 8,* W. Montagna and F. Hu (eds.), pp. 361–377, Pergamon Press, Oxford, 1967.

Race, R.R., and Sanger, R.: *Blood Groups in Man*, 4th ed., Blackwell, Oxford, 1962.

Radochońska, A.: Effect of the gene mosaic (*Ms*) on growth rate, weight of organs and hair structure in mouse. *Genet. Pol.* **11:** 257–274, 1970.

Raisz, L.G., Simmons, H.A., Gworek, S.C., and Eilon, G.: Studies on congenital osteopetrosis in microphthalmic mice using organ cultures: Impairment of bone resorption in response to physiologic stimulators. *J. Exp. Med.* **145:** 857–865, 1977.

Rauch, H., and Yost, M.T.: Phenylalanine metabolism in dilute-lethal mice. *Genetics* **48:** 1487–1495, 1963.

Rawles, M.E.: The development of melanophores from embryonic mouse tissues grown in the coelom of chick embryos. *Proc. Nat. Acad. Sci. USA* **26:** 673–680, 1940.

Rawles, M.E.: Origin of pigment cells from the neural crest in the mouse embryo. *Physiol. Zool.* **20:** 248–266, 1947.

Rawles, M.E.: Origin of melanophores and their role in development of color patterns in vertebrates. *Physiol. Rev.* **28:** 383–408, 1948.

Rawles, M.E.: Origin of the mammalian pigment cell and its role in the pigmentation of hair. In *Pigment Cell Growth*, M. Gordon (ed.), pp. 1–12, Academic Press, New York, 1953.

Reams, W.M., Jr.: Morphogenesis of pigment cells in the connective tissue of the PET mouse. *Ann. N.Y. Acad. Sci.* **100:** 486–495, 1963.

Reams, W.M., Jr.: Regulation of pigment cell colonization of PET mouse leg muscles. *Anat. Rec.* **155:** 89–96, 1966.

Reams, W.M., Jr.: Pigment cell population pressure within the skin and its role in the pigment cell invasion of extraepidermal tissues. In *The Pigmentary System, Advances in Biology of Skin, Vol. 8,* W. Montagna and F. Hu (eds.), pp. 489–501, Pergamon Press, Oxford, 1967.

Reams, W.M., Jr., and Schaeffer, M.A.: Effects of single dose X-irradiation on the pigment cell system of the PET mouse leg skin. *J. Invest. Dermatol.* **50:** 297–300, 1968.

Reams, W.M., Jr., Shervette, M.S., and Dorman, W.H.: Refractoriness of mouse dermal melanocytes to hormones. *J. Invest. Dermatol.* **50:** 338–339, 1968.

Reams, W.M., Jr., Salisbury, R.L., Earnhardt, J.T., and Howard, V.H.: Miniature melanocytes. A specific cell type in murine epidermis. In *Pigment Cell*, V. Riley (ed.), Vol. 3, pp. 220–227, Karger, Basel, 1976.

Reed, S.C.: Determination of hair pigments. III. Proof that expression of the black-and-tan gene is dependent upon tissue organization. *J. Exp. Zool.* **79:** 337–346, 1938.

Reed, S.C., and Henderson, J.M.: Pigment cell migration in mouse epidermis. *J. Exp. Zool.* **85:** 409–418, 1940.

Reed, S.C., and Sander, G.: Time of differentiation of hair pigments in the mouse. *Growth* **1:** 194–200, 1937.

Reynolds, J.: The epidermal melanocytes of mice. *J. Anat.* **88:** 45–58, 1954.

Rittenhouse, E.: Genetic effects on fine structure and development of pigment granules in mouse hair bulb melanocytes. I. The *b* and *d* loci. *Develop. Biol.* **17:** 351–365, 1968a.

Rittenhouse, E.: Genetic effects on fine structure and development of pigment granules in mouse hair bulb melanocytes. II. The *c* and *p* loci, and *ddpp* interaction. *Develop. Biol.* **17:** 366–381, 1968b.

Rittenhouse, E.: Effects of four radiation-induced lethal alleles at the albino locus on the fine structure of melanin granules in the mouse. *J. Invest. Dermatol.* **54:** 96, 1970 (Abstract).

Roberts, E.: A new mutation in the house mouse (*Mus musculus*). *Science* **74:** 569, 1931.

Roberts, E., and Quisenberry, J.H.: Linkage of the genes for nonyellow (*y*) and pink-eye (p_2) in the house mouse (*Mus musculus*). *Amer. Nat.* **69:** 181–183, 1935.

Roberts, R.C., Falconer, D.S., Bowman, P., and Gauld, I.K.: Growth regulation in chimeras between large and small mice. *Nature* (*London*) **260:** 244–245, 1976.

Robertson, G.G.: An analysis of the development of homozygous yellow mouse embryos. *J. Exp. Zool.* **89:** 197–231, 1942a.

Robertson, G.G.: Increased viability of homozygous yellow mouse embryos in new uterine environments. *Genetics* **27:** 166–167, 1942b (Abstract).

Robinson, R.: Mosaicism in mammals. *Genetica* **29:** 120–145, 1957.

Robinson, R.: Sable and umbrous mice. *Genetica* **29:** 319–326, 1959.

Roderick, T.H.: Private communication. *Mouse News Letter* **49:** 33, 1973.

Roderick, T.H.: Private communication. *Mouse News Letter* **52:** 38, 1975.

Roderick, T.H., Davisson, M.T., and Lane, P.W.: Private communication. *Mouse News Letter* **55:** 18, 1976.

Rothman, I.K., Zanjani, E.D., Gordon, A.S., and Silber, R.: Nucleoside deaminase: An enzymatic marker for stress erythropoiesis in the mouse. *J. Clin. Invest.* **49:** 2051–2067, 1970.

Rovee, D.T., and Reams, W.M., Jr.: An experimental and descriptive analysis of the melanocyte population in the venter of the PET mouse. *Anat. Rec.* **149:** 181–190, 1964.

Rowe, D.W., McGoodwin, E.B., Martin, G.R., Sussman, M.D., Grahn, D., Faris, B., and Franzblau, C.: A sex-linked defect in the cross-linking of collagen and elastin associated with the mottled locus in mice. *J. Exp. Med.* **139:** 180–192, 1974.

Rowe, D.W., McGoodwin, E.B., Martin, G.R., and Grahn, D.: Decreased lysyl oxidase activity in the aneurysm-prone, mottled mouse. *J. Biol. Chem.* **252:** 939–942, 1977.

Ruscetti, F.W., Boggs, D.R., Torok, B.J., and Boggs, S.S.: Reduced blood and marrow neutrophils and granulocytic colony-forming cells in *Sl/Sl^d^* mice. *Proc. Soc. Exp. Biol. Med.* **152:** 398–402, 1976.

Russell, E.S.: A quantitative histological study of the pigment found in the coat-color mutants of the house mouse. I. Variable attributes of the pigment granules. *Genetics* **31:** 327–346, 1946.

Russell, E.S.: A quantitative histological study of the pigment found in the coat-color mutants of the house mouse. II. Estimates of the total volume of pigment. *Genetics* **33:** 228–236, 1948.

Russell, E.S.: A quantitative histological study of the pigment found in the coat-color mutants of the house mouse. III. Interdependence among the variable granule attributes. *Genetics* **34:** 133–145, 1949a.

Russell, E.S.: A quantitative histological study of the pigment found in the coat-color mutants of the house mouse. IV. The nature of the effects of genic substitution in five major allelic series. *Genetics* **34:** 146–166, 1949b.

Russell, E.S.: Analysis of pleiotropism of the *W*-locus in the mouse: Relationship between the effects of *W* and W^v substitution on hair pigmentation and on erythrocytes. *Genetics* **34:** 708–723, 1949c.

Russell, E.S.: Review of the pleiotropic effects of *W*-series genes on growth and differentiation. In *Aspects of Synthesis and Order of Growth, 13th Symp. Soc. Study Develop. and Growth*, D. Rudnick (ed.), pp. 113–126, Princeton Univ. Press, New Jersey, 1954.

Russell, E.S.: Abnormalities of erythropoiesis associated with mutant genes in mice. In *Regulation of Hematopoiesis*, A.S. Gordon (ed.), pp. 649–675, Appleton, New York, 1970.

Russell, E.S.: Private communication, *Mouse News Letter* **50:** 43, 1974.

Russell, E.S., and Bernstein, S.E.: Blood and blood formation. In *Biology of the Laboratory Mouse*, 2nd ed., E.L. Green (ed.), pp. 351–372, McGraw-Hill, New York, 1966.

Russell, E.S., and Bernstein, S.E.: Influence of non-H-2 genetic factors on success in implantation therapy of *W* series hereditary anemias of the mouse. *Transplantation* **5:** 142–153, 1967.

Russell, E.S., and Bernstein, S.E.: Proof of whole-cell implant therapy of *W*-series anemia. *Arch. Biochem. Biophys.* **125:** 594–597, 1968.

Russell, E.S., and Bernstein, S.E.: Private communication. *Mouse News Letter* **50:** 44, 1974.

Russell, E.S., and Fekete, E.: Analysis of *W*-series pleiotropism in the mouse: Effect of W^vW^v substitution on definitive germ cells and on ovarian tumorigenesis. *J. Nat. Cancer Inst.* **21:** 365–381, 1958.

Russell, E.S., and Fondal, E.L.: Quantitative analysis of the normal and four alternative degrees of an inherited macrocytic anemia in the house mouse. I. Number and size of erythrocytes. *Blood* **6:** 892–905, 1951.

Russell, E.S., and Keighley, G.: The relation between erythropoiesis and plasma erythropoietin levels in normal and genetically anaemic mice during prolonged hypoxia or after whole-body irradiation. *Brit. J. Haematol.* **22:** 437–452, 1972.

Russell, E.S., and Lawson, F.A.: Selection and inbreeding for longevity of a lethal type. *J. Hered.* **50:** 19–25, 1959.

Russell, E.S., and McFarland, E.C.: Erythrocyte populations in fetal mice with and without two hereditary anemias. *Fed. Proc.* **24:** 240, 1965 (Abstract).

Russell, E.S., and McFarland, E.C.: Analysis of pleiotropic effects of *W* and *f* genic substitutions in the mouse. *Genetics* **53:** 949–959, 1966.

Russell, E.S., Coulombre, J.L., and Fekete, E.: Contributions of studies on gonad development and function and of hair spotting pattern to analysis of the *W*-series pleiotropism. *Genetics* **37:** 621, 1952 (Abstract).

Russell, E.S., Snow, C.M., Murray, L.M., and Cormier, J.P.: The bone marrow in inherited macrocytic anemia in the house mouse. *Acta Haematol.* **10:** 247–259, 1953.

Russell, E.S., Murray, L.M., Small, E.M., and Silvers, W.K.: Development of embryonic mouse gonads transferred to the spleen: Effects of transplantation combined with genotypic autonomy. *J. Embryol. Exp. Morphol.* **4:** 347–357, 1965a.

Russell, E.S., Smith, L.J., and Lawson, F.A.: Implantation of normal blood forming tissue in radiated genetically anemic hosts. *Science* **124:** 1076–1077, 1956b.

Russell, E.S., Lawson, F., and Schabtach, G.: Evidence for a new allele at the *W*-locus of the mouse. *J. Hered.* **48:** 119–123, 1957.

Russell, E.S., Bernstein, S.E., Lawson, F.A., and Smith, L.J.: Long-continued function of normal blood-forming tissue transplanted into genetically anemic hosts. *J. Nat. Cancer Inst.* **23:** 557–566, 1959.

Russell, E.S., Bernstein, S.E., McFarland, E.C., and Modeen, W.R.: The cellular basis of differential radiosensitivity of normal and genetically anemic mice. *Radiat. Res.* **20:** 677–694, 1963.

Russell, E.S., Thompson, M.W., and McFarland, E.C.: Analysis of effects of *W* and *f* genic substitutions on fetal mouse haematology. *Genetics* **58:** 259–270, 1968.

Russell, L.B.: Private communication. *Mouse News Letter* **23:** 58–59, 1960.

Russell, L.B.: Genetics of mammalian sex chromosomes. *Science* **133:** 1795–1803, 1961.

Russell, L.B.: Genetic and functional mosaicism in the mouse. In *Role of Chromosomes in Development, 23rd Symp. Soc. Study of Develop. and Growth*, M. Locke (ed.), pp. 153–181, Academic Press, New York, 1964.

Russell, L.B.: Private communication. *Mouse News Letter* **32:** 68–69, 1965.

Russell, L.B.: Definition of functional units in a small chromosomal segment of the mouse and its use in interpreting the nature of radiation-induced mutations. *Mutat. Res.* **11:** 107–123, 1971.

Russell, L.B.: Validation of the *in vivo* somatic mutation method in the mouse as a prescreen of germinal point mutations. *Arch. Toxicol* **38:** 75–85, 1977.

Russell, L.B.: Analysis of the albino-locus region of the mouse. II. Mosaic mutants. *Genetics* **91:** 141–147, 1979.

Russell, L.B., and Maddux, S.C.: Private communication. *Mouse News Letter* **30:** 48, 1964.

Russell, L.B., and Major, M.H.: A high rate of somatic reversion in the mouse. *Genetics* **41:** 658, 1956 (Abstract).

Russell, L.B., and Major, M.H.: Radiation-induced presumed somatic mutations in the house mouse. *Genetics* **42:** 161–175, 1957.

Russell, L.B., and Raymer, G.D.: Analysis of the albino-locus region of the mouse. III. Embryological analysis of lethals. *Genetics*, in press.

Russell, L.B., and Russell, W.L.: A study of the physiological genetics of coat color in the mouse by means of the dopa reaction in frozen sections of skin. *Genetics* **33:** 237–262, 1948.

Russell, L.B., and Saylors, C.L.: Induction of paternal sex chromosome losses by irradiation of mouse spermatozoa. *Genetics* **47:** 7–10, 1962.

Russell, L.B., and Woodiel, F.N.: A spontaneous mouse chimera formed from separate fertilization of two meiotic products of oogenesis. *Cytogenetics* **5:** 106–119, 1966.

Russell, L.B., McDaniel, M.N.C., and Woodiel, F.N.: Crossing-over within a "locus" of the mouse. *Genetics* **48:** 907, 1963 (Abstract).

Russell, L.B., Russell, W.L., and Kelly, E.M.: Analysis of the albino-locus region of the mouse. I. Origin and viability. *Genetics* **91:** 127–139, 1979.

Russell, W.L.: Splotch, a new mutation in the house mouse *Mus musculus*. *Genetics* **32:** 102, 1947 (Abstract).

Russell, W.L.: The effect of radiation dose rate and fractionation on mutation in mice. In *Repair from Genetic Radiation Damage*, F.H. Sobels (ed.), pp. 205–217, Pergamon Press, London, 1963.

Russell, W.L.: Evidence from mice concerning the nature of the mutation process. *Proc. 11th Int. Congr. Genet.* **2:** 257–264, 1965.

Russell, W.L., and Gower, J.S.: Offspring from transplanted ovaries of fetal mice homozygous for a lethal gene (*Sp*) that kills before birth. *Genetics* **35:** 133–134, 1950 (Abstract).

Russell, W.L., and Russell, E.S.: Investigation of the sterility-producing action of the W^v gene in the mouse by means of ovarian transplantation. *Genetics* **33:** 122–123, 1948 (Abstract).

Russell, W.L., Russell, E.S., and Brauch, L.R.: Problems in the biochemistry and physiological genetics of pigmentation in mammals. *Spec. Publ. N.Y. Acad. Sci.* **4:** 447–453, 1948.

Sakurai, T., Ochiai, H., and Takeuchi, T.: Ultrastructural change of melanosomes associated with agouti pattern formation in mouse hair. *Develop. Biol.* **47:** 466–471, 1975.

Sarvella, P.A.: Pearl, a new spontaneous coat and eye color mutation in the house mouse. *J. Hered.* **45:** 19–20, 1954.

Sarvella, P.A., and Russell, L.B.: Steel, a new dominant gene in the house mouse. *J. Hered.* **47:** 123–128, 1956.

Schaible, R.H.: Private communication. *Mouse News Letter* **23:** 31, 1960.

Schaible, R.H.: Private communication. *Mouse News Letter* **24:** 38, 1961.

Schaible, R.H.: Development of piebald spotting patterns in the mouse and chick. *Amer. Zool.* **2:** 444, 1962 (Abstract).

Schaible, R.H.: Developmental genetics of spotting patterns in the mouse. Ph.D. Thesis, Iowa State University, Ames, 1963a.

Schaible, R.H.: Private communication. *Mouse News Letter* **28:** 39–40, 1963b.

Schaible, R.H.: Private communication. *Mouse News Letter* **29:** 48, 1963c.

Schaible, R.H.: Clonal distribution of melanocytes in piebald-spotted and variegated mice. *J. Exp. Zool.* **172:** 181–199, 1969.

Schaible, R.H.: Comparative effects of piebald-spotting genes on clones of melanocytes in different vertebrate species. In *Pigmentation Its Genesis and Biologic Control*, V. Riley (ed.), pp. 343–357, Appleton-Century-Crofts, New York, 1972.

Schaible, R.H., and Gowen, J.W.: Delimitation of coat pigment areas in mosaic and piebald mice. *Genetics* **45:** 1010, 1960 (Abstract).

Schlager, G., and Dickie, M.M.: Spontaneous mutation rates at five coat-color loci in mice. *Science* **151:** 205–206, 1966.

Schlager, G., and Dickie, M.M.: Spontaneous mutations and mutation rates in the house mouse. *Genetics* **57:** 319–330, 1967.

Schlager, G., and Dickie, M.M.: Natural mutation rates in the house mouse. Estimates for five specific loci and dominant mutations. *Mutat. Res.* **11:** 89–96, 1971.

Schneider, L.C., Hollinshead, M.B., and Lizzack, L.S.: Tooth eruption induced in grey lethal mice using parathyroid hormone. *Arch. Oral Biol.* **17:** 591–594, 1972.

Schultz, W.: Schwarzfärbung weisser Haare durch Rasier und die Entwicklungsmechanik der Farben von Haaren und Federn. I. *Abh. Arch. f. Entw-mech.* **41:** 535–557, 1915.

Schumann, H.: Die entstehung der Scheckung bei der Mausen mit weisser Blesse. *Develop. Biol.* **2:** 501–515, 1960.

Schwarz, E., and Schwarz, H.K.: The wild and commensal stocks of the house mouse *Mus musculus* Linnaeus. *J. Mammal.* **24:** 59–72, 1943.

Searle, A.G.: A lethal allele of dilute in the house mouse. *Heredity* **6:** 395–401, 1952.

Searle, A.G.: Private communication. *Mouse News Letter* **25:** 33, 1961.

Searle, A.G.: Private communication. *Mouse News Letter* **34:** 28, 1966.

Searle, A.G.: *Comparative Genetics of Coat Colour in Mammals*, Logos Press, London, 1968a.

Searle, A.G.: An extension series in the mouse. *J. Hered.* **59:** 341–342, 1968b.

Searle, A.G.: Private communication. *Mouse News Letter* **38:** 22, 1968c.,

Searle, A.G.: The use of pigment loci for detecting reverse mutations in somatic cells of mice. *Arch. Toxicol.* **38:** 105–108, 1977.

Searle, A.G., and Beechey, C.V.: Private communication. *Mouse News Letter* **42:** 28, 1970.

Searle, A.G., and Beechey, C.V.: Private communication. *Mouse News Letter* **50:** 39–40, 1974.

Searle, A.G., and Truslove, G.M.: A gene triplet in the mouse. *Genet. Res.* (*Camb.*) **15:** 227–235, 1970.

Searle, A.G., Beechey, C.V. and Palmer, A.: Private communication. *Mouse News Letter* **51:** 20, 1974.

Seiji, M., Fitzpatrick, T.B., and Birbeck, M.S.C.: The melanosome: A distinctive subcellular particle of mammalian tyrosinase and the site of melanogenesis. *J. Invest. Dermatol.* **36:** 243–252, 1961.

Seller, M.J.: Donor haemoglobin in anaemic mice of the *W*-series transplanted with haematopoietic tissue from an unrelated donor. *Nature* (*London*) **212:** 81–82, 1966.

Seller, M.J.: Erythrocyte chimerism after injection of spleen cells into anemic mice of the *W*-series. *Science* **155:** 90–91, 1967.

Seller, M.J.: Transplantation of anemic mice of the *W*-series with haemopoietic tissue bearing marker chromosomes. *Nature* (*London*) **220:** 300–301, 1968.

Seller, M.J.: The establishment of tolerance in adult anaemic mice of the W-series treated with anti-lymphocyte serum and allogeneic haemopoietic cells. *Clin. Exp. Immunol.* **6:** 639–643, 1970.

Seller, M.J.: The presence of donor-type immunoglobulins in anaemic mice of the *W*-series transplanted with allogeneic foetal liver cells. *Immunology* **24:** 249–252, 1973.

Seller, M.J., and Polani, P.E.: Experimental chimerism in a genetic defect in the house mouse *Mus musculus*. *Nature* (*London*) **212:** 80–81, 1966.

Seller, M.J., and Polani, P.E.: Transplantation of allogeneic haemopoietic tissue in adult anaemic mice of the *W* series using antilymphocyte serum. *Lancet* **i:** 18–21, 1969.

Shaklai, M., and Tavassoli, M.: Ultrastructural analysis of haemopoiesis in W/W^v anaemic mice. *Brit. J. Haematol.* **40:** 383–388, 1978.

Sheridan, W.: The dominant effects of a recessive lethal in the mouse. *Mutat. Res.* **5:** 323–328, 1968.

Sidman, R.L., and Pearlstein, R.: Pink-eyed dilution (*p*) gene in rodents: Increased pigmentation in tissue culture. *Develop. Biol.* **12:** 93–116, 1965.

Sidman, R.L., Green, M.C., and Appel, S.H.: *Catalog of the Neurological Mutants of the Mouse*, Harvard Univ. Press, Cambridge, 1965.

Silberberg, R.: Epiphyseal growth and osteoarthrosis in blotchy mice. *Exp. Cell. Biol.* **45:** 1–8, 1977.

Silver, A.F., Fleischmann, R.D., and Chase, H.B.: The fine structure of the melanocytes of the adult mouse hair follicle during their amelanotic phase (telogen and early anagen). *Amer. J. Anat.* **150:** 653–658, 1977.

Silvers, W.K.: Pigment cells: Occurrence in hair follicles. *J. Morphol.* **99:** 41–55, 1956.

Silvers, W.K.: Melanoblast differentiation secured from different mouse genotypes after transplantation to adult mouse spleen or to chick embryo coelom. *J. Exp. Zool.* **135:** 221–238, 1957.

Silvers, W.K.: An experimental approach to action of genes at the agouti locus in the mouse. II. Transplants of newborn *aa* ventral skin to a^ta, A^wa and *aa* hosts. *J. Exp. Zool.* **137:** 181–188, 1958a.

Silvers, W.K.: An experimental approach to action of genes at the agouti locus in the mouse. III. Transplants of newborn A^w—,*A*—, a^t-skin to A^y—, A^w—, *A*— and *aa* hosts. *J. Exp. Zool.* **137:** 189–196, 1958b.

Silvers, W.K.: Origin and identity of clear cells found in hair bulbs of albino mice. *Anat. Rec.* **130:** 135–144, 1958c.

Silvers, W.K.: Genes and the pigment cells of mammals. *Science* **134:** 368–373, 1961.

Silvers, W.K.: Tactics in pigment-cell research. In *Methodology in Mammalian Genetics*, W.J. Burdette (ed.), pp. 323–343, Holden-Day, San Francisco, 1963.

Silvers, W.K.: Agouti locus: Homology of its method of operation in rats and mice. *Science* **149:** 651–652, 1965.

Silvers, W.K., and Russell, E.S.: An experimental approach to action of genes at the agouti locus in the mouse. *J. Exp. Zool.* **130:** 199–220, 1955.

Snell, G.D.: Inheritance in the house mouse, the linkage relations of short ear, hairless, and naked. *Genetics* **16:** 42–74, 1931.

Snell, G.D.: Gene and chromosome mutations. In *Biology of the Laboratory Mouse*, G.D. Snell (ed.), pp. 234–247, Blakiston, Philadelphia, 1941.

Snell, G.D., Dickie, M.M., Smith, P., and Kelton, D.E.: Linkage of loop-tail, leaden, splotch and fuzzy in the mouse. *Heredity* **8:** 271–273, 1954.

Sô, M., and Imai, Y.: The types of spotting in mice and their genetic behaviour. *J. Genet.* **9:** 319–333, 1920.

Sô, M., and Imai, Y.: On the inheritance of ruby eye in mice. *Japan. J. Genet.* **4:** 1–9, 1926.

Southard, J.L.: Private communication. *Mouse News Letter* **51:** 23, 1974.

Southard, J.L., and Eicher, E.M.: Private communication. *Mouse News Letter* **56:** 40, 1977.

Southard, J.L., and Green, M.C.: Private communication. *Mouse News Letter* **45:** 29, 1971.

Steele, M.S.: Private communication. *Mouse News Letter* **50:** 52, 1974.

Stelzner, K.F.: Private communication. *Mouse News Letter* **31:** 40–41, 1964.

Stelzner, K.F.: Private communication. *Mouse News Letter* **34:** 41, 1966.

Stevens, W.L.: Accuracy of mutation rates. *J. Genet.* **43:** 301–307, 1942.

Straile, W.E.: A study of the hair follicle and its melanocytes. *Develop. Biol.* **10:** 45–70, 1964.

Strong, L.C., and Hollander, W.F.: Two non-allelic mutants resembling "*W*" in the house mouse. *J. Hered.* **44:** 41–44, 1953.

Styrna, J.: Survival of *Ms*/-males in two lines of mice selected for a different expression of the gene mosaic (*Ms*) in heterozygous females. *Genet. Pol.* **16:** 213–219, 1975.

Styrna, J.: Private communication. *Mouse News Letter* **55:** 19, 1976.

Styrna, J.: Private communication. *Mouse News Letter* **57:** 23, 1977a.

Styrna, J.: Analysis of causes of lethality in mice with the *Ms* (mosaic) gene. *Genet. Pol.* **18:** 61–79, 1977b.

Sutherland, D.J.A., Till, J.E., and McCulloch, E.A.: A kinetic study of the genetic control of hemopoietic progenitor cells assayed in culture and *in vivo*. *J. Cell. Physiol.* **75:** 267–274, 1970.

Swank, R.T., and Brandt, E.J.: Turnover of kidney β-glucuronidase in normal and Chediak-Higashi (beige) mice. *Amer. J. Pathol.* **92:** 755–772, 1978.

Sweet, H.O., and Lane, P.W.: Private communication. *Mouse News Letter* **57:** 19, 1977.

Sweet, S.E., and Quevedo, W.C., Jr.: Role of melanocyte morphology in pigmentation of mouse hair. *Anat. Rec.* **162:** 243–254, 1968.

Takeuchi, T: Regulatory function of the agouti locus in the mouse melanocyte. *J. Invest. Dermatol.* **54:** 98, 1970 (Abstract).

Taylor, R.F., and Farrell, R.K.: Light and electron microscopy of peripheral blood neutrophils in a killer whale affected with Chediak–Higashi syndrome. *Fed. Proc.* **32:** 822, 1973. (Abstract).

Theriault, L.L., and Hurley, L.S.: Ultrastructure of developing melanosomes in C57 black and pallid mice. *Develop. Biol.* **23:** 261–275, 1970.

Thiessen, D.D., Lindzey, G., and Owen, K.: Behavior and allelic variations in enzyme activity and coat color at the *c* locus of the mouse. *Behav. Genet.* **1:** 257–267, 1970.

Thompson, M.W., McCulloch, E.A., Siminovitch, L., and Till, J.E.: The cellular basis for the defect in haemopoiesis in flexed-tailed mice. I. The nature and persistence of the defect. *Brit. J. Haematol.* **12:** 152–160, 1966.

Thorndike, J., Trigg, M.J., Stockert, R., Gluecksohn-Waelsch, S., and Cori, C.F.: Multiple biochemical effects of a series of x-ray induced mutations at the albino locus in the mouse. *Biochem. Genet.* **9:** 25–39, 1973.

Till, J.E., Siminovitch, L., and McCulloch, E.A.: The effect of plethora on growth and differentiation of normal hemopoietic colony-forming cells transplanted in mice of genotype W/W^v. *Blood* **29:** 102–113, 1967.

Tost, M.: Cataracta hereditaria mit Mikrophthalmus beir der Hausmaus. *Z. Mensch. Vererb. Konst.* **34:** 593–600, 1958.

Travassoli, M., Ratzen, R.J., Maniatis, A., and Crosby, W.H.: Regeneration of hemopoietic stroma in anemic mice of Sl/Sl^d and W/W^v genotype. *J. Reticuloendothelial Soc.* **13:** 518–526, 1973.

Trigg, M.J., and Gluecksohn-Waelsch, S.: Ultrastructural basis of biochemical effects in a series of lethal alleles in the mouse. *J. Cell Biol.* **58:** 549–563, 1973.

Truslove, G.M.: A new allele at the patch locus in the mouse. *Genet. Res.* (*Camb.*) **29:** 183–186, 1977.

VanWoert, M.H., Nicholson, A.R., and Cotzias, G.C.: Functional similarities between the cytoplasmic organelles of melanocytes and the mitochondria of hepatocytes. *Nature* (*London*) **208:** 810–811, 1965.

Vassalli, J.D., Granelli-Piperno, A., Griscelli, G., and Reich, E.: Specific protease deficiency in polymorphonuclear leukocytes of Chediak–Higashi syndrome and beige mice. *J. Exp. Med.* **147:** 1285–1290, 1978.

Veneroni, G., and Bianchi, A.: Correcting the genetically determined sterility of W^vW^v male mice. *J. Embryol. Exp. Morphol.* **5:** 422–427, 1957.

Vlahakis, G., and Heston, W.E.: Increase of induced skin tumors in the mouse by the lethal yellow gene (A^y). *J. Nat. Cancer Inst.* **31:** 189–195, 1963.

von Lehmann, E.: Private communication. *Mouse News Letter* **48:** 23, 1973.

von Lehmann, E.: Private communication. *Mouse News Letter* **50:** 26–27, 1974.

Wabik, B.: Effect of homo- and heterozygous genetic background on the survival rate of male mice with lethal *mosaic* (*Ms*) gene. *Genet. Pol.* **12:** 545–556, 1971.

Walker, D.G.: Counteraction to parathyroid therapy in osteopetrotic mice as revealed in the plasma calcium level and ability to incorporate H^3-proline into bone. *Endocrinology* **79:** 836–842, 1966.

Walker, D.G.: Congenital osteopetrosis in mice cured by parabiotic union with normal siblings. *Endocrinology* **91:** 916–920, 1972.

Walker, D.G.: Osteopetrosis cured by temporary parabiosis. *Science* **180:** 875, 1973.

Walker, D.G.: Bone resorption restored in osteopetrotic mice by transplants of normal bone marrow and spleen cells. *Science* **190:** 784–785, 1975a.

Walker, D.G.: Spleen cells transmit osteopetrosis in mice. *Science* **190:** 785–787, 1975b.

Walker, D.G.: Control of bone resorption by hematopoietic tissue. The induction and reversal of congenital osteopetrosis in mice through use of bone marrow and splenic transplants. *J. Exp. Med.* **142:** 651–663, 1975c.

Wallace, M.E.: A case of mutual reduction of dominance: Observed in *Mus musculus*. *Heredity* **7:** 435–437, 1953.

Wallace, M.E.: A mutation or a crossover in the house mouse. *Heredity* **8:** 89–105, 1954.

Wallace, M.E.: Private communication. *Mouse News Letter* **27:** 22, 1962.

Wallace, M.E.: Pseudoallelism at the agouti locus in the mouse. *J. Hered.* **56:** 267–271, 1965.

Wallace, M.E.: Private communication. *Mouse News Letter* **42:** 20, 1970.

Wallace, M.E.: Private communication. *Mouse News Letter* **44:** 18, 1971.

Wallace, M.E.: Private communication. *Mouse News Letter* **47:** 23, 1972.

Welshons, W.J., and Russell, L.B.: The Y-chromosome as the bearer of male determining factors in the mouse. *Proc. Nat. Acad. Sci. USA* **45:** 560–566, 1959.

Werneke, F.: Die pigmentierung der Farbenrassen von *Mus musculus* und inre Beziehung zur Vererbung. *Arch. Entw. Mech. Org.* **42:** 72–106, 1916.

Weston, J.A.: A radioautographic analysis of the migration of trunk neural crest cells in the chick. *Develop. Biol.* **6:** 279–310, 1963.

Wiktor-Jedrzejczak, W., Sharkis, S., Ahmed, A., Sell, K.W., and Santos, G.W.: Theta-sensitive cells and erythropoiesis: Identification of a defect in W/W^v anemic mice. *Science* **196:** 313–315, 1977.

Wilson, F.D., and O'Grady, L.: Some observations of the hematopoietic status *in vivo* and *in vitro* on mice of genotype Sl/Sl^d. *Blood* **48:** 601–608, 1976.

Windhorst, D.B., and Padgett, G.: The Chediak–Higashi syndrome and the homologous trait in animals. *J. Invest. Dermatol.* **60:** 529–537, 1973.

Windhorst, D.B., Zelickson, A.S., and Good, R.A.: Chediak-Higashi syndrome: Hereditary gigantism of cytoplasmic organelles. *Science* **151:** 81–83, 1966.

Windhorst, D.B., Zelickson, A.S., and Good, R.A.: A human pigmentary dilution on a heritable subcellular structural defect—the Chediak–Higashi syndrome. *J. Invest. Dermatol.* **50:** 9–18, 1968.

Winston, H., and Lindzey, G.: Albinism and water escape performance in the mouse. *Science* **144:** 189–191, 1964.

Witkop, C.J., Jr., King, R.A., and Creel, D.J.: The abnormal albino animal. In *Pigment Cell*, V. Riley (ed.), Vol. 3, pp. 201–210, Karger, Basal, 1976.

Wolf, N.S.: Dissecting the hematopoietic microenvironment. I. Stem cell lodgement and commitment, and the proliferation and differentiation of erythropoietic descendants in the Sl/Sl^d mouse. *Cell Tissue Kinet.* **7:** 89–98, 1974.

Wolfe, H.G.: Private communication. *Mouse News Letter* **26:** 35, 1962.

Wolfe, H.G.: Two unusual mutations affecting pigmentation in the mouse. In *Proc. XI Int. Congr. Genet. Vol. 1,* S.J. Geerts (ed.), pp. 251, Pergamon Press, New York, 1963 (Abstract).

Wolfe, H.G.: Private communication. *Mouse News Letter* **37:** 43–44, 1967.

Wolfe, H.G.: Genetic influence on gonadotropic activity in mice. *Biol. Reprod.* **4:** 161–173, 1971.

Wolfe, H.G., and Coleman, D.L.: *Mi*-spotted: A mutation in the mouse. *Genet. Res. (Camb.)* **5:** 432–440, 1964.

Wolfe, H.G., and Coleman, D.L.: *Pigmentation.* In *Biology of the Laboratory Mouse*, 2nd ed., E.L. Green (ed.), pp. 405–425, McGraw-Hill, New York, 1966.

Wolfe, H.G., Erickson, R.P., and Schmidt, L.C.: Effects on sperm morphology by alleles at the pink-eyed locus in mice. *Genetics* **85:** 303–308, 1977.

Wolff, G.L.: Growth of inbred yellow (A^ya) and non-yellow (*aa*) mice in parabiosis. *Genetics* **48:** 1041–1058, 1963.

Wolff, G.L.: Body composition and coat color correlation in different phenotypes of "viable yellow" mice. *Science* **147:** 1145–1147, 1965.

Wolff, G.L.: Differential growth of hepatoma-susceptible liver induced by gene x genome interaction. *Cancer Res.* **30:** 1722–1725, 1970a.

Wolff, G.L.: Genetic influences on response to castration of liver growth and hepatoma formation. *Cancer Res.* **30:** 1726–1730, 1970b.

Wolff, G.L.: Stimulation of growth of transplantable tumors by genes which promote spontaneous tumor development. *Cancer Res.* **30:** 1731–1735, 1970c.

Wolff, G.L.: Genetic modification of homeostatic regulation in the mouse. *Amer. Nat.* **105:** 241–252, 1971.

Wolff, G.L.: Influence of maternal phenotype on metabolic differentiation of agouti locus mutants in the mouse. *Genetics* **88:** 529–539, 1978.

Wolff, G.L., and Bartke, A.: Decreased survival of embryos in yellow ($A^{y}a$) female mice. *J. Hered.* **57:** 14–17, 1966.

Wolff, G.L., and Flack, J.D.: Genetic regulation of plasma corticosterone concentration and its response to castration and allogeneic tumor growth in the mouse. *Nature (London)* **232:** 181–182, 1971.

Wolff, G.L., and Pitot, H.C.: Response of hepatic enzyme activities to castration and sarcoma 37 growth in different mouse genotypes. *J. Nat. Cancer Inst.* **49:** 405–413, 1972a.

Wolff, G.L., and Pitot, H.C.: Variation of hepatic malic enzyme capacity with hepatoma susceptibility in mice of different genotypes. *Cancer Res.* **32:** 1861–1863, 1972b.

Wolff, G.L., and Pitot, H.C.: Influence of background genome on enzymatic characteristics of yellow ($A^{y}/-, A^{vy}/-$) mice. *Genetics* **73:** 109–123, 1973.

Wolff, G.L., and Richard, C.A., Jr.: Response of serum insulin concentration to tumor growth in different genetic systems. *Hormone Metab. Res.* **2:** 68–71, 1970.

Wolff, G.L., Galbraith, D.B., Domon, O.E., and Row, J.M.: Phaeomelanin synthesis and obesity in mice: Interaction of the viable yellow (A^{vy}) and sombre (e^{so}) mutations. *J. Hered.* **69:** 295–298, 1978.

Wolff, S., Dale, D.C., Clark, R.A., Root, R.K., and Kimball, H.R.: The Chediak–Higashi syndrome: Studies of host defenses. *Ann. Int. Med.* **76:** 293–306, 1972.

Wood, B.C., and Miner, G.: Private communication. *Mouse News Letter* **40:** 32, 1969.

Woolf, L.I.: Inherited metabolic disorders: Errors of phenylalanine metabolism. *Adv. Clin. Chem.* **6:** 97–230, 1963.

Woolley, G.W.: "Misty", a new coat color dilution in the mouse. *Mus musculus. Amer. Nat.* **75:** 507–508, 1941.

Woolley, G.W.: "Misty", a new coat color dilution in the mouse, *Mus musculus. Genetics* **28:** 95–96, 1943 (Abstract).

Woolley, G.W.: Misty dilution in the mouse. *J. Hered.* **36:** 269–270, 1945.

Wright, S.: The physiological genetics of coat color of the guinea pig. *Biol. Symp.* **6:** 337–355, 1942.

Yntema, C.L., and Hammond, W.S.: The origin of intrinsic ganglia of trunk and viscera from vagal neural crest in the chick embryo. *J. Comp. Neurol.* **101:** 515–541, 1954.

Zannoni, V.G., Weber, W.W., Van Valen, P., Rubin, A., Bernstein, R., and La Du, B.N.: Phenylalanine metabolism and "phenylketonuria" in dilute-lethal mice. *Genetics* **54:** 1391–1399, 1966.

Zelickson, A.S., Windhorst, D.B., White, J.G., and Good, R.A.: The Chediak–Higashi syndrome: Formation of giant melanosomes and the basis of hypopigmentation. *J. Invest. Dermatol.* **49:** 575–581, 1967.

Zimmerman, A.A., and Becker, S.W., Jr.: Melanoblasts and melanocytes in fetal negro skin. *Illinois Monogr. Med. Sci.* **6,** No. 3, University of Illinois Press, Urbana, 1959.

Zimmerman, K.: Some results of genetical analysis in populations of wild rodents. *Proc. VII Int. Genet. Congr.* pp. 332, 1941 (Abstract).

Author Index

Subject Index

Biology of the Mouse Histocompatibility-2 Complex
Principles of Immunogenetics Applied to a Single System
Jan Klein, University of Texas, Southwestern Medical School

"In spite of the importance of the H-2 system, the literature on it is usually complicated and hard to comprehend because of complex technical details and a somewhat confusing terminology. Many immunologists, mammalian geneticists, and cell and membrane biologists have long felt a need for a clear and systematic account of the subject. This requirement is met excellently by Klein's book. The author illuminates every seemingly esoteric aspect of the subject by defining phenomena in simple, lucid language, and the organization is clear and systematic. . . . Its grand scale, considerable depth, and clear, systematic outline make this book an excellent summary that should be useful to both beginners and experts in the field as well as to interested outsiders."
—*Science*

1974/xii, 620 pp./69 illus. 76 tables/cloth
ISBN 0-387-06733-7

The Major Histocompatibility Systems in Man and Animals
Edited by **D. Gotze,** The Wistar Institute, Philadelphia

Three quarters of a century ago, an albino mouse refused to accept tumorous tissue from a Japanese waltzing mouse. This obstinate behavior on the part of the albino mouse marked the beginning of the science of immunogenetics. It has since been shown that this rejection phenomenon is controlled by several genes, the most important of which are designated the Major Histocompatibility Gene Complex.

The Major Histocompatibility System in Man and Animals provides a comprehensive survey of the data accumulated on this central issue in cellular immurology. The effects this gene complex has upon gene action and regulation are discussed, as are the evolution of this complex and the role it plays in the pathogensis of disease in man and animals.

1976/xii, 404 pp./23 fig./cloth